Plasma Technology for Hyperfunctional Surfaces

Edited by
Hubert Rauscher, Massimo Perucca, and
Guy Buyle

Related Titles

Kawai, Y., Ikegami, H., Sato, N., Matsuda, A., Uchino, K., Kuzuya, M., Mizuno, A. (eds.)

Industrial Plasma Technology

Applications from Environmental to Energy Technologies

2010

ISBN: 978-3-527-32544-3

Heimann, R. B.

Plasma Spray Coating

Principles and Applications

2008

ISBN: 978-3-527-32050-9

Hippler, R., Kersten, H., Schmidt, M., Schoenbach, K. H. (eds.)

Low Temperature Plasmas

Fundamentals, Technologies and Techniques

2008

ISBN: 978-3-527-40673-9

d'Agostino, R., Favia, P., Kawai, Y., Ikegami, H., Sato, N., Arefi-Khonsari, F. (eds.)

Advanced Plasma Technology

2008

ISBN: 978-3-527-40591-6

Smirnov, B. M.

Plasma Processes and Plasma Kinetics

580 Worked-Out Problems for Science and Technology

2007

ISBN: 978-3-527-40681-4

Plasma Technology for Hyperfunctional Surfaces

Food, Biomedical and Textile Applications

Edited by
Hubert Rauscher, Massimo Perucca, and Guy Buyle

WILEY-VCH

WILEY-VCH Verlag GmbH & Co. KGaA

The Editor

Dr. Hubert Rauscher
Institute for Health and Consumer
Protection
European Commission
Joint Research Centre
Via E. Fermi 2749
21027 Ispra (VA)
Italy

Dr. Massimo Perucca
Environment Park S.p.A.
Clean NT Lab
Via Livorno 60
10144 Torino
Italy

Dr. Guy Buyle
Centexbel
Technologiepark 7
9052 Zwijnaarde
Belgium

■ All books published by Wiley-VCH are
carefully produced. Nevertheless, authors,
editors, and publisher do not warrant the
information contained in these books,
including this book, to be free of errors.
Readers are advised to keep in mind that
statements, data, illustrations, procedural
details or other items may inadvertently be
inaccurate.

Library of Congress Card No.: applied for

**British Library Cataloguing-in-Publication
Data**
A catalogue record for this book is available
from the British Library.

**Bibliographic information published by
the Deutsche Nationalbibliothek**
The Deutsche Nationalbibliothek lists this
publication in the Deutsche Nationalbib-
liografie; detailed bibliographic data are
available on the Internet at
http://dnb.d-nb.de.

© 2010 WILEY-VCH Verlag GmbH & Co.
KGaA, Weinheim

Typesetting Laserwords Private Limited,
Chennai
Printing and Binding betz-druck GmbH,
Darmstadt
Cover Design Adam Design, Weinheim

Printed in the Federal Republic of Germany
Printed on acid-free paper

ISBN: 978-3-527-32654-9

Contents

Preface *XV*
List of Contributors *XIX*
List of Contacts *XXIII*

Part I   Introduction to Plasma Technology for Surface Functionalization *1*

1   Introduction to Plasma and Plasma Technology *3*
         Massimo Perucca
1.1   Plasma: the Fourth State of Matter *3*
1.2   Historical Highlights *4*
1.3   Plasma Fundamentals *6*
1.3.1   Free Ideal Gas *7*
1.3.2   Interacting Gas *8*
1.3.3   The Plasma as a Fluid *11*
1.3.4   Waves in Plasmas *12*
1.3.5   Relevant Parameters that Characterize
         the State of Plasma *14*
1.4   Classification of Technological Plasmas *17*
1.4.1   Hot (Thermal) Plasmas and Their
         Applications *18*
1.4.2   Cold (Nonthermal) Plasmas and Their Applications *19*
1.5   Reactive Plasmas *22*
1.5.1   Elementary Plasma–Chemical Reactions *22*
1.5.2   Elastic Scattering and Inelastic Thomson Scattering: Ionization
         Cross-section *24*
1.5.3   Molecular Ionization Mechanisms *25*
1.5.4   Stepwise Ionization by Electron Impact *26*
1.6   Plasma Sheaths *28*
1.7   Summary *31*
         References *31*

Plasma Technology for Hyperfunctional Surfaces. Food, Biomedical and Textile Applications.
Edited by Hubert Rauscher, Massimo Perucca, and Guy Buyle
Copyright © 2010 WILEY-VCH Verlag GmbH & Co. KGaA, Weinheim
ISBN: 978-3-527-32654-9

2   **Plasma Systems for Surface Treatment**   *33*
        Guy Buyle, Joachim Schneider, Matthias Walker, Yuri Akishev, Anatoly Napartovich, and Massimo Perucca
2.1   Introduction   *33*
2.2   Low Pressure Plasma Systems   *34*
2.2.1   Microwave Systems   *35*
2.2.1.1   Introduction   *35*
2.2.1.2   Standard Microwave System for Textile Treatment   *36*
2.2.1.3   Example: Duo-Plasmaline–a Linearly Extended Plasma Source   *36*
2.2.1.4   Electron Cyclotron Resonance Heated Plasmas   *40*
2.2.2   Capacitively Coupled Systems   *43*
2.2.2.1   Introduction   *43*
2.2.2.2   Capacitive Coupled Plasma for Biomedical Applications   *44*
2.2.3   Physical Vapor Deposition Plasma: LARC®   *45*
2.2.3.1   Background   *45*
2.2.3.2   Cathodic Arc PVD Systems   *45*
2.2.3.3   Example: Treatment of Food Processing Tools by LARC® PVD System   *48*
2.3   Atmospheric Pressure Plasma Systems   *49*
2.3.1   Corona-type Surface Treatment   *51*
2.3.1.1   Standard Corona Treatment   *51*
2.3.1.2   Controlled Atmosphere Corona Treatment–Aldyne Treatment   *52*
2.3.1.3   Liquid Deposition   *52*
2.3.2   Remote Surface Treatment   *54*
2.3.2.1   Plasma Sources Used for Modeling   *55*
2.3.2.2   Example: AcXys Plasma Jet   *57*
2.4   Summary   *58*
        Acknowledgment   *59*
        References   *59*

3   **Plasma-surface Interaction**   *63*
        Domenico D'Angelo
3.1   Introduction   *63*
3.2   Polymer Etching   *65*
3.3   Plasma Grafting   *66*
3.4   Chemical Kinetics   *68*
3.4.1   Chain Polymerization   *68*
3.4.2   Plasma Polymerization   *70*
3.5   Example: Plasma Polymerization   *71*
3.5.1   Plasma Polymerization of HEMA   *72*
3.5.1.1   Theoretical Background   *72*
3.5.1.2   Example: Polymerization of HEMA on PET Fabric   *73*
3.5.2   Plasma Polymerization of HDMSO   *75*
3.6   Conclusion   *76*
        References   *77*

4   **Process Diagnostics by Optical Emission Spectroscopy**   *79*
        Giacomo Piacenza
4.1   Introduction   *79*
4.2   Optical Emission Spectroscopy   *79*
4.2.1   Theory of Optical Emission   *80*
4.2.2   Spectroscopy   *82*
4.2.3   OES Bench and Set-up   *83*
4.3   Optical Absorption Spectroscopy   *85*
4.3.1   Actinometry   *86*
4.4   Laser Induced Fluorescence (LIF)   *87*
4.5   Conclusion   *88*
        References   *88*

5   **Surface Analysis for Plasma Treatment Characterization**   *91*
        Amandine David, Yves de Puydt, Laurent Dupuy, Séverine Descours,
        Françoise Sommer, Minh Duc Tran, and Jocelyn Viard
5.1   Introduction to Surface Characterization Techniques   *91*
5.2   X-ray Photoelectron Spectroscopy (XPS) or Electron Spectroscopy
        for Chemical Analysis (ESCA)   *94*
5.2.1   Principles of XPS   *95*
5.2.2   XPS Core Level Chemical Shift   *96*
5.2.3   Quantitative Analysis   *97*
5.2.4   Quantitative Analysis of Nitrogen Plasma-Treated Polypropylene   *98*
5.2.5   Angle-Resolved XPS Depth Profiling and Surface Sensitivity
        Enhancement by Grazing Angle XPS Detection   *100*
5.2.6   Determination of Thin Coating Thickness by Angle-Resolved
        XPS   *100*
5.2.7   Mapping   *104*
5.2.8   Summary of XPS   *105*
5.3   Static Secondary Ion Mass Spectrometry by Time of Flight
        (ToF-SSIMS)   *106*
5.3.1   Principles of ToF-SSIMS   *106*
5.3.1.1 Secondary Ion Emission   *107*
5.3.1.2 Static and Dynamic Modes   *107*
5.3.1.3 Molecular SIMS   *107*
5.3.2   Applications of ToF-SSIMS   *107*
5.3.2.1 Spectrometry Mode   *108*
5.3.2.2 Secondary Ion Imaging   *108*
5.3.2.3 Depth Profiling   *108*
5.3.2.4 Data Treatment by Multivariate Methods: Multi-Ion SIMS   *108*
5.3.2.5 Examples   *109*
5.3.2.5.1 Poly(ethylene terephthalate) Tissue   *109*
5.3.2.5.2 Polypropylene Packaging   *109*
5.3.2.5.3 SiO_x Barrer Coating on PET   *111*
5.3.2.5.4 Anti-UV Additive qualification on PET Films   *112*

5.4	Atomic Force Microscopy	*114*
5.4.1	Operating Modes in AFM	*114*
5.4.1.1	Contact Mode	*115*
5.4.1.1.1	Constant Force Mode	*115*
5.4.1.2	Resonant Modes	*117*
5.4.1.2.1	The Contact –No Contact Mode	*118*
5.4.1.2.2	Phase Contrast Mode	*118*
5.4.1.3	Other Modes	*119*
5.4.2	Summary and Outlook	*119*
5.5	Scanning Electron Microscopy (SEM)	*121*
5.5.1	Principles of SEM	*121*
5.5.2	Imaging in SEM	*122*
5.5.3	New Generation of SEM	*122*
5.5.4	Chemical Analysis	*123*
5.5.5	Sample Preparation and Applications	*124*
5.6	Transmission Electron Microscopy (TEM)	*124*
5.6.1	Principles of TEM	*124*
5.6.2	Resolution	*126*
5.6.3	Image Contrast	*126*
5.6.4	Chemical Analysis	*126*
5.6.5	Typical Applications of TEM	*127*
5.6.6	Sample Requirements	*127*
5.7	Contact Angle Measurement	*129*
5.7.1	Surface Energy Calculation	*130*
5.7.1.1	Owens and Wendt Model for Surface Energy Calculation	*130*
5.7.1.2	Good and Van Oss Model for Surface Energy Calculation	*131*
5.8	Conclusions	*132*
	References	*132*

Part II   Hyperfunctional Surfaces for Textiles, Food and Biomedical Applications *133*

6	**Tuning the Surface Properties of Textile Materials**	*135*
	Guy Buyle, Pieter Heyse, and Isabelle Ferreira	
6.1	Introduction	*135*
6.1.1	Potential Impact of Plasma on the Textile Industry	*135*
6.1.2	Plasma Basics	*137*
6.1.3	Fundamental Advantage of Plasma Processing	*138*
6.1.4	Classification of Plasmas from the Textile Viewpoint	*138*
6.1.4.1	Pressure-based	*140*
6.1.4.2	Substrate-based	*141*
6.2	Plasma Treatment of Textile Materials	*142*
6.2.1	Overview of Functionalizations	*142*
6.2.2	Effect of Plasma Treatment on Textile Substrates	*143*
6.2.2.1	Interaction of Active Plasma Species with a Surface	*143*

6.2.2.2   Basic Plasma Effect on Substrate   *143*
6.2.2.3   Aging   *144*
6.3   Integration of Plasma Processes into the Textile Manufacturing Chain   *146*
6.3.1   Fiber Level   *147*
6.3.2   Filament Level   *148*
6.3.3   Yarn Level   *149*
6.3.3.1   Natural Materials   *149*
6.3.3.1.1   Cotton   *149*
6.3.3.1.2   Wool   *149*
6.3.3.1.3   Other Natural Fibers   *149*
6.3.3.2   Non-natural Materials   *150*
6.3.4   Fabric Level   *150*
6.3.4.1   Woven Textiles   *151*
6.3.4.1.1   Natural Materials   *151*
6.3.4.1.2   Non-natural Materials   *152*
6.3.4.2   Knitted Textiles   *152*
6.3.4.3   Non-wovens   *153*
6.3.5   Intermediate/Finished Textile Material   *154*
6.4   Specific Requirements for the Textile Industry   *155*
6.4.1   Chemical Composition   *155*
6.4.2   Surface Cleanliness   *155*
6.4.3   Three-dimensional Structure of Textiles   *156*
6.4.4   Large Surface Area   *157*
6.4.5   Moisture Regain and Air Adsorption   *158*
6.5   Case Studies   *158*
6.5.1   Assessing the Surface Energy of Textiles   *158*
6.5.1.1   Introduction to Methods for Evaluating the Surface Energy and Wetting of Textiles   *159*
6.5.1.1.1   Wilhelmy Method   *159*
6.5.1.1.2   Washburn Method   *160*
6.5.1.2   Evaluation of Methods for Measuring Hydrophilic Properties   *161*
6.5.1.2.1   Wilhelmy Method   *161*
6.5.1.2.2   Washburn Method   *162*
6.5.1.2.3   Summary of Evaluation   *163*
6.5.1.3   Tests and Standards for Evaluating Hydrophobic/Oleophobic Properties   *163*
6.5.1.3.1   Water Repellency: Spray Test   *164*
6.5.1.3.2   Water/Alcohol Repellency   *165*
6.5.1.3.3   Oil Repellency   *166*
6.5.2   Hydrophilic Properties Imparted by Plasma   *167*
6.5.2.1   Plasma Experiments at Low Pressure   *167*
6.5.2.1.1   First Screening of Precursors   *168*
6.5.2.1.2   Aging of the Samples   *169*
6.5.2.2   Plasma Experiments at Atmospheric Pressure (Aldyne System)   *170*

6.5.3 Hydrophobic/Oleophobic Properties Imparted by Plasma *171*
6.5.3.1 Preliminary Experiments *171*
6.5.3.2 Washing Durability *172*
6.5.3.3 Abrasion Durability *173*
6.5.3.4 Summary of Oleophobic Properties *174*
6.6 Transferring Plasma Technology to Industrial Processes *174*
6.6.1 Textile Sector Related Issues *175*
6.6.2 Fundamental Aspects Regarding Industrialization *176*
6.7 Summary *177*
References *178*

7 Preventing Biofilm Formation on Biomedical Surfaces *183*
Virendra Kumar, Hubert Rauscher, Frédéric Brétagnol, Farzaneh
Arefi-Khonsari, Jerome Pulpytel, Pascal Colpo, and François Rossi
7.1 Bacterial Adhesion to Biomaterials: Biofilm Formation *183*
7.1.1 'Biofilm' and Its Implications in the Biomedical Field *184*
7.1.2 Mechanism for Bacterial Adhesion to Surfaces *184*
7.1.3 Biofilm Formation – a Multistep Process *186*
7.1.4 Factors Influencing Biofilm Formation *187*
7.1.4.1 Role of the Conditioning Film *187*
7.1.4.2 Material Surface Characteristics *188*
7.1.4.3 Micro-organism Characteristics *190*
7.1.4.4 Environmental Factors *191*
7.2 Biofilm Prevention Strategies *192*
7.2.1 Pre-surgery Precautionary Approach *192*
7.2.2 Antimicrobial-releasing Biomaterials *193*
7.2.3 Surface-engineering Approach *193*
7.2.3.1 High Surface Energy Approach *194*
7.2.3.2 Low Surface Energy Approach *195*
7.2.3.3 Surfaces with Bound Tethered Antimicrobial Agents *196*
7.2.4 'Antibiofilm' Approach *197*
7.3 Role of Plasma Processing in Biofouling Prevention *198*
7.3.1 Plasma Surface Functionalization *199*
7.3.2 Plasma-Induced Grafting *199*
7.3.3 Plasma Polymerization *200*
7.3.4 Plasma Sterilization *201*
7.4 Case Study: Plasma-deposited Poly(ethylene oxide)-like Films
for the Prevention of Biofilm Formation *202*
7.4.1 PEO Films and Plasma Deposition *202*
7.4.2 Plasma Polymerization by Continuous Wave Plasma *203*
7.4.2.1 Retention of the PEO Character and Film Stability *203*
7.4.2.2 Protein Adsorption *205*
7.4.2.3 Cell Attachment and Proliferation *206*
7.4.2.4 Aging *208*
7.4.3 Plasma Polymerization in Pulsed Mode *208*

7.4.4   Sterilization of PEO-like Films   *210*
7.4.5   Composite Films: Ag Nanoparticles in a PEO-like Matrix   *211*
7.4.5.1   Synthesis of Ag Nanoparticles and Deposition on Surfaces   *212*
7.4.5.2   Composite AgNP/PEO Surfaces and Their Antibacterial Activity   *213*
7.5   Summary   *216*
         References   *217*

8   **Oxygen Barriers for Polymer Food Packaging**   *225*
         Joachim Schneider and Matthias Walker
8.1   Introduction   *225*
8.2   Fundamentals of Gas Diffusion through Polymers   *225*
8.2.1   Diffusion, Solubility, and Permeability of Polymers   *227*
8.2.2   Diagnostic Methods   *230*
8.2.3   Barrier Concepts   *233*
8.3   Case Study: Plasma Deposition of SiO_x Barrier Films on Polymer
         Materials Relevant for Packaging Applications   *234*
8.3.1   Materials and Measurements   *234*
8.3.1.1   Selection of Two-dimensional and Three-dimensional Polymer
         Substrates   *234*
8.3.1.2   Measurement of the Steady-state O_2 Particle Flux   *235*
8.3.1.3   Measurement of the Coating Thickness   *235*
8.3.2 SiO_x Barrier Films on PET Foil   *236*
8.3.2.1 SiO_x Barrier Films Deposited from O_2: HMDSO Gas Mixtures   *236*
8.3.2.1.1 O_2 Permeation Measurements: Determination of the Diffusion
         Coefficient   *237*
8.3.2.1.2 O_2 Permeation Measurements: Variation of the O_2: HMDSO Gas
         Mixture Ratio   *238*
8.3.2.1.3 FTIR Analysis: Chemical Composition of the Surface of the SiO_x
         Barrier Films Deposited from Different O_2: HMDSO Gas
         Mixtures   *239*
8.3.2.2 SiO_x Barrier Films Deposited from O_2 : HMDSN Gas Mixtures   *243*
8.3.2.2.1 O_2 Permeation Measurements: Variation of the O_2: HMDSN Gas
         Mixture Ratio   *243*
8.3.2.2.2 FTIR Analysis: Comparing Best Performing SiO_x Barrier Films
         Deposited from O_2 : HMDSO and from O_2 : HMDSN Gas
         Mixtures   *245*
8.3.2.2.3 O_2 Permeation Measurements: Variation of the Film Thickness   *246*
8.3.3 SiO_x Barrier Films on PP Foil   *247*
8.3.3.1   ECR Plasma Source: Comparing the Barrier Properties of SiO_x Films
         Deposited on PP and on PET Foil by Variation of the O_2 : HMDSO Gas
         Mixture Ratio   *247*
8.3.3.2   Duo-Plasmaline Plasma Source: SiO_x Barrier Films Deposited from
         O_2: HMDSN Gas Mixtures   *249*
8.3.4   ECR Plasma Deposition of SiO_x Barrier Films on Polymer Trays
         Designed for Food Packaging   *251*

8.3.4.1   ECR Plasma Deposition of SiO$_x$ Barrier Films Without Directed Gas
           Supply and Customized Magnet Configuration: Variation of the
           Plasma Deposition Time and of the Distance between Sample and
           Plasma   *252*

8.3.4.2   Achieving Industrially Relevant Plasma Deposition Times by Directed
           Gas Supply and Customized Magnet Configuration   *255*

8.4   Conclusions   *258*
           Acknowledgments   *259*
           References   *259*

9   **Anti-wear Coatings for Food Processing**   *263*
           Maddalena Rostagno and Federico Cartasegna

9.1   Introduction   *263*

9.2   Recent Developments in PVD Coatings   *264*

9.3   Coatings Trends and Market Share   *267*

9.4   Coatings Application in the Food Processing Sector   *268*

9.5   Coating Requirements in the Food Sector   *269*

9.5.1   Wear Resistance   *270*

9.5.2   Coefficient of Friction (COF)   *271*

9.6   Selection of Methodologies for Effective Characterization of Coatings
           for the Food Sector   *271*

9.6.1   Chemical and Structural Characterization   *273*

9.6.1.1   Scanning Electron Microscopy (SEM)   *273*

9.6.1.1.1 Application to Anti-wear Coatings for Food Processing Tools   *273*

9.6.1.2   Energy Dispersive X-ray Spectrometry (EDX)   *274*

9.6.1.2.1 Application to Anti-wear Coatings   *274*

9.6.1.3   Calotest and Optical Microscopy (OM)   *275*

9.6.1.3.1 Application to Anti-wear Coatings for Food Processing Tools   *276*

9.6.2   Mechanical Characterization   *276*

9.6.2.1   Hardness   *276*

9.6.2.1.1 Application to Anti-wear Coatings for Food Processing Tools   *277*

9.6.2.2   Pin-on-disk   *279*

9.6.2.2.1 Application to Anti-wear Coatings for Food Processing Tools   *280*

9.6.3   Atoxicity and Corrosion Characterization   *280*

9.6.3.1   Food Compatibility: Heavy Metals Release   *280*

9.6.3.2   Food Compatibility: Oxidation Test   *280*

9.6.3.3   Salt Spray Test   *280*

9.7   Case Studies: Development and Characterization of Ceramic Coatings
           for Food Processing Applications   *281*

9.7.1   Relevant Substrates and Functionalities Required for Cutting
           Applications   *281*

9.7.2   Technical Analysis and Choice of the Proper Coating Chemistry and
           Technique   *282*

9.7.3   Coating Development   *285*

9.7.4   Case Study: PVD Coating of Saw Blades   *288*

9.7.5   Case Study: PVD Coating of Hammers for Food Treatment   *291*
9.8   Conclusions   *294*
           References   *294*

10   **Physics and Chemistry of Nonthermal Plasma at Atmospheric Pressure Relevant to Surface Treatment**   *295*
           Yuri Akishev, Anatoly Napartovich, Michail Grushin, Nikolay Trushkin, Nikolay Dyatko, and Igor Kochetov
10.1   Introduction   *295*
10.2   Discharge Modeling   *297*
10.2.1   Full Kinetic Models and Reduced Model for Technological Plasma   *297*
10.2.2   Electron Kinetics   *299*
10.2.3   Plasma Chemistry   *301*
10.2.4   Experimental UV, Optical, and Near Infra-red Emission Spectra   *302*
10.2.4.1   Air-based Discharges   *302*
10.2.4.2   Nitrogen-based Discharges   *306*
10.2.4.3 CF_4-based Discharges   *309*
10.2.5   Influence of Impurities on Composition of Gas Activated by Nonthermal Plasma   *310*
10.3   Kinetic Model for Chemical Reactions on a Polypropylene Surface in Atmospheric Pressure Air Plasma   *314*
10.3.1   Description of Kinetic Model   *314*
10.3.1.1   Description of Chemical Reaction Modeling   *314*
10.3.1.2   Description of Surface Concentration Modeling   *320*
10.3.1.2.1 Abstraction of H Atoms from H-sites by OH Radicals   *320*
10.3.1.2.2 Abstraction of H Atoms from H-sites by Alkoxy Radicals   *321*
10.3.1.2.3 Chain Backbone Scission Due to Interaction of Alkoxy Radicals with the Polymer Backbone   *321*
10.3.2   Results of Modeling and Comparison with Experimental Data   *321*
10.4   Conclusions   *328*
           Acknowledgement   *328*
           References   *328*

           Part III   Economical, Ecological, and Safety Aspects   *333*

11   **Economic Aspects**   *335*
           Elisa Aimo Boot
11.1   Market Analysis: an Overview   *335*
11.1.1   Textile Market Analysis   *335*
11.1.1.1   General   *335*
11.1.1.2   Technical Textiles   *336*
11.1.1.3   Hydrophobic and Oleophobic Textile Market   *336*
11.1.2   Biomedical Market Perspective   *337*
11.1.3   Food Packaging Market Potential   *339*

11.2 Case Study: Up-Scaling of the Plasma Treatment of Hammers for Meat Milling *340*

11.2.1 Analysis of the Reference Scenario *341*

11.2.2 Analysis of Scenario 2 – Outsourcing *341*

11.2.3 Analysis of Scenario 3 – In-house *342*

11.2.4 Investment and Operating Cost *343*

11.2.5 Comparative Analysis of All Three Scenarios *344*

11.2.6 Final Considerations *345*

References *346*

12 Environment and Safety *347*

Massimo Perucca and Gabriela Benveniste

12.1 Introduction to LCA *347*

12.2 Environmental Impact of Traditional Surface Processing: the Reason for Developing Innovative Solutions Supported by Dedicated LCA *350*

12.3 LCA Applied to Plasma Surface Processing: Case Studies *353*

12.3.1 Scope, Functional Unit, and System Boundaries *354*

12.3.2 Life Cycle Inventory (LCI) and Hypothesis *356*

12.3.3 Inventory Data and Results *360*

12.3.3.1 The Anti-corrosion Process *361*

12.3.3.2 Textile Processes *364*

12.3.3.2.1 Total Energy Requirement *364*

12.3.3.2.2 Output of the Oleophobic PET Processes *366*

12.3.3.2.3 Output of the Hydrophobic PET/Cotton Processes *367*

12.3.4 Impact Assessment *369*

12.3.5 Sensitivity Analysis *371*

12.3.5.1 Managing Uncertainties *371*

12.3.5.2 Example 1: General Sensitivity Analysis for the LCA Study of the Textile Processes *371*

12.3.5.3 Example 2: Design of Plasma Processes via LCA *375*

12.3.6 Concluding Considerations on LCA Study *375*

12.4 Process Safety for the Working Environment *378*

12.4.1 Atmospheric Pressure Plasma Unit: Standard Configuration *379*

12.4.2 Devising Safe Processes for Industrial Applications Maintaining the Semi-continuous Feeding *381*

12.4.3 Final Considerations on Process Safety *388*

References *389*

Index *391*

Preface

The principal aim of this book is the promotion and dissemination of knowledge on plasma technology, underlining its technical applicability, economic sustainability, and minimal environmental impact. This is illustrated via plasma processes that are implemented in traditional or innovative industrial applications in the textile, food packaging and/or processing and biomedical sector. A further objective of this book is to provide selected application examples and case studies deriving from the research, development and technology transfer experienced within ACTECO, a project supported by the European Commission under the 6th Framework Programme. The project provided environmentally friendly, economically sustainable solutions for specific surface functionalities (hyperfunctional surfaces).

This book promotes a broader perspective in the exploitation of plasma technology by thoroughly evaluating the competitive advantages and limitations leading to a new concept of eco-design. In this view, components and products are engineered starting from their functional needs and specifications, rather than from traditional material choice. In this framework, hyperfunctional surfaces, through sustainable dry plasma processing may represent a powerful technique to provide added value via dedicated surface finishing while, at the same time, preserving the beneficial physico-chemical characteristics of the bulk material. Additionally, plasma surface processing can overcome the need for complex composite materials or materials whose specific bulk chemical composition is actually only required at the very surface (e.g., for wear and oxidation resistance of steels).

As a matter of fact, the performance of materials used in major industrial applications in the field of health, food, textile, and environment depends very strongly on the physico-chemical properties of the surfaces. For instance, the very functioning of several biomedical devices is linked to the ability of their surfaces to repel proteins and to avoid biofilm formation. Likewise, textiles for clothing and technical applications are a major target for finishing techniques because imparting, for example, durable hydrophobicity, hydrophilicity, or oleophobicity is a major challenge for several applications. Improved recyclable and/or biodegradable food packaging for a longer shelf life can be realized via more advanced surface barrier properties, while efficient and safe food processing benefits from components whose surfaces are treated against wear, corrosion, and heavy metal migration.

Plasma Technology for Hyperfunctional Surfaces. Food, Biomedical and Textile Applications.
Edited by Hubert Rauscher, Massimo Perucca, and Guy Buyle
Copyright © 2010 WILEY-VCH Verlag GmbH & Co. KGaA, Weinheim
ISBN: 978-3-527-32654-9

In general, providing breakthrough competitive and innovative solutions requires a radical new vision for the development of the technological fields involved. Such a new vision should stem from cutting edge scientific knowledge (e.g., from latest progress in nanosciences and nanotechnologies) and be followed by the transfer of exploitable content into up-scalable, industrial solutions. However, the feasibility of a specific surface treatment on the laboratory scale does not necessarily imply its applicability as an industrial process.

For this reason, this book provides a thorough analysis of the developments made for several applications in the form of case studies, thus delivering the stepping stones for wider, more industrial take-up. In particular, the selected examples illustrate that controlling the surface properties has a major impact on the eco-efficiency of the industrial sectors concerned via the reduction of energy and water consumption. Industrial solutions are presented to provide control of adhesion, barrier properties, and wear resistance of materials.

In summary, this book suggests tools and basic knowledge to support the development of novel, knowledge-based added value products and processes, also in traditional industries, less dependent on research and technological development. Its content has been selected to stimulate process design based on eco-innovation and eco-efficiency criteria. Additionally, this book considers modern general demands on novel industrial processes, meaning that the book not only discusses state-of-the-art approaches but also presents a discussion of economic, ecological, and safety issues related to plasma surface processing.

This book consists of three parts. The first part starts with an introduction to plasma technology through plasma fundamentals (Chapter 1) and includes plasma sources (Chapter 2), plasma-surface interactions (Chapter 3), plasma diagnostics (Chapter 4), and surface characterization techniques (Chapter 5).

The second part covers applications studied within ACTECO and, therefore, covers the three domains (food, biomedical, and textile) targeted within the project. These fields currently experience some of the most innovative applications of surface processing by plasma. We will discuss how plasma treatment can be used to tune the surface properties of textiles (Chapter 6), prevent biofilm formation on biomedical surfaces (Chapter 7), provide oxygen barriers for food packaging (Chapter 8) and obtain anti-wear coatings in food processing machines (Chapter 9). A comprehensive theoretical approach is provided to model the interactions of nonthermal atmospheric pressure plasma with surfaces (Chapter 10).

The third part is dedicated to the technical and economic aspects of plasma technology. It includes an analysis of the market potential as well as the economic impact arising from the introduction of plasma technology into the textile, food, and biomedical sectors (Chapter 11). This part concludes with a discussion of environmental and safety issues related to plasma surface treatments (Chapter 12). This includes a comparative life cycle analysis to assess the eco-efficiency of surface plasma functionalization with respect to traditional surface treatment and an assessment of plasma processing safety in terms of process reliability for environmental working conditions as well as the potential local impact due to emissions.

Within this book we have tried to achieve a sufficient cohesion and self consistency. Internal referencing among chapters, although written by different authors, is provided to enable the reader to browse through the content via different pathways, even starting from different points, according to different interests, needs, and background.

This philosophy was followed when putting the manuscript together. Clearly, we cannot list all these chapter interconnections in the preface but we want to highlight an example as a possible suggestion for a path through the book. One of the applications mentioned in the book regards tuning of the surface energy of textiles by plasma treatment. Starting from the specific application discussed in Chapter 6 the reader may move up-stream to Chapter 2 in order to find out details related to plasma systems used for textile processing. Further on the reader may explore the related economic and ecological aspects connected to these processes by visiting Chapters 11 and 12, respectively. Furthermore, additional information can be found in Chapter 5, which illustrates some of the surface characterization techniques employed to assess the physico-chemical changes induced by plasma treatment. A similar approach may be followed for the other main applications dealt with in this book. Nevertheless, the reader may follow a more orthodox approach by sequentially going through each chapter, which provides a more general perspective of the topics treated. The sequential approach is particularly recommended to readers completely new to plasma surface functionalization.

This work has been partially funded by the European Commission in the 6th Research Framework programme through the integrated project ACTECO for small and medium enterprises (IP 515859-2), contract number NMP-CT-2005-515859, launched on 1 May 2005 and ended on 30 April 2009. The full project title is: 'Eco efficient activation for hyperfunctional surfaces'; this highlights its main focus: addressing the use of plasma technology for efficient and effective surface functionalization and activation (*http://www.acteco.org/*).

The consortium, whose composition evolved during the course of the project, consisted of several partners that can be grouped into different categories. The first are end-users within the different application areas: food related (Diad s.r.l., Tops Foods), biomedical applications (PlasMATec, Covidien-Sofradim), and textile companies (Jovertex, Creat-Chargeurs, Luxilon). Another group of companies were the plasma technology providers related to atmospheric as well as low pressure plasma (Muegge, CPI, AcXys, Dow Corning Plasma Solutions, Environment Park, and Europlasma, the project coordinator). Also high-tech SME companies dedicated to surface analysis formed part of the project (CSMA, Biophy, Biomatech). Further, several research centers and universities supported the R&D activities (UPMC, TRINITI, USTUTT-IPF, IFTH, EC-JRC, Centexbel). ACTECO also included a partner to perform market studies (Nodal) and sector associations covering the three targeted domains via IVLV, Clubtex, and Eucomed. A full list with the contact details of the project partners, according to the situation at the end of the project, can be found at the end of the preface.

We would like to thank all ACTECO partners, whose valuable contributions throughout the project created a synergy that made ACTECO a success story. Without them, their precious work carried out within the project and their input during the writing of the manuscript, this book could not have been written.

We tried our best to design and write a book that is useful for people already working in plasma technology as well as for those whose focus is more on one of the application fields discussed. Moreover, suggestions may be found for the application of plasma technology in industrial sectors not explicitly treated here.

Plasma Technology for Hyperfunctional Surfaces: Food, Biomedical and Textile Applications addresses industry professionals, researchers, academic teachers and PhD students specializing in the field of plasma physics and chemistry, as well as people entering the field of plasma surface treatments and technical staff involved in economic sustainability and ecology. Our intention is that also policy makers in the field of clean, environmentally friendly, and economically efficient technological innovations will find useful information here on trends and potentials of plasma surface engineering.

Hubert Rauscher
Massimo Perucca
Guy Buyle

List of Contributors

Yuri Akishev
State Research Center of the
Russian Federation
Troitsk Institute for Innovation
and Fusion Research
Pushkovykh st., domain 12
Troitsk 142190
Moscow region
Russia

Farzaneh Arefi-Khonsari
Université Pierre et Marie Curie
Laboratoire de Génie des
Procédés Plasmas et Traitement
de Surface
ENSCP
11 rue Pierre et Marie Curie
75231 Paris cedex 05
France

Gabriela Benveniste
Environment Park S.p.A.
Clean NT Lab
Via Livorno 60
10144 Torino
Italy

Elisa Aimo Boot
Environment Park S.p.A.
Clean NT Lab
Via Livorno 60
10144 Torino
Italy

Frédéric Brétagnol
European Commission
Joint Research Centre
Institute for Health and
Consumer Protection
Via E. Fermi 2749
21027 Ispra (VA)
Italy

Guy Buyle
Centexbel
Technologiepark 7
9052 Zwijnaarde
Belgium

Federico Cartasegna
Environment Park S.p.A.
Clean NT Lab
Via Livorno 60
10144 Torino
Italy

Pascal Colpo
European Commission
Joint Research Centre
Institute for Health and
Consumer Protection
Via E. Fermi 2749
21027 Ispra (VA)
Italy

Plasma Technology for Hyperfunctional Surfaces. Food, Biomedical and Textile Applications.
Edited by Hubert Rauscher, Massimo Perucca, and Guy Buyle
Copyright © 2010 WILEY-VCH Verlag GmbH & Co. KGaA, Weinheim
ISBN: 978-3-527-32654-9

Domenico D'Angelo
Environment Park S.p.A.
Clean NT Lab
Via Livorno 60
10144 Torino
Italy

Amandine David
Biophy Research
Actipôle Saint Charles
131 Av. de l'Etoile
13710 Fuveau
France

Séverine Descours
Biophy Research
Actipôle Saint Charles
131 Av. de l'Etoile
13710 Fuveau
France

Laurent Dupuy
Biophy Research
Actipôle Saint Charles
131 Av. de l'Etoile
13710 Fuveau
France

Nikolay Dyatko
State Research Center of the
Russian Federation
Troitsk Institute for Innovation
and Fusion Research
Pushkovykh st., domain 12
Troitsk 142190
Moscow region
Russia

Isabelle Ferreira
Institut Français du Textile et de
L'Habillement–IFTH
Direction Régionale Rhône-Alpes
PACA
Avenue Guy de Collongue
69134 Ecully Cedex
France

Michail Grushin
State Research Center of the
Russian Federation
Troitsk Institute for Innovation
and Fusion Research
Pushkovykh st., domain 12
Troitsk 142190
Moscow region
Russia

Pieter Heyse
Centexbel
Technologiepark 7
9052 Zwijnaarde
Belgium

Igor Kochetov
State Research Center of the
Russian Federation
Troitsk Institute for Innovation
and Fusion Research
Pushkovykh st., domain 12
Troitsk 142190
Moscow region
Russia

Virendra Kumar
Université Pierre et Marie Curie
Laboratoire de Génie des
Procédés Plasmas et Traitement
de Surface
ENSCP
11 rue Pierre et Marie Curie
75231 Paris cedex 05
France

Anatoly Napartovich
State Research Center of the
Russian Federation
Troitsk Institute for Innovation
and Fusion Research
Pushkovykh st., domain 12
Troitsk 142190
Moscow region
Russia

Massimo Perucca
Environment Park S.p.A.
Clean NT Lab
Via Livorno 60
10144 Torino
Italy

Giacomo Piacenza
Environment Park S.p.A.
Clean NT Lab
Via Livorno 60
10144 Torino
Italy

Jerome Pulpytel
Université Pierre et Marie Curie
Laboratoire de Génie des
Procédés Plasmas et Traitement
de Surface
ENSCP
11 rue Pierre et Marie Curie
75231 Paris cedex 05
France

Yves de Puydt
Biophy Research
Actipôle Saint Charles
131 Av. de l'Etoile
13710 Fuveau
France

Hubert Rauscher
European Commission
Joint Research Centre
Institute for Health and
Consumer Protection
Via E. Fermi 2749
21027 Ispra (VA)
Italy

François Rossi
European Commission
Joint Research Centre
Institute for Health and
Consumer Protection
Via E. Fermi 2749
21027 Ispra (VA)
Italy

Maddalena Rostagno
Diad s.r.l.
St., Della Praia 12/C
0090 Buttigliera Alta (TO)
Italy

Joachim Schneider
Institut für Plasmaforschung der
Universität Stuttgart
Pfaffenwaldring 31
70569 Stuttgart
Germany

Françoise Sommer
Biophy Research
Actipôle Saint Charles
131 Av. de l'Etoile
13710 Fuveau
France

Minh Duc Tran
Biophy Research
Actipôle Saint Charles
131 Av. de l'Etoile
13710 Fuveau
France

Nikolay Trushkin
State Research Center of the
Russian Federation
Troitsk Institute for Innovation
and Fusion Research
Pushkovykh st., domain 12
Troitsk 142190
Moscow region
Russia

Jocelyn Viard
Biophy Research
Actipôle Saint Charles
131 Av. de l'Etoile
13710 Fuveau
France

Matthias Walker
Institut für Plasmaforschung der
Universität Stuttgart
Pfaffenwaldring 31
70569 Stuttgart
Germany

List of Contacts

Europlasma (Belgium)
www.europlasma.be
Mr. Filip Legein
email: filip.legein@europlasma.
be
tel: +32-55-303205
fax: +32-55-318753

AcXys Technologies (France)
www.acxys.com
Mr. Thierry Sindzingre
email: thierry.sindzingre@acxys.
com
tel: +33-476-756079
fax: +33-476-759275

Muegge Electronics (Germany)
www.muegge.de
Mr. Horst Muegge
email: hmuegge@muegge.de
tel: +49-6164-930736
fax: +49-6164-930793

Biophy Research (France)
www.biophyresearch.com
Ms. Françoise Sommer
email:
fsommer@biophyresearch.
com
tel: +33-442-538326
fax: +33-442-538319

CSMA - CERAM Surface and Materials Analysis (United Kingdom)
www.csma.ltd.uk
Mr. Alan Paul
email: alanpaul@ceram.com
tel: +44-1782-764440
fax: +44-1782-412331

Jovertex (Spain)
www.jover.es
Mr. Miguel Jover Perez
email: mjover@jover.es
tel: +34-965-590507
fax: +34-965-500402

Luxilon (Belgium)
www.luxilon.be
Mr. Herbert De Breuck
email:
herbert.debreuck@luxilon.be
tel: +32-3-3263388
fax: +32-3-3263324

Biomatech (France)
www.biomatech.fr
Ms. Rosy Eloy
email: r.eloy@biomatech.fr
tel: +33-478-079234
fax: +33-472-240812

Plasma Technology for Hyperfunctional Surfaces. Food, Biomedical and Textile Applications.
Edited by Hubert Rauscher, Massimo Perucca, and Guy Buyle
Copyright © 2010 WILEY-VCH Verlag GmbH & Co. KGaA, Weinheim
ISBN: 978-3-527-32654-9

Covidien – Sofradim Production (France)
www.covidien.com
Mr. Olivier Lefranc
email: olivier.lefranc@covidien.com
tel: +33-474-089000
fax: +33-474-089230

Diad (Italy)
www.diadsrl.com
Ms. Maddalena Rostagno
email: maddalena.rostagno@diadgroup.com
tel: +39-347-3302727
fax: +39-011-9319173

Tops Foods (Belgium)
www.topsfoods.com
Mr. Rudy Tops
email: Rudy.Tops@topsfoods.com
tel: +32-14-285560
fax: +32-14-226150

Environment Park (Italy)
www.envipark.com
Mr. Massimo Perucca
email: massimo.perucca@envipark.com
tel: +39-011-2257523
fax: +39-011-2257221

Nodal (France)
www.nodal.fr
Mr. Benoit Rivollet
email: benoit.rivollet@nodal.fr
tel: +33-140-027555
fax: +33-140-027544

IFTH (France)
www.ifth.org
Ms. Isabelle Ferreira
email: iferreira@ifth.org
tel: +33-472-861655
fax: +33-478-433966

Centexbel (Belgium)
www.centexbel.eu
Mr. Guy Buyle
email: gbu@centexbel.be
tel: +32-9-2204151
fax: +32-9-2204955

European Commission – Joint Research Centre (Belgium)
ec.europa.eu/dgs/jrc
Mr. Hubert Rauscher
email: hubert.rauscher@jrc.ec.europa.eu
tel: +39 -0332-785128
fax: +39-0332-785787

University of Stuttgart – Institut für Plasmaforschung (Germany)
www.ipf.uni-stuttgart.de
Mr. Matthias Walker
email: walker@ipf.uni-stuttgart.de
tel: +49-711-6852156
fax: +49-711-6853102

University of Pierre et Marie Curie (France)
www.enscp.fr/labos/LGPPTS/
Ms. Farzaneh Arefi-Khonsari
email: farzi-arefi@enscp.fr
tel: +33-146-334283
fax: +33-143-265813

Clubtex (France)
www.clubtex.com
Ms. Edith Degans
email: contact@clubtex.com
tel: +33-320-994612
fax: +33-320-994613

Eucomed (Belgium)
www.eucomed.org
Mr. John Brennan
email: John.Brennan@eucomed.
be
tel: +32-2-7759232
fax: +32-2-7713909

**IVLV – Industrievereiniging für
Lebensmitteltechnologie and
Verpackung (Germany)**
www.ivlv.de
Mr. Rainer Brandsch
email: rainer.brandsch@ivlv.org
tel: +49-89-1490090
fax: +49-89-14900980

**TRINITI – State Research Center
of the Russian Federation, Troitsk
Institute for Innovation and Fusion
Research (Russian Federation)**
www.triniti.ru
Mr. Yuri Akishev
email: akishev@triniti.ru
tel: +7-095-3345236
fax: +7-095-3345776

Mat PlasMATec (Germany)
www.mat-dresden.de
Mr. Andreas Mucha
email: mucha@mat-dresden.de
tel: +49-351-207720
fax: +49-351-2077222

**CPI – Coating Plasma Industrie
(France)**
www.cpi-plasma.com
Mr. Tran Minh Duc
email: tranminh@cpi-plasma.
com
tel: +33-442-538311
fax: +33-442-538329

Part I
Introduction to Plasma Technology for Surface Functionalization

Plasma Technology for Hyperfunctional Surfaces. Food, Biomedical and Textile Applications.
Edited by Hubert Rauscher, Massimo Perucca, and Guy Buyle
Copyright © 2010 WILEY-VCH Verlag GmbH & Co. KGaA, Weinheim
ISBN: 978-3-527-32654-9

1
Introduction to Plasma and Plasma Technology

Massimo Perucca

1.1
Plasma: the Fourth State of Matter

The term *plasma* was first used by Lewi Tonks and Irving Langmuir in 1929 [1] to describe a collection of charged particles in their studies of oscillations in the inner region of an electrical discharge. Later, the definition was broadened to define a state of matter ('the fourth state of matter') in which a significant number of atom and/or molecules are electrically charged or ionized with the fundamental characteristic of exhibiting a collective behavior due to the long-range Coulomb interactions.

A rough but comprehensive definition of plasma is that of an ensemble of charged, excited, and neutral species, including some or all of the following: electrons, positive and negative ions, atoms, molecules, radicals, and photons. On average a plasma is electrically neutral, because any charge unbalance would result in electric fields that would tend to move the charges in such way as to neutralize charges of opposite sign.

It is estimated that more than 99.9% of the apparent universe is in the plasma state: gaseous nebulae, interstellar gas [2], stars, including our sun [3–5], have extremely high surface temperatures (from 2000 to 22 000 K) and they consist entirely of plasma. Our planet is not an exception to the presence of the plasma state. The earth's atmosphere (in the altitude range between 90 and 500 km, the thermosphere), is continuously bombarded by intense cosmic rays and solar wind radiation and as a consequence its components become electrically charged species leading to the formation of an atmospheric shell, called the '*ionosphere*' [6–9]. The solar UV radiation is almost completely absorbed by the ionosphere, producing the electrically charged particles (especially electrons) which are deflected and funneled by the magnetic field of the earth to the poles, which results in the northern hemisphere in the formation of spectacular 'northern lights' known as *Aurora Borealis* [10]. Lightning developed during thunderstorms is a natural plasma state as well. It is estimated that about 100 cloud–ground and cloud–cloud lightning strikes happen every second over the whole earth. Lightning is an intense transient electric discharge of extremely long path-length (often many kilometers).

Plasma Technology for Hyperfunctional Surfaces. Food, Biomedical and Textile Applications.
Edited by Hubert Rauscher, Massimo Perucca, and Guy Buyle
Copyright © 2010 WILEY-VCH Verlag GmbH & Co. KGaA, Weinheim
ISBN: 978-3-527-32654-9

Even though the mechanism of electric field development in clouds is incompletely understood, it is suggested that this phenomenon is associated with the freezing of water. It was found that in the absence of ice the field build-up is slow and lightning is rare. Colliding ice particles are believed to generate electric-field growth and their separation under gravity probably contributes to the development of charge transfer. The plasma state can be produced in laboratories and by raising the energy content of matter regardless of the nature of the energy source. Thus plasmas can be generated by mechanical (close to adiabatic compression), thermal (electrically heated furnaces), chemical (exothermic reactions), radiant (high energy electromagnetic and particle radiations, e.g., electron beams) and electromagnetic (arcs, coronas, direct current (DC), and radio frequency (RF), microwave (MW), electron cyclotron resonance (ECR) discharges) energies, and by the combination of them, as in an explosion, in which mechanical and thermal energies are present.

1.2
Historical Highlights

In the mid-nineteenth century the Czech medical scientist, Johannes Purkinje (1787–1869) used the Greek word plasma (which means 'moldable substance' or 'jelly') to denote the clear fluid which remains after removal of all the corpuscular material in blood.

Half a century later, in 1927, the Nobel prize winning American chemist Irving Langmuir first used this term to describe an ionized gas. Langmuir was reminded of the way blood plasma carries red and white corpuscles by the way an electrified fluid carries electrons and ions.

Langmuir, along with his colleague Lewi Tonks, was investigating the physics and chemistry of tungsten-filament lightbulbs, with a view to finding a way to greatly extend the lifetime of the filament. In the process, he developed the theory of plasma sheaths, the boundary layers which form between ionized plasmas and solid surfaces. He also discovered that certain regions of a plasma discharge tube exhibit periodic variations of the electron density, which we nowadays term *Langmuir waves*. In the 1920s and 1930s a few isolated researchers began the study of what is now called *plasma physics*. In the 1940s Hannes Alfvén developed a theory of hydromagnetic waves (now called *Alfvén waves*) and proposed that these waves would be important in astrophysical plasmas.

The creation of the hydrogen bomb generated a great deal of interest in controlled thermonuclear fusion as a possible power source for the future, since 1952. At first, this research was carried out secretly, and independently, by the United States, the Soviet Union, and Great Britain. But the 'International conference on the peaceful uses of atomic energy', held in Geneva in 1955 [11, 12], ratified the beginning of the studies on peaceful use of nuclear fusion. The constitution of the International Atomic Energy Agency (IAEA) was almost contemporaneous (1957). Nowadays the agency works with its member states and multiple partners worldwide to promote safe, secure, and peaceful nuclear technologies.

Fusion progress was slow through most of the 1960s, but by the end of that decade the empirically developed Russian tokamak configuration began producing plasmas with parameters far better than the lackluster results of the previous two decades. By the 1970s and 1980s many tokamaks with progressively improved performance were constructed and at the end of the twentieth century fusion break-even had nearly been achieved in tokamaks [13].

In the 1970s a new application of plasma physics appeared, and has developed as critical technique for the fabrication of the tiny, complex integrated circuits used in modern microelectronic devices. This application is now of great economic importance for industry, indeed plasma processing has been an enabling technology for the broader applications of microelectronics which has led to the actual information and communication society with radical changes in worldwide living habits.

Plasma processing is a technology used in a large number of industries, and whilst semiconductor device fabrication for computers is perhaps the best known, it is equally important in other sectors such as automotive, textile, food packaging, biomedical, polymers, and solar energy. A common theme in the applications is plasma treatment of surfaces. Plasma is an environmentally friendly process technology, producing an extremely low level of industrial by-products, especially when compared to more traditional wet chemical treatments.

The principal applications of plasma treatments concern the processes that induce a limited and selected transformation of the outermost surface layer (nanometric scale).

Many fundamental processes take place during surface treatment of a material: the surface undergoes bombardment by fast electrons, ions, and free radicals, combined with the continued electromagnetic radiation emission in the UV-Vis spectrum enhancing chemical-physical reactions in order to obtain the desired functional and aspect properties.

Many properties and functionalities can be obtained by plasma treatment, and they depend on the application: the plasma can be used to change the surface wettability which can be changed from hydrophilic to hydrophobic and vice versa, to enhance the barrier characteristics, adhesion, dye-ability, printability, or the oleophobicity.

For example, plasma treatment of textile fabrics and yarns (see Chapter 6) is investigated as an alternative to wet chemical fabric treatment and pre-treatment processes, for example, shrink resistance, water repellent finishing, or improvement of dye-ability, which tend to alter the mechanical properties of the fabric and are environmentally hazardous [14, 15].

In the food sector the application of plasma processing is interesting for many purposes, in particular to provide barrier effects on homopolymeric packaging for their recyclability (see Chapter 8) or on biodegradable films (polylactic or starch based compounds) to attain an extended shelf life by product preservation [16]. Food processing also requires surfaces with enhanced functionalities in terms of wear resistance, chemical inertness, controlled surface energy for adhesion control, and barriers against migration of heavy metals; whose properties are efficiently achieved by deposition of coatings via plasma processing (see Chapter 9) or by surface plasma treatment.

In the biomedical sector plasma technology is used for cold sterilization [17] of instruments and prostheses as well as many thermolabile materials used in the biomedical technology sector for its particular advantages, including its moderate or negligible impact on substrate materials and use on nontoxic compounds). Low pressure plasma (LPP) processing is also being investigated for the production of non-fouling surfaces to prevent the formation of biofilms with promising results (as detailed in Chapter 7)

1.3
Plasma Fundamentals

Plasma is sometimes considered as the fourth state of matter, an expression that was first coined by Crooks in 1879 to describe the ionized medium created in a gas discharge. Though this attribute can be somehow misleading, it highlights the unique feature of the plasma phase. The concept of the fourth state of matter results from the idea that phase transitions occur by progressively providing energy to the matter, such as the one from the solid state to liquid up to the gas state.

A further 'phase transition' may be thought of as the one from the gas state to the plasma state, even if this state is reached gradually by providing more and more energy to the system. Therefore this process cannot be rigorously defined as a 'phase transition', because it lacks the signature of all real phase transitions: that is, an abrupt change (discontinuity) in the order parameter defining the thermodynamic phase.

Plasma can be seen as a particular ionized gas, which retains some unique features which distinguish it from an (ideal) gas. One of the main differences is that plasma particles do interact, because of the electromagnetic coupling between charged particles and electric and magnetic collective perturbations which constitute the plasma itself.

Another important difference from an ideal gas is found in nonthermal plasmas (NTPs) in which quite different temperatures are associated with different species, due to the out-of-equilibrium thermodynamic regime. Indeed, in cold plasmas, neutrals and ions may be at ambient temperature whereas electrons may reach temperatures close to 10^5 K. These plasma systems are of particular interest for molecular processing and surface treatment, because cold plasmas often produce 'activated' states of matter, capable of enhancing plasma–surface chemical reactions and physical processing which cannot be achieved under thermodynamic equilibrium conditions. Therefore molecular and surface micro/nano-structures can be fabricated by plasma processing in a way which is different from any other available method.

In the following sections we do not claim to provide a rigorous derivation of plasma fundamentals, which is available in several reference textbooks [18–21]; we rather aim at providing a comprehensive overview of the basic concepts and to offer some helpful tools to better understand the plasma state and the different plasma regimes determining the processing conditions for surface functionalization.

1.3.1
Free Ideal Gas

An ideal gas is an ensemble of non-interacting rigid and negligible-size particles, each one undergoing frequent elastic scattering with other particles and with the physical boundaries of the system. Real gases (containing many particles) may be described in good approximation as ideal gases if they have a low density. Due to frequent (elastic) collisions the ideal gases obey the Maxwell–Boltzmann velocity distribution statistics, that is, the particles' velocities (or energies) are distributed according to the statistical curve depicted in Figure 1.1, which is named after the two physicists Maxwell and Boltzmann.

For ideal gases the kinetic theory of gases provides a very good theoretical description of microscopic and macroscopic physical state variables fully describing the gas behavior. The kinetic theory of gases must be modified if there are strong interactions or, more in general, where the conditions for an ideal gas no longer hold. For completeness sake we must say that strong interactions are indeed not the only cause for a non-ideal gas regime. Quantum correlation responsible for gas degeneracy can be another sufficient reason to give up the kinetic theory of gases. Anyway it is useful to recall the basis of the kinetic theory of gases in order to have a starting point from which we can move toward a satisfactory description of plasma physics.

If we think of the gas as an ensemble of N particles in a cubic box with volume V and edge length l, a single particle moving along a box direction, say x, will undergo scattering by another particle or the box (perpendicular) surface. If the particle collides onto the surface, its momentum change will be: $\Delta p_x = 2mv_x$, where m

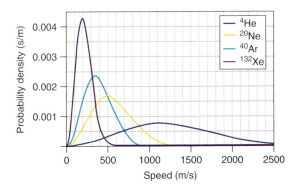

Figure 1.1 Maxwell–Boltzman velocity distribution of some noble gases for a given system temperature (298.15 K). The x axis represents the velocity and the y axis reports the relative particles probability density, that is, the fraction of particles per infinitesimal speed interval.

is the particle mass and v_x its velocity. The force exerted by the particle will be then: $F_x = \Delta p_x / \Delta t$. By considering the motion of the particle along the whole box edge, we can consider the time during which momentum is changed as $\Delta t = l/v_x$. The force exerted by the *i*-th particle then results in: $F_i = 2mv_x^2/l$. By summing up over all the particles (assumed to have the same mass) hitting the surface we may estimate the overall force exerted on the surface, and remembering that pressure is $p = F/A$, we get (Equation 1.1):

$$F_{tot} = \sum_{i=1}^{N} F_i = \sum_{i=1}^{N} \frac{2mv_i^2}{l}; \quad p_{tot} = \sum_{i=1}^{N} p_i = \sum_{i=1}^{N} \frac{2mv_i^2}{l^3} = \sum_{i=1}^{N} \frac{2mv_i^2}{V} \quad (1.1)$$

In order to evaluate the overall pressure on each face we can consider the velocity components along the three axes and assume that the particles' velocities are equally distributed on average along each direction (Pascal principle). The total pressure on the six faces of the box is then defined as:

$$p = N\frac{2m\langle v^2 \rangle}{6V} = N\frac{m\langle v^2 \rangle}{3V}, \quad \text{with } \langle v^2 \rangle = \frac{1}{N}\sum_{i=1}^{N} v_i^2 \quad (1.2)$$

or the average square velocity along one axis. By comparison of Equation (1.2), known as the *Joule–Clausius formula*, derived by microscopic considerations with the ideal gas law $pV = Nk_BT$, ($k_B \approx 1,381 \cdot 10^{-23}$ J/K) valid for macroscopic system quantities, it is possible to connect macroscopic physical variables with microscopic ones. In particular the temperature reads $T = m\langle v^2 \rangle/(3 k_B)$, and the total kinetic energy of the system is $\varepsilon_{tot} = 1/2\ Nm\langle v^2 \rangle = 3/2Nk_BT$, whence we obtain the (average) single particle kinetic energy $\varepsilon = 3/2\ k_BT$ (*m* being the particle mass, k_B the Boltzman constant, and *N* the number of particles).

1.3.2
Interacting Gas

When considering a plasma – even if it is only partially ionized – it is necessary to take into account long-range particle interactions in the ensemble. Indeed, electrically charged particles feel electric Coulomb forces whose range is long compared with the characteristic scale length of the plasma. When charged species are in motion they generate electric currents inducing internal magnetic and electric fields which, in turn, interact with external fields. This dramatically complicates the modeling of such a system, which is self-consistent through a strong coupling between its dynamics and the electromagnetic configuration. Although intrinsically complex, this coupling allows one to control to some extent the dynamics of an ensemble of charged particles by biasing its collective behavior with external electromagnetic fields. A typical example of such plasma controlling by electromagnetic fields is its spatial confinement, which turns out to be very useful for technological applications and particularly for fusion.

Whereas an ideal gas (see Section 1.3.1) expands indefinitely in free space, a plasma can be temporarily confined in a small spatial region by applying suitable electromagnetic fields. In order to describe the plasma while taking into account the particle interactions a new theoretical framework is needed.

Electromagnetic interactions are described by the Maxwell equations. Relevant fields entering the equations are **E** and **B**, the electric and magnetic field in vacuum, respectively.

Here, vacuum does not mean that the equations hold only for the low pressure regime, but that the relevant fields are outside the medium, that is, the particle ensemble. We do not need to describe fields inside the matter, because we are only interested in gas dynamics. In order to properly describe the dynamics of a population of many particles we need a statistical approach. The statistical information can be encoded in the distribution function $f(v, r, t)$ describing the population time evolution in phase space. It depends on velocity (v, or momentum $p = mv$), position (r), and time (t), and gives the statistical probability of the particles' velocity (momentum) and position at each time step. In other words, the distribution function contains the information on how many particles have momentum p and position r. For short range particle interactions the Boltzmann equation provides a good model for plasma dynamics, describing the time evolution of the distribution function of a particle population subject to external forces:

$$\frac{\partial f_\alpha}{\partial t} + v \cdot \frac{\partial f_\alpha}{\partial r} + \frac{\partial f_\alpha}{\partial v} \cdot \frac{F}{m} = \frac{\partial f_\alpha}{\partial t}\bigg|_{coll} \tag{1.3}$$

In Equation 1.3 quantities in bold character are regarded as vectors (or space differential operators) and the subscript α refers to the different species to which the equation can apply, for example, electrons or ions. The first term is the free evolution term, the second one represents the diffusion term from regions of higher density to lower density regions, the third one represents the drift term (from higher to lower density regions in reciprocal space). The right hand side (RHS) of the equation is the term taking into account the collisions between particles. In a plasma the relevant force **F** is represented by the Lorentz force (Equation 1.4):

$$F = q_\alpha \cdot (E + v \times B) \tag{1.4}$$

with q_α the electrical charge of particle type α. Different from the kinetic theory of ideal gases, the collisional term includes both elastic and inelastic scattering. Another important difference from the kinetic theory is that the Boltzmann equation holds also in the nonequilibrium regime. The Boltzmann equation is easily derived from the Liouville theorem ($Df/Dt = 0$) as a consequence of matter and energy conservation, once the distribution function f is well defined.

Despite its intrinsic complexity due to the collective behavior of particles in plasma systems, some approximations of this collisional term have been proposed, leading to different forms of the Boltzmann transport equation which are valid in specific plasma regimes.

For example, for fully ionized plasmas (e.g., in thermal plasmas), the suitable collision term leads to the Fokker–Planck equation based on the fundamental assumption that long range Coulomb interactions produce large angle deflections of particle trajectories due to the rapid succession of multiple collisions with distant particles. In the Fokker–Planck model the collision operator reads:

$$\frac{\partial f}{\partial t}\bigg|_{coll} = \sum_i \frac{\partial}{\partial v_i}\left\langle \frac{\Delta v_i}{\Delta t}\right\rangle f(\boldsymbol{v}, t) + \frac{1}{2}\sum_{i,j}\frac{\partial^2}{\partial v_i \partial v_j}\left[\left\langle \frac{\Delta v_i \Delta v_j}{\Delta t}\right\rangle f(\boldsymbol{v}, t)\right] \qquad (1.5)$$

In Equation 1.5 the $< >$ symbols denote the average change in velocity moments of the distribution function per time unit and the indexes i and j refer to the space coordinates. This form of the collisional term takes into account nonequilibrium thermodynamic behavior at first expansion order. The first term on the RHS takes into account acceleration or slowing down of a group of particles (for instance due to effects similar to friction). The second term on the RHS accounts for diffusion effects due to spreading of the distribution function.

For partially or weakly ionized plasmas, which are of greater interest for industrial plasma processing, short range interaction of ionized plasma particles with the neutral background dominates the other mechanisms such as diffusion and conductivity and a suitable formal representation of the two-body collisional operator is provided by Equation 1.6:

$$\frac{\partial f_\alpha}{\partial t}\bigg|_{coll} = \sum_\gamma n_\gamma \int d\mathbf{V}\int d\Omega \frac{d\sigma}{d\Omega}|\boldsymbol{v}-\boldsymbol{v}_1|\left[f_\alpha(\mathbf{r}, \boldsymbol{v}', t)f_\gamma(\mathbf{r}, \boldsymbol{v}'_1, t) - f_\alpha(\mathbf{r}, \boldsymbol{v}, t)f_r(\mathbf{r}, \boldsymbol{v}_1, t)\right]$$

$$(1.6)$$

where primed velocities refer to velocities of scattered particles and $\boldsymbol{v}, \boldsymbol{v}_1$ represent velocities of two particles approaching each other before scattering. $d\Omega$ is the solid angle element in velocity space, α and γ refer to the incident and target particle species and σ is the scattering cross-section, which quantifies the particles collision probability. However, the Boltzmann equation with the above collision operator is practically intractable for its intrinsic complexity and nonlinearity. For weakly ionized plasmas simplifying assumptions, which yield more tractable equations are possible, provided that we consider the time evolution of the distribution function as mainly determined by:

- average fields of charged particles included in a self consistent form;
- applied external electromagnetic fields, included in the Lorentz force (see Equation 1.4);
- collisions dominated by charged species with the neutral background.

The last condition assumes, contrary to the Fokker–Planck theory, that the collision duration is much shorter than the time lap in-between two collisions. Along with the previous assumptions the Boltzmann transport equation may be further simplified by using a mean-free-transit time (τ). This τ is the average time between collisions and is independent on the particle velocity. Hence, the collision integral reduces to Equation (1.7):

$$\frac{\partial f}{\partial t}\bigg|_{coll} = \frac{f_o - f(\mathbf{r}, \boldsymbol{v}, t)}{\tau(|\vec{v}|)} \approx \frac{f_o - f(\mathbf{r}, \boldsymbol{v}, t)}{\tau} \qquad (1.7)$$

The modified Boltzmann equation (sometimes referred to as the *Krook model*), becomes

$$\frac{\partial f_\alpha}{\partial t} + v \frac{\partial f_\alpha}{\partial r} + \frac{q_\alpha}{m}(E + v \times B)\frac{\partial f_\alpha}{\partial v} = -\frac{f_\alpha(r, v, t)}{\tau_\alpha} \tag{1.8}$$

Equation 1.8 accounts for the distribution function f_α evolution of the α-species (e.g., electrons, positive and negative ions) toward relaxation to the equilibrium distribution on local scale, which should be chosen as a local Maxwellian distribution function. Conservation of particles is assumed and the collision operator is considered as source or sink of particles. A final simplification named after Lorentz considers only the electrons as diffusing particles having collisions in a background of heavy particles described by an equilibrium statistical distribution.

1.3.3
The Plasma as a Fluid

Starting from the statistical treatment it is possible to derive macroscopic plasma quantities by averaging the distribution function and providing the (mass, momentum, and energy) continuity equations. These equations coupled with the Maxwell equations for electromagnetic fields provide a self consistent plasma model. The system of Equations 1.9 represents the plasma in the two-fluid theory. Electrons and ions are considered as two conducting fluids and are coupled through momentum transfer collisions (by a suitable definition of the collisional operator) and through Maxwell equations. In this representation we assume isotropy (the pressure tensor reduces to a scalar) and adiabatic conditions, providing energy conservation (Equation 1.9c).

$$\frac{\partial n_\alpha}{\partial t} + \nabla \cdot (n_\alpha u_\alpha) = 0 \tag{1.9a}$$

$$m_\alpha n_\alpha \frac{\partial u_\alpha}{\partial t} + m_\alpha n_\alpha u_\alpha \cdot \nabla u_\alpha = q_\alpha m_\alpha (E + u_\alpha \times B)$$
$$- \nabla p_\alpha - \sum_\beta m_\alpha n_\alpha v_{\alpha\beta}(u_\alpha - u_\alpha) \tag{1.9b}$$

$$\frac{\partial [p_\alpha \cdot (m_\alpha n_\alpha)^\gamma]}{\partial t} + u_\alpha \nabla \cdot [p_\alpha \cdot (m_\alpha n_\alpha)^\gamma] = 0 \tag{1.9c}$$

$$\nabla \cdot E = \frac{1}{\varepsilon_0} \cdot \sum_\alpha q_\alpha n_\alpha \tag{1.9d}$$

$$\nabla \cdot B = 0 \tag{1.9e}$$

$$\nabla \times B = \frac{1}{\varepsilon_0}\frac{\partial E}{\partial t} + \mu_0 \sum_\alpha q_\alpha n_\alpha u_\alpha \tag{1.9f}$$

$$\nabla \times E = -\frac{\partial B}{\partial t} \tag{1.9g}$$

The subscript α (or β) refers to the particle (conducting fluid) species.

A simplified and modified set of equations is provided by the three-fluid model (Equation 1.10) introduced in Chapter 10 .The model is simplified in the sense that only the continuity equations of the three different species (electrons and both negative and positive ions) are included and only the electric field is taken into account, through the Poisson equation, which understands plasma quasi-neutrality conditions (negative and positive charges are balanced).

$$\partial n_e/\partial t + div n_e w_e = (v_i - v_a)n_e + v_d n_n, \tag{1.10a}$$

$$\partial n_p/\partial t + div n_p w_p = v_i n_e, \tag{1.10b}$$

$$\partial n_n/\partial t + div n_n w_n = v_a n_e - v_d n_n, \tag{1.10c}$$

$$div E = e(n_p - n_e - n_n)/\varepsilon_0, \tag{1.10d}$$

In Equation 1.10 the indexes e, p, and n refer to electrons, positive, and negative ions, respectively, n_p, n_e, and n_n are the positive ion, electron, and negative ion number densities derived by the respective distribution functions integrated in the space domain, w_p, w_e, and w_n their drift velocities, v_i, v_a, and v_d are the ionization, attachment, and detachment frequencies, e is the electronic charge and ε_0 is the permittivity of free space. The electron drift velocity and kinetic coefficients are to be determined from solving the electron Boltzmann equation Equation 1.8 for $\alpha = e$, the ion drift velocities are calculated using the known ion mobilities (see Equation 1.29 below).

1.3.4
Waves in Plasmas

Waves occurring in plasmas are described starting from the study of particle dynamics either by a statistical approach or through the fluid models. Complete description of waves in plasmas has been made extensively in dedicated works [22]. Waves in cold plasmas of interest for technological applications are described by considering the plasma as a dispersive dielectric medium (accounting for plasma inhomogeneities and anisotropies) for which the appropriate dielectric tensor may be defined.

Restricting our description of plasma waves relevant to applications described in this book (see Chapter 2),we consider the case of magnetized plasmas, in which it is possible to describe the cold plasma as collisionless by taking into account only electrons (ions and neutrals species are considered as an almost motionless particle background). For the sake of simplicity the external magnetic field is directed along the z axis and regarded as uniform and stationary (B_0).

Assuming harmonic time dependence of time varying quantities (n_α, p_α, u_α, E, B) in the two-fluid model (Equation 1.9), and combining the last two Maxwell equations (Equation 1.9f,g) by taking the curl of Equation 1.9g it is possible to derive the wave equation (Equation 1.11):

$$\nabla \times \nabla \times E = \frac{\omega^2}{c^2}\varepsilon \cdot E \tag{1.11}$$

where ε is represents the dielectric tensor. By taking the 3D Fourier transform of the same quantities with respect to space dependence, the vectorial Equation 1.12 leading to the dispersion relation is obtained:

$$-k \times k \times \hat{E} = \frac{\omega^2}{c^2}\varepsilon \cdot \hat{E} \tag{1.12}$$

which can be written more explicitly as (Equation 1.13):

$$\begin{pmatrix} k_z^2 & 0 & -k_x k_z \\ 0 & k_x^2 + k_z^2 & 0 \\ -k_x k_z & 0 & k_x^2 \end{pmatrix} \begin{pmatrix} \hat{E}_x \\ \hat{E}_y \\ \hat{E}_z \end{pmatrix} = \frac{\omega^2}{c^2} \cdot \begin{pmatrix} \varepsilon_\perp & -j \cdot \varepsilon_\times & 0 \\ -j \cdot \varepsilon_\times & \varepsilon_\perp & 0 \\ 0 & 0 & \varepsilon_= \end{pmatrix} \begin{pmatrix} \hat{E}_x \\ \hat{E}_y \\ \hat{E}_z \end{pmatrix} \quad (1.13)$$

The first term in parentheses of the RHS of Equation 1.13 is the explicit form of the plasma dielectric tensor.

For a non-collisional homogeneous plasma immersed in a steady uniform magnetic field the resulting tensor components read:

$$\varepsilon_\perp = 1 - \frac{\omega_{pe}^2}{\omega^2 - \omega_{ce}^2}$$

$$\varepsilon_\times = \frac{\omega_{ce}}{\omega} \frac{\omega_{pe}^2}{\omega^2 - \omega_{ce}^2} \quad (1.14)$$

$$\varepsilon_= = 1 - \frac{\omega_{pe}^2}{\omega^2}$$

with $\omega_{ce} = eB_0/m_e$ is defined as the electron cyclotron frequency, which corresponds to the gyration frequency of electrons along the magnetic field lines, $\omega_{pe} = (e^2 n/\varepsilon_0 m)^{1/2}$ is the electron plasma frequency, corresponding to undriven small amplitude electron oscillations. In the case of collisional plasma the plasma dielectric tensor components are given by (Equation 1.15):

$$\varepsilon_\perp = 1 - \frac{\omega - jv_m}{\omega} \frac{\omega_{pe}^2}{(\omega - jv_m)^2 - \omega_{ce}^2}$$

$$\varepsilon_\times = \frac{\omega_{ce}}{\omega} \frac{\omega_{pe}^2}{(\omega - jv_m)^2 - \omega_{ce}^2} \quad (1.15)$$

$$\varepsilon_= = 1 - \frac{\omega_{pe}^2}{\omega(\omega - jv_m)}$$

where v_m represents the electron–neutral collision frequency. The last terms in Equation 1.14 and in Equation 1.15, corresponding to a direction parallel to the magnetic field (and transverse to the electric field) also represent the plasma dielectric constants in a nonmagnetized cold collisionless and collisional plasma, respectively. This simple case of transverse modes ($k \cdot E = 0$) leads to the following dispersion relation:

$$\omega^2 = \omega_{p,e}^2 + c^2 k^2 \quad (1.16)$$

Equation 1.16 for the dielectric permittivity exhibits a phenomenon called *cut-off*. Cut-off occurs when ε goes to zero. That is, when (Equation 1.17)

$$\omega^2 = \omega_{p,e}^2 = \frac{e^2 n_e}{\varepsilon_0 m_e} \quad (1.17)$$

which corresponds to the critical electron density $n_c = m_e \varepsilon_0 \omega^2 / e^2$. For MWs with a frequency of $f = 2.45\,\text{GHz}$, (see Chapters 2 and 8) wave reflection occurs at the critical electron density $n_c = 7.5 \times 10^{16}\,\text{m}^{-3}$. This is quite crucial for MW-generated plasmas in which a MW source may produce a plasma with a maximum electron density equal to n_c.

1.3.5
Relevant Parameters that Characterize the State of Plasma

From the above theoretical framework it is possible to nail down some useful physical quantities and parameters which help in characterizing the plasma state and different plasma regimes.

In particular, for systems in thermodynamic equilibrium such as high temperature plasmas (thermal plasmas) a plasma temperature may be defined starting from the particle velocity (Maxwell) distribution. However also nonequilibrium plasmas can be described by parameters provided that there is local equilibrium, or that at least the equilibrium within each particle species is maintained. This means that it is possible to define different equilibrium regimes for each plasma species even if the whole system is not in equilibrium. In this case different temperatures may be defined for different particle populations, for cold plasmas typically electrons, ions, and neutrals. Thermal equilibrium within each species is due to the inefficient collisions among different particle populations. In this regime mutual collisions among species (e.g., electron–ion or electron–neutral) do not lead to significant energy losses (or gains) for any of the species involved in the interactions but collisions among particles of the same species are efficient in terms of energy exchange, providing fast thermalization and attainment of the equilibrium regime. A typical case is that in which electron, ion, and neutral temperatures may be defined separately. For each plasma species assumed to be in thermodynamic equilibrium a Maxwell distribution is associated in velocity space (obtained by integrating the particle distribution function over Euclidian space) and the selected plasma species temperature is related to the average kinetic energy of the particles derived from their velocity distribution function second order momentum; therefore ion (i) and electron (e) temperatures are defined as: $T_i = m\langle v_i^2 \rangle/(3\, k_B)$ and $T_e = m\langle v_e^2 \rangle/(3\, k_B)$.

Other characteristic physical parameters in space and time (frequency) domains are useful to characterize plasmas and for their classification.

The *Debye length* (λ_D) represents the distance over which the electric field of each charge carrier (usually electrons) is screened, representing the interaction range of single charged particles. It is given by Equation 1.18:

$$\lambda_D = \sqrt{\frac{\varepsilon_0\, k_B T_e}{n_e e^2}} \tag{1.18}$$

where n_e, T_e, and e are the electron density, temperature, and charge ($e = 1{,}602 \cdot 10^{-19}$ C), respectively, k_B the Boltzmann constant, and ε_0 the vacuum electric permeability ($\varepsilon_0 = 8{,}859 \cdot 10^{-12} F/m$).

If non-negligible size particles are considered, another useful parameter of gas dynamics is the so-called *mean free path*, representing the average distance traveled by particles in rectilinear motion between subsequent collisions within the gas box (see the free ideal gas model of Section. 1.3.1). Where d is the classical particle diameter, T and p the gas temperature and pressure, the mean free path is evaluated by (Equation 1.19):

$$\lambda = \frac{k_B T}{\sqrt{2\pi}\, d^2 p} \tag{1.19}$$

The *plasma parameter* (g) is a nondimensional parameter defined through the Debye length and the plasma (equilibrium) density: it is the measure of the number of particles present in a Debye sphere (Equation 1.20):

$$g = \frac{1}{n_0 (\lambda_D)^3} \tag{1.20}$$

The plasma parameter has to be small for a many body system to be treated in the so-called '*plasma approximation*'. This means that the average potential energy of a particle is much less than its kinetic energy. Departure from this limit implies that the particles interaction energy to becomes more relevant and that the plasma may not be treated as an ideal gas (this holds typically in the case of highly charged particles density per Debye sphere).

The *thermal De Broglie wavelength* Λ is the average particle de Broglie length defined as Equation 1.21:

$$\Lambda = \sqrt{\frac{h^2}{2\pi\, m k_B T}} \tag{1.21}$$

where h is the Plank constant ($h \approx 6.626 \times 10^{-34}$ Js). To consider the plasma as a classical system the following relations must be satisfied: $\Lambda \ll \sqrt[3]{n_0}$; $\Lambda \ll e^2/(k_B T)$, where the last term on the RHS in the two relations are the mean spacing among plasma particles and the distance of closest approach in Coulomb interactions, respectively. If the first inequality is violated, the binary collision among near neighbor particles can no longer be treated classically. If the second approximation relation does not hold, the system cannot be described by the Maxwell–Boltzmann statistics, since degeneracy occurs and a quantum statistics has to be adopted, either Fermi–Dirac (fermion gas) or Bose–Einstein (boson gas).

Concerning the frequency domain, the plasma frequency may be defined as:

$$\omega_{p,\alpha} = \sqrt{\frac{n_\alpha q_\alpha^2}{\varepsilon_0\, m_\alpha}} \tag{1.22}$$

When Equation (1.22) concerns the electron species ($\alpha = e$, where here q_e represents the electron charge) the *electron plasma frequency* is the rate of electron free oscillations (seen as a non-collisional plasma slab) over a background of almost steady still positive ions, when the system is perturbed by a small amplitude electric impulse. The plasma frequency may also be seen as a measure of the electron density and it is the parameter that influences the transmission (or damping) of specified frequency external electromagnetic waves. A similar quantity may be defined for ions when $\alpha = i$ and $q_i = Ze$ is the total ion charge, providing the *ion plasma frequency*.

In magnetized low density plasmas charged particle motion is characterized by spinning around the magnetic field lines, and the ion and electron *gyrofrequency* is defined as Equation (1.23):

$$\omega_{c,\alpha} = \frac{q_\alpha B}{m_\alpha} \tag{1.23}$$

where B denotes the intensity of the magnetic field. For $\alpha = e$ and $\alpha = i$ the electron and ion gyrofrequencies are defined, respectively. Due to the electron-to-mass ratio and to typical reactor dimensions, in plasma processing the electron gyrofrequency is of great interest, for example, in ECR plasmas (see Chapters 2 and 8). The corresponding gyration radius or gyroradius can be written as Equation (1.24):

$$r_c = \frac{v_{\perp,\alpha}}{|\omega_{c,\alpha}|} \tag{1.24}$$

where $v_{\perp,\alpha}$ represents the particle(s) of type-α speed component perpendicular to the magnetic field.

Regarding characteristic velocity parameters, it is worth mentioning the drift velocity of the guiding center of particles gyrating around magnetic field lines. The $E \times B$ *drift velocity* is due to the presence of perpendicular components of electric and magnetic fields in the plasma (Equation 1.25):

$$v_{ExB} = \frac{E \times B}{B^2} \tag{1.25}$$

The $E \times B$ drift motion is of particular importance, for example, in the confinement of secondary electrons in magnetron sputtering sources as well as for the control of the arc spot on arc- physical vapor deposition (PVD) targets (see Chapter 2).

Considering electrostatic waves in plasmas the *adiabatic electron sound speed* accounts for the propagation of waves in plasma parallel to the electric field (which is possible in plasma but not in vacuum or in a conventional dielectric material) due to an exchange between thermal and electric energy (γ being the ratio of specific heats in Equation 1.26):

$$c_{e,\gamma} = \sqrt{\frac{\gamma k_B T_e}{m_e}} \tag{1.26}$$

In magnetized plasmas the Alfvén velocity represents the phase speed of transverse waves produced by oscillation of ions through a perpendicular magnetic field whose exerted force acts as a restoring force (Equation 1.27):

$$v_A = \frac{B}{\sqrt{\mu_0 n_0 m_i}} = \frac{\omega_{c,i}}{\omega_{p,i}} \cdot c \tag{1.27}$$

where n_i and m_i are the ion number density and mass, B is the magnetic field strength, μ_0 the vacuum magnetic permeability, and c the speed of light.

Regarding the *transport parameters*, when assuming only electron-neutral collisions and for low electric field frequencies (e.g., low frequency electric field perturbations) the *plasma resistivity* is defined as Equation (1.28):

$$\rho = \frac{m_e v_{m,e}}{n_e e^2} = \frac{v_{m,e}}{\varepsilon_0 \omega_p^2} \tag{1.28}$$

where $v_{m,e}$ is the electron collision frequency, n_e is the electron density, ω_p is plasma frequency, m_e is the electron mass. For kinetic transport the mobility (μ) and diffusion coefficients (D) of the species α through a background of neutral (steady) particles are given by Equation (1.29):

$$\mu_\alpha = \frac{q_\alpha}{m_\alpha v_{m,\alpha}}; D_\alpha = \frac{kT_\alpha}{m_\alpha v_{m,\alpha}} \tag{1.29}$$

where T_α and $v_{m\alpha}$ are the species temperature and collision frequency with neutrals.

Another kinetic transport parameter relevant for electrically driven plasmas is the *ambipolar diffusion coefficient* (Equation 1.30) accounting for the diffusion process induced by the presence of static electric field:

$$D_A = \frac{\mu_i D_e + \mu_e D_i}{\mu_i + \mu_e} \tag{1.30}$$

where the subscripts i and e refer to ion and electron species, respectively. For magnetically driven and magnetized plasmas electron diffusion perpendicular to the magnetic field are described by the relative directional mobility and diffusion coefficients (Equation 1.31):

$$\mu_\perp = \frac{\mu_e}{1 + (\omega_{c,e}/v_{m,e})^2}; D_\alpha = \frac{kT_e}{m_\alpha v_{m,e}} \cdot \frac{1}{1 + (\omega_{c,e}/v_{m,e})^2} \tag{1.31}$$

Finally, we report Loschmidt's number, corresponding to the particle number density at standard pressure and temperature: $n_0 \approx 2.6868 \times 10^{25}\ m^{-3}$, useful to gauge the variety of plasma densities in LPPs such as encountered in PVD or plasma enhanced chemical vapor deposition (PECVD) reactors. In atmospheric pressure plasma (APP) regimes, examples are plasma jets, dielectric barrier discharge (DBD), or plasma coronas.

1.4
Classification of Technological Plasmas

A first broad classification of plasmas may be done in terms of their thermodynamic properties: thermal plasmas (TPs) and non thermal plasmas (NTPs) also regarded as plasmas in thermodynamic equilibrium and nonequilibrium plasmas, respectively. The first class includes high pressure hot plasmas characterized by pressures (p) exceeding 10^3 Pa with high electron temperatures (T_e) in the order of 10^4 K and higher. In these systems the collision frequency among different plasma species is efficient: the collision frequency is high with respect to the particles transit time on the plasma scale length and allows the electrons to lose energy in favor of the ion species, providing thermalization of different particle species to the thermodynamic equilibrium temperature. The energy equipartition principle holds and energy content is evenly distributed among vibration, rotational, and translation energies. The ionization degree (number of ions over total plasma particles) is close or equal to 100%.

On the contrary cold NTPs are in a thermodynamic nonequilibrium state in which it is sometimes possible to define different temperatures for the different plasma species. The typical situation in plasma processing is to find relatively hot electrons with temperatures of the order of 10^4 K ($T_e \sim 10^4$K) and cold ions and neutrals, often found at almost ambient temperature ($T_i \sim T_n \sim 10^4$K). Nonthermal cold plasmas are associated with low degrees of ionization in the range 10^{-4}–10^{-1}.

1.4.1
Hot (Thermal) Plasmas and Their Applications

Hot plasmas, such as electrical arcs, plasma jets of rocket engines, plasmas generated by thermonuclear reactions, and so on, have an extremely high energy content, which induces fragmentation of all organic molecules to atomic levels. As a consequence, these plasmas can only be used to generate extremely high caloric energy or to modify thermally stable inorganic materials (metals, metal oxides, etc.). The thermal plasma is obtained by generating an arc discharge in a gas submitted to electric fields of varied frequency. The bundle of gas ionized at very high temperature is able to remove, fuse or to thermally modify a material. The bundle can be compared to a tool, it is easily controllable and not in direct contact with the treated surface. The applications of thermal plasmas depend on temperature, gaseous reagents and tiny particles injected into the plasma (plasma spraying, synthesis) or exposed to plasma in the form of 'bulk materials' (fusion and refining in metallurgy).

The potential applications of thermal plasma processing technology cover a wide range of activities, such as the extraction of metals, the refining/alloying of metals/alloys, the synthesis of fine ceramic powders, spray coatings, and the consolidation and destruction of hazardous waste. In particular cases the thermal plasma finds applications in complicated chemical processes, for instance fast quenching chemistry or synthesis of nanoparticles.

In metal melting and remelting, the plasma is used primarily as an effective source of process heat, making use of the anode heat transfer characteristics of an arc between a cathode and the metal. The relatively long characteristic process times (from 0.1 second to minutes) reduce the importance of instability effects. In plasma cutting and welding, the use of a plasma is more economical than using a laser or an electron beam which may provide higher power flux densities. New approaches are driven by improvement of the product quality and process reliability. Examples are the expanded use of pulsing the weld current and of sensors for feedback control in automated welding.

Plasma sprayed coating has evolved extensively during the last 20 years, except for the basic design of the plasma gun (nozzle), which has not been changed significantly. The essential part of the gun is the nozzle that consists of a cone-shaped cathode located within a cylindrical anode, which usually extends beyond the cathode. Reactive or inert gases or mixtures of them traversing the space located between the electrodes are 'instantly' ionized and as a result the plasma state is generated. For coating purposes, powders can be injected into the plasma jet at desired locations relative to the nozzle to control the caloric energy absorption of the materials for deposition and the pathways of the plasma-borne particles and droplets. The coating particles (powders) introduced into the jet are instantly molten and the resulting droplets are deposited and cooled on the target surfaces usually leading to strongly bound – though porous – coatings. In plasma waste treatment, the major advantages of using thermal plasmas are the fast heating rates, the high processing temperatures allowing the formation of stable vitrified slugs, and the low off-gas flow rates. Off-gas cleaning is a major economic factor in any waste processing installation, and the costs scale down with increasing gas flows. The major issue is the economics of the specific process, and all new developments have been directed toward improving the economics either by combining plasma processes with conventional incinerators to make use of the heating value of the wastes, or by using the waste heat to obtain a useful co-product.

1.4.2
Cold (Nonthermal) Plasmas and Their Applications

In common perception, plasmas are hot gases that emit light and conduct electricity. Indeed, plasmas often contain energetic electrons (at $E \cong 1 \text{ eV} = 1.1604 \times 10^4 \text{K}$ or higher) that in turn transfer their energy to neutral molecules and excite radiating transitions. However, not all plasmas are hot. Cold plasmas, including low-pressure DC and RF discharges (silent discharges), discharges from fluorescent (e.g., neon) illuminating tubes, DBDs may be found both at low pressure or atmospheric pressure. Cold LPPs for surface processing are found in the range between 10^{-6} and 1 Pa, with a typical Debye length, corresponding to the electromagnetic screening distance, of 10^{-5} m, much less than the typical plasma scale length in the order of 10^{-1}–1 m. Cold LPPs have neutrals densities between $10^{-5}n_0$ and $10^{-2}n_0$ (n_0, being the Loschmidt number, as defined in Section 1.3.3), while cold APPs such as DBDs are characterized by electron number densities of the order of n_0 with an energy range of 1–10 eV.

Electrons, which are small and light particles, cannot heat the large and heavy molecules very efficiently, so in many cases the background gas remains almost at room temperature. In such nonequilibrium systems (often called *nonthermal plasmas* or *cold plasmas*), the complex plasma chemistry is driven by electrons. They perform ionization, necessary to sustain the plasma; in addition, they are responsible for atomic/molecular excitation, dissociation, and production of radicals and metastable molecular states. The result is an active gaseous medium that can be safely used without thermal damage to the surrounding materials. Such exceptional nonequilibrium chemistry is the base of plasma applications in lighting technology, exhaust gas treatment, and material processing. There are several methods to generate cold plasmas. When charged particles are in the minority, heating of neutral molecules is limited. This leads to diffuse plasmas where the fraction of ionized species is below $10^{-7}n_0$ and pressures reach 10^3 Pa. The effect of low pressure is double: in a rare gas ionization events are scarce, which keeps the charge density low. Moreover, the frequency of elastic collisions between electrons and atoms/molecules is low, so electrons do not have much chance to convey their energy to the gas. LPPs are of great value in fundamental research as well as in plasma technology, but they have many serious drawbacks. These plasmas must be maintained in massive vacuum reactors, in which the chemistry is optimally controlled, but their operation is costly due to long pump down times, energy requirement, as well as the reactor maintenance burden. The access for observation or sample treatment is limited, and there are limitations for the materials to be treated because of degassing problems. Therefore, one of the recent trends focuses on developing new plasma sources, which operate at atmospheric pressure, and target to retain the properties of low-pressure media. Also these approaches are characterized by positive features such as substrate accessibility, high throughput, continuous or semi-continuous processing, but limitations are also found in the shapes of substrate to be treated, less precise control of plasma chemistry due to plasma pollution, different physical and chemical regimes posing limitations to surface processing such as to plasma polymerization of high boiling point precursors.

NTPs may be generated for processing purposes by using different principles where the energy input comes from diverse sources.

In **micro-plasmas** gas heating occurs in the plasma volume, and the energy is carried away by thermal diffusion/convection to the outside. If the plasma has a small volume and a relatively large surface, gas heating is limited.

Coronas are gas discharges where the electrode geometry controls and confines the ionization processes of gases in a high-field ionization environment, in the absence of insulating surfaces or when the dielectric surfaces are far away from the discharge zone. Corona discharges are often called *negative, positive, bipolar, AC, DC,* or high frequency (HF) coronas, according to the polarity of the stressed electrodes, to whether one or both positive and/or negative ions are involved into the current conduction, and to the nature of the driving field. What makes corona discharges unique in comparison to other plasmas is the presence of a large low field drift region located between the ionization zone and the passive (low field) electrode. Ions and electrons penetrating the drift space will undergo neutralization, excitation, and recombination reactions involving both electrons and neutral and charged molecular and atomic species. However, because of multiple inelastic collision processes in atmospheric pressure environments the charged active species escaping from the ionization zone (electrons, ions) will have energies lower than the ionization energies, and as a consequence, neutral chemistry (free radical chemistry) will characterize the drift region. According to various electrode configurations, point-to-plane, wire-cylinder, and wire-to-plane corona discharges can be identified.

Dielectric barrier discharges. These plasmas are typically generated between parallel metal plates, which are covered by a thin layer of dielectric or highly resistive material. Usually they are driven by a HF electric current (in the kHz range), but it is also possible to obtain a DBD by simple transformation of 50 Hz/220 V network voltage to about 10 kV to 40 kHz electric input. The dielectric layer plays an important role in suppressing the current (sparking due to streamers): the cathode/anode layer is charged by incoming positive ions/electrons, which reduces the electric field and hinders charge transport toward the electrode. DBDs have typically low ionization degrees and currents in the order of mA. Besides, the electrode plates are quite large ($\sim$0.1 m^2, in some cases with a large aspect ratio of 10 : 1) and the distance between them usually does not exceed the millimeter range. Thus, DBD has a large surface-to-volume ratio, which promotes diffusion losses and maintains a low gas temperature (at most a few tens of degrees above the ambient). The only serious drawback of a DBD is its limited flexibility. Since the distance between the plates must be kept small, treatment of large and irregular (3D) samples is impossible, at least with conventional planar electrodes.

In **electron cyclotron resonance ion sources** (ECRISs), the plasma is confined in a special magnetic field configuration where an axial magnetic field is produced by, for example, two solenoid coils (magnetic mirror). Superimposed is a radial magnetic field usually produced by a permanent multipole magnet. This geometry leads to a minimum-B-structure, that is, from the geometrical middle the magnetic field increases in all directions. Electrons confined in that magnetic field gyrate around the magnetic field lines with the cyclotron frequency ($\omega_{c,e}$). The microwave energy is radiated into the plasma and electrons can be heated resonantly when the microwave frequency equals the cyclotron frequency. Every time an electron passes the resonance region it can gain 1–2 keV energy. Electrons from ECRIS can have energies up to several kiloelectronvolts and therefore a good magnetic plasma confinement is required. Ions are not accelerated due to their high mass and are confined electrostatically by the space charge of the plasma electrons. When electrons leave the plasma through the loss cone of the magnetic mirror, ions can follow and can then be extracted from the ion source by applying a high voltage. The maximum obtainable charge state depends on the confinement time of the ions and on the energy of the plasma electrons. These can be varied by tuning the ion source parameters like gas pressure, microwave power, magnetic field strength, and so on.

Microwave plasmas are sustained by microwave energy dissipated into the reaction media by coaxial cables or by waveguides in the case of higher powers. The physical dimensions of coaxial cables (cross-sections) and waveguides are selected according to the microwave frequency. Most materials efficiently absorb or reflect microwaves, and as a result microwave energy cannot be transported using conventional cables. Microwave discharges are more difficult to sustain under low-pressure conditions ($<10^3$ Pa). In a collisionless condition the energy gained by an electron during one cycle is too small to produce ionization. In collisional plasmas at constant power density and electric field, the average (RF) microwave power transferred from the driving field has a maximum value when the collision frequency equals the driving frequency. The absorption of microwave power depends on the collision frequency of the electrons which is controlled by the atomic and molecular species. At comparable plasma parameters, RF discharges most often fill the entire reactor, whose dimensions are usually smaller than the wavelength of the RF field (13.56 MHz corresponds roughly to 22 m). Microwave plasmas exhibit a strong peaking in field intensity at the coupling to the microwave cavity that diminishes gradually with increasing distance from the coupling, rather than being deposited throughout the discharge.

1.5
Reactive Plasmas

In this section we particularly focus on the reactivity of NTPs and the capability of promoting plasma–chemical reactions through intermediate steps in a way to exploit the plasma state as a very efficient environment as compared to conventional conditions where chemical reactions usually take place. The free electric charges – electrons and ions – make a plasma electrically conductive, internally interactive and strongly responsive to electromagnetic fields. Plasma is widely used in practice, in particular for surface processing, and NTPs offer three major features that are attractive for industrial applications:

- The temperature of at least some plasma components and the energy density can significantly exceed those in conventional chemical technologies (e.g., energetic electrons), abating the activation energy thresholds.
- Plasmas are able to produce very high concentrations of energetic and chemically active species (electrons, ions, atoms and radicals, excited and metastable states, and photons with different wavelengths which can interact with the processing surface).
- Cold plasmas are far from thermodynamic equilibrium, providing extremely high concentrations of the chemically active species while keeping the bulk temperature as low as room temperature, thus not affecting the properties of material to be treated other than the ones of the functionalized surface.

These plasma features allow significant intensification of traditional chemical processes, a dramatic increase of their efficiency and often successful promotion of chemical reactions which in conventional chemistry would require significant energy input and the use of additional chemical compounds such as catalysts. Moreover, plasma treatments are by definition dry processing methods, thus avoiding the use of water and solvents, minimizing emissions and the overall environmental burden, as discussed in detail in Chapter 12.

1.5.1
Elementary Plasma–Chemical Reactions

To reach the required degree of reactivity in a plasma not all particles need to be ionized; a common condition in plasma chemistry is that the gases are only partially ionized. The ionization degree in the conventional plasma-chemical systems spans a range of seven orders of magnitude (10^{-7}–10^{-14}). When the ionization degree is close to unity, such a plasma is called *completely ionized plasma*, which is often the case in high temperature thermal plasmas. When the ionization degree is low, the plasma is called *weakly ionized plasma*, which is the main focus of plasma chemistry in the present context.

The total yield of the plasma–chemical processes is due to synergistic contributions of numerous different elementary reactions taking place simultaneously in a discharge system. The sequence of transformations of the initial chemical substances and electric energy into products and thermal energy is usually referred to as the mechanism of *the plasma–chemical process*. Elementary reaction rates are determined by the microkinetic characteristics of individual reactive collisions (like, for example, collision cross-sections, i.e., elementary reaction probabilities) as well as by relevant kinetic distribution functions (see Section 1.3), like the electron energy distribution function (EEDF) and ion energy distribution function (IEDF), respectively defined generically as f_α in the previous sections, and by other distribution functions such as that of excited molecular states. Formally reactions rates (k) are calculated by means of the reaction cross-sections and interacting particle speeds as Equation (1.32):

$$k = \langle v \cdot \sigma \rangle \tag{1.32}$$

where the $<.>$ symbol stands for an averaged quantity over the distribution function and σ is the reaction cross-section, while v is the relative velocities of the reacting particles.

Indeed, the elementary reaction rate is actually the result of integration of the reactive collision cross-section over the relevant distribution function and it is characterized by the energy or excitation state of the reactant.

The key process to sustain the plasma discharge and therefore to allow plasma-chemical reactions is ionization, which means conversion of neutral atoms or molecules into *electrons* and *positive ions*. Thus, ionization is the first processes to be considered.

In quasi-neutral plasmas, typically employed in surface processing, the number density of electron and positive ions species are comparable or equal $(n_e \approx n_i)$ in case of high electron affinity of heavy particle species negative ions are also effectively formed and give rise to 'electronegative' plasmas. Mainly responsible for ionization are inelastic effective collisions. Therefore we dwell briefly on a quantitative description of collision phenomena principles and dynamics to get the main understanding of these processes.

Due to continuous impact of natural energetic cosmic rays on the neutral gas some free electrons are continuously generated and always available. When external (intense) electric fields are applied, an ionization avalanche process is started, providing more and more available free electrons. The latter are the first species gaining kinetic energy from electric fields, because of their low mass and high mobility with respect to other species such as ions. Those energetic free electrons transfer energy to all other plasma components, providing energy for ionization, excitation, dissociation, and other plasma chemical processes. The rates of such processes depend on how many electrons have enough energy to do the job. This can be quantitatively described by means of the EEDF, which is the probability density $f(\varepsilon)$ for an electron for having an energy ε. The EEDF strongly depends on the applied electric field and the gas composition in the plasma (especially in nonthermal discharges) and often can be very far from the equilibrium distribution.

1.5.2
Elastic Scattering and Inelastic Thomson Scattering: Ionization Cross-section

Electron–electron, electron–ion, and ion–ion scattering processes are so-called Coulomb collisions. Their cross-sections are quite high with respect to those of collision with neutral partners, but they are much less frequent in a discharge with a low degree of ionization. An important feature of Coulomb collisions is the strong dependence of their cross-section on the kinetic energy of the colliding particles. This can be demonstrated by a simple analysis, where two particles (considered as rigid spheres) have the same charge and, for the sake of simplicity, one collision partner is considered to be at rest. A scattering event takes place if the Coulomb interaction energy ($U \sim q^2/b$, where b is the impact parameter) is comparable to the kinetic energy ε of a moving particle. Then, the impact parameter $b \sim q^2/\varepsilon$ and the ionization reaction cross action σ can be estimated as $\sigma = \pi b^2$, in the classical hard spheres approximation.

In order to sustain a plasma and to provide its chemical reactivity, continuous ionization is necessary. Electron collisions with the background neutral species and ions provide the mechanism to determine these conditions when collisions are said to be non-elastic and a certain amount of collisional energy is spent to directly ionize or excite and subsequently ionize the molecules.

Starting with the Rutherford formula for differential cross-section from (classical) collision particle dynamics it is possible to derive the Thomson ionization cross-section:

$$\sigma_{iz}(\varepsilon) = \pi \left(\frac{e^2}{4\pi\varepsilon_0}\right)^2 \frac{1}{\varepsilon}\left(\frac{1}{\varepsilon_{iz}} - \frac{1}{\varepsilon}\right) \tag{1.33}$$

Equation 1.33 is valid for $\varepsilon > \varepsilon_{iz}$ and for $\varepsilon \leq \varepsilon_{iz}$ the ionization cross-section is identically zero: $\sigma_{iz}(\varepsilon) \equiv 0$.

When considering typical scattering cross-sections at room temperature ($\sim 293\,\mathrm{K} = 3.39 \times \cdot10^{-2}\mathrm{eV}$) it is straightforward to realize that there is a gap of three orders of magnitude with respect to the scattering cross-section of electrons at a temperature of $1\,\mathrm{eV}$ ($\sim 1.16 \times 10^{-4}\mathrm{K}$), typical for electric discharges. Beside this we recall that for a charged particle scattering on a neutral molecule which has a permanent dipole moment (interaction energy $U \sim 1/r^2$) and an induced dipole moment (interaction energy $U \sim 1/r^4$), the ionization cross-sections are $\sigma_{ix}(\varepsilon) \sim 1/\varepsilon$ and $\sigma_{iz}(\varepsilon) \sim 1/\varepsilon^{1/2}$, respectively. Similar considerations may be made for electrons.

Moreover, energy transfer during an elastic collision (in the hard sphere scattering approximation) is possible only as a transfer of kinetic energy. However the average fraction γ of kinetic energy (Equation 1.34), transferred from one particle of mass m (m_e for electron mass) to another particle of mass M (m_i for ion mass), is equal to:

$$\gamma = \frac{2\,m_e m_i}{(m_e + m_i)^2} \tag{1.34}$$

For elastic collisions of electrons with heavy neutrals or ions, $m_e \ll m_i$ and, hence, $\gamma \sim 2\,m_e/m_i$, which means that the fraction of transferred energy is very small ($\gamma \sim 10^{-4}$). In particular, this explains why the direct impact ionization due to a collision of an incident electron with a valence electron of an atom predominates (here $\gamma = 0.5$).

1.5.3
Molecular Ionization Mechanisms

Nondissociative and dissociative ionization of molecules by direct electron impact is presented here as an example for the complex mechanisms that occur in reactive plasmas. Such a process can be written, for the case of diatomic molecules AB respectively as:

$$e + AB \longrightarrow AB^+ + e + e$$
$$e + AB \longrightarrow A + B^+ + e + e \qquad (1.35)$$
$$e + AB \longrightarrow A + B^* + e + e$$

Here, AB denotes a diatomic molecule, AB^+ an ionized diatomic molecule, B^+ an ionized atom as product of molecular dissociation, B^* an atom in an excited state after molecular dissociation.

The first of the listed processes (Equation 1.35) takes place when the electron energy does not greatly exceed the ionization potential. Some peculiarities of molecule ionization by electron impact can be seen from the illustrative potential energy curves for AB and AB^+, shown in Figure 1.2, for collisions having a threshold energy ε_{iz}. For collisions having threshold energies higher than ionization, molecule dissociation occurs with excitation or further ionization of products (represented in states labeled as c and b, respectively in Figure 1.2).

The fastest internal motion of atoms inside molecules is their molecular vibration. But even molecular vibrations have a typical time period of 10^{-14}–10^{-13} s, which is much longer than the interaction time between the plasma electrons and the molecules: $a_0/v_e \sim 10^{-16} - 10^{-15}$ s (where a_0 is the atomic unit of length and v_e is the mean electron velocity). This means that all kinds of electronic excitation processes under consideration, which are induced by electron impact, are much faster than the atomic motion inside the molecules. As a result, all the atoms inside a molecule can be considered to be frozen during the process of electronic transition, stimulated by electron impact. This fact is known as the *Frank–Condon principle*. The nondissociative ionization process (Equation 1.35) usually results in the formation of a vibrationally excited ion $(AB^+)^*$ and requires a little more energy than the corresponding atomic ionization. When the electron energy is relatively high and substantially exceeds the ionization potential, dissociative ionization can take place:

$$e + AB \longrightarrow A + B^+ + e + e \qquad (1.36)$$

This ionization process (Equation 1.36) corresponds to electronic excitation into a repulsive state of the ion, $(AB^+)^*$, followed by a decay of this molecular ion. The energy threshold for the dissociative ionization is essentially greater than that for the nondissociative situation.

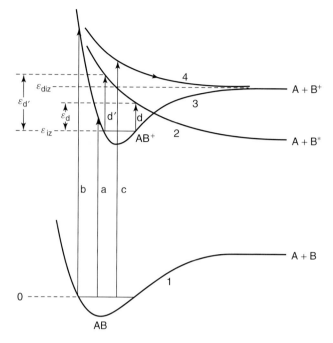

Figure 1.2 Qualitative representation of potential energy curves for ionization of molecules AB and AB$^+$ by electron impact. Energy input to a molecule by collisions may be comparable to or exceed the ionization energy ε_{iz}. In the case of comparable energy (a) the molecule AB is ionized into the AB$^+$ state; if higher energy is transferred to the molecule by electron impact, dissociation may occur (b and c).

In reactive plasmas many other similar two- and three-body processes, including association, dissociation, recombination, attachment, detachment, excitation transfer processes of increasing complexity may also occur.

1.5.4
Stepwise Ionization by Electron Impact

The efficiency of the ionization and excitation mechanisms associated with specific molecules (regarded as precursors in plasma processing) strongly influences the plasma reactivity.

However, the mechanisms previously described do not account for bridging the three orders of magnitude gap missing for the cross-section to explain why such efficient ionization processes sustain plasma discharges. To understand this and to see why the plasma state may provide such efficient reaction pathways we consider the molecular stepwise ionization process by electron impact.

When the plasma density and, therefore, the concentration of excited neutrals is high enough, the energy (ε_{iz}) necessary for ionization can be provided in two different ways. First, as in the case of direct ionization, it can be provided by the energy of plasma electrons. Second, the high energy of preliminary electronic excitation of the neutrals can be converted in the ionization process, which is called *stepwise ionization*. If the level of electronic excitation is high enough, stepwise ionization is much faster and more efficient than direct ionization, because the statistical weight of electronically excited neutrals is greater than that of free plasma electrons. Hence, the probability that particle collisions are effective is higher for excitation than for direct ionization. Furthermore, the probability that the subsequent electron collisions are effective for ionization is higher, because the energy difference between the excited state and the unbound electron state is lower. We can consider the ionization process $e + A \rightarrow A^+ + e + e$, as the inverse to the three-body recombination event: $A^+ + e + e \rightarrow A^* + e \rightarrow A + e$, realized through a set of excited states. For quasi-equilibrium or thermal plasmas we may apply the principle of detailed equilibrium and consider the ionization process $e + A \rightarrow A^+ + e + e$ as if going through the set of electronically inverse excited states of the three-body recombination process, in this interpretation the ionization is regarded as a stepwise process.

For the ionization process we may define a rate coefficient k_{iz} (Equation 1.37), derived by the summation of partial rate coefficients $k_{iz,i}$, corresponding to the j-th electronically excited state, over all states of excitation, taking also into account their concentration:

$$k_{iz} = \sum_{j}^{n} k_{iz,j} \frac{N_j(\varepsilon_j)}{N_0} \tag{1.37}$$

Further, we assume that the electron excited states in the target atoms, radicals, or molecules, are in quasi equilibrium with plasma electrons, and apply the Boltzmann statistics describing the electronically excited states as having a defined electron temperature T_e, and by taking into account the statistical weights (g_j) of the states, their number density (N_j), Equation (1.38), and their energy (ε_n), become linked (note that the subscript '$_0$' refers to the ground state):

$$N_j = \left(\frac{g_j}{g_0} \right) N_0 \cdot e^{\left(-\frac{\varepsilon_j}{k_b T_e} \right)} \tag{1.38}$$

From statistical thermodynamics $g_j = 2 g_{iz} j^2$, where g_{iz} is the statistical weight of an ion.

Assuming effective (inelastic) electron neutral excitation collision, the energy transfer from the impinging electron to an electron bound to the neutral target particle undergoing excitation is about T_e. This means that excited particles with an energy of about $\varepsilon_j = \varepsilon_{iz} - T_e$ make the major contribution to the sum of Equation (1.37). Taking into account that $\varepsilon_j \sim 1/j^2$, the number of states with an energy of about $\varepsilon_j = \varepsilon_{iz} - T_e$ and a ionization potential of about $\varepsilon_{iz} = k_B T_e$ is of order j. Thus, from Equations (1.33) and 1.38 we can derive the reaction rates at each step:

$$k_{iz} = \left(\frac{g_{iz}}{g_0} \right) j^3 \cdot \langle \sigma_{iz} \cdot v \rangle \cdot e^{\left(-\frac{\varepsilon_{iz}}{k_b T_e} \right)} \tag{1.39}$$

By substituting the thermal velocity and ionization cross-section as a result, considering that from quantum mechanics $j^2 \approx \frac{z^2 e^4}{(4\pi\varepsilon_0)^2 \hbar^2 \varepsilon_{iz}^2}$, the stepwise ionization rate coefficient can be expressed as:

$$k_{iz} \approx \left(\frac{g_{iz}}{g_0}\right) \frac{1}{(4\pi\varepsilon_0)^5} \frac{m e^{10}}{\hbar^3 T_e^3} \cdot e^{\left(-\frac{\varepsilon_{iz}}{k_b T_e}\right)} \tag{1.40}$$

Finally, comparing the stepwise- (Equation 1.40) and direct-ionization rate $(\bar{k}_{iz}(T_e))$ coefficients we obtain:

$$\frac{k_{iz}(T_e)}{\bar{k}_{iz}(T_e)} \approx \left(\frac{g_{iz} a_0^2}{g_0 \sigma_0}\right) \cdot \left(\frac{e^4 m}{(4\pi\varepsilon_0)^2 \hbar^2 T_e^2}\right)^{\frac{7}{2}} \approx \left(\frac{\varepsilon_{iz}}{T_e}\right)^{\frac{7}{2}} \tag{1.41}$$

From Equation (1.41) it is possible to argue that for $\sigma_0 \approx a_0$ and for $\varepsilon_{iz}/T_e \approx 10$, as it is the case for a typical discharge, the stepwise ionization can be 10^3–10^4 times faster than the direct one, as needed to sustain a plasma discharge.

1.6
Plasma Sheaths

Confined plasmas which enter into contact with materials are of particular interest for technical applications. A description of physical and chemical interactions between plasma and material surfaces is provided in Chapter 3 for processing applications. To understand the effective plasma regime at the plasma–substrate interface, the features of plasma sheaths are introduced in the following.

Plasmas which are quasi neutral ($n_i \approx n_e$), develop positively charged boundary layers called *sheaths* when approaching the material confining surfaces. This is due to the higher thermal velocity of electrons, $v_{th,e} = (2 k_B T_e/m_e)^{1/2}$, exceeding that of ions, $v_{th,i} = (2 k_B T_i/m_i)^{1/2}$, by two orders of magnitude because of the disproportioned mass ratio $m_e/m_i \ll 1$. Considering a simple (ideal) configuration of a plasma slab between two identical parallel grounded surfaces (walls), due to charge neutrality, the electric field would be zero everywhere. However, since in this configuration electrons are not confined by any electric field, due to their higher mobility they are rapidly lost to the walls, causing an abrupt change of charge concentration. Therefore at the plasma–wall interface charge neutrality is no longer satisfied and the electrical potential is found to be positive in the plasma (due to lack of negative charged particles) and rapidly decreasing within the plasma sheath space-domain, and approaching zero close to the walls within a few Debye lengths. The onset of this natural charge unbalance and consequent generation of potential barrier provides a self-confining mechanism for electrons that are then pulled into the plasma by the electric field directed from the plasma to the walls. Under these conditions ions approaching the plasma sheath are accelerated to the walls, causing ion bombardment. An important aspect of plasma sheaths is that their typical scale length is much smaller than the plasma spatial extension.

In fact, the sheath structure is more complex than just that of a boundary layer. Indeed, to satisfy the continuity of ion flux through the sheath, a pre-sheath between the plasma and the sheath has to be assumed. Furthermore, sheath features are strongly dependent both on boundary conditions (applied external voltages: continuous, oscillating high/low voltage) and on the plasma characteristics (i.e., presence of electronegative ions, ion temperature, ionization degree).

For simplification, we consider a non-collisional plasma with Maxwellian electron temperature T_e, assuming cold ions such that ($T_e \gg T_i$) and quasi neutrality. Under these conditions at the plasma–sheath interface, considering a one-dimensional space dependence of quantities (1D-system) the momentum (ion-flux) and energy conservation along with the Maxwell Equation 1.9d read:

$$
\begin{cases}
n_i(x)u(x) = n_{i,s}(x)u_s & \text{(1.42a)} \\[2mm]
\dfrac{1}{2}m_i u^2(x) = \dfrac{1}{2}m_i u_s^2 - e\varphi(x) & \text{(1.42b)} \\[2mm]
\nabla^2\varphi(x) = \dfrac{e}{\varepsilon_0}(n_e - n_i) & \text{(1.42c)}
\end{cases}
$$

where u is the ion speed and n_i and n_e the ion and electron densities in the plasma, respectively, n_{is} and u_s are ion density and speed within the sheath, where ionization is assumed to be absent.

Given the electrons thermal equilibrium, the electron density is expressed by the Boltzmann relation $n_e(x) = n_{es}\exp[-e\varphi(x)/k_B T_e]$, with n_{es} the electron density within the sheath and k_B the Boltzmann constant. By solving Equations 1.42a and 1.42b with respect to u we find a relation for n_i as a function of $\varphi(x)$; which together with the former Boltzmann equation for n_e may be introduced into Equation 1.42c to give:

$$
\frac{d^2\varphi(x)}{dx^2} = \frac{en_s}{\varepsilon_0}\left\{\exp[-e\varphi(x)/k_B T_e] - \left[1 - \frac{e\varphi(x)}{\varepsilon_s}\right]^{-\frac{1}{2}}\right\}
\tag{1.43}
$$

where $\varepsilon_s = 1/2m_i u_s^2$ and $n_s = n_{is} = n_{es}$ are the kinetic energy and density of ions within the plasma sheath. Equation (1.43) is the basic nonlinear equation governing the sheath potential and the ion and electron densities; it allows for a first exact integral obtained by multiplying Equation (1.43) by $d\varphi/dx$ and then integrating with respect to x with the field free plasma boundary conditions at $x = 0 : \varphi(0) = 0; [d\varphi/dx]_0 = 0$; thus yielding:

$$
\frac{1}{2}\left[\frac{d\varphi(x)}{dx}\right]^2 = \frac{en_s}{\varepsilon_0}\left\{k_B T_e\exp[-e\varphi(x)/k_B T_e] - k_B T + 2\varepsilon_s\left[1 - \frac{e\varphi(x)}{\varepsilon_s}\right]^{\frac{1}{2}} - 2\varepsilon_s\right\}
$$

$$
\tag{1.44}
$$

Equation 1.44 may be integrated numerically. However, analytical considerations provide some fundamental information on sheath features. The RHS of Equation (1.44) is positive-defined yielding to the consideration that the ion density within the sheath must always be larger than the electron density. Furthermore, by expanding up to the second order in $\varphi(x)$ we get the relation $e\varphi(x)^2/k_B T_e - e\varphi(x)^2/2\varepsilon_s \geq 0$, which may be satisfied for any value of x only if $2\varepsilon_s \geq k_B T_i$, leading to the Bohm sheath criterion (Equation 1.45):

$$u_s \geq u_B \equiv \sqrt{\frac{k_B T_e}{m_i}} \tag{1.45}$$

From this condition the need is derived for an intermediate region between plasma (bulk) and plasma sheath in which ions are accelerated by electric an field to reach the speed u_s: this region is named *presheath*. Given the structural complexity of the plasma–wall interface matching (analytical and numerical) computational techniques are needed to provide solutions in specific configuration conditions and plasma regimes.

However, configurations in which a simple analytical solution is possible are still representative in some plasma regimes and often applied in surface function-alization processes. This is the case of high negative voltage biasing of one of the electrodes connected to the substrate to be plasma-treated or plasma-coated (see for instance PVD processes dealt with in Chapters 2 and 9) imply driving high negative voltages to one of the electrodes. In this case the sheath voltage is high and the related energy is large compared with the electron thermal energy, $e\varphi(x)^2 \gg k_B T_e$, allowing Equation (1.44) to be simplified by neglecting the exponential term on the RHS. This is the case of the so called *matrix sheath* in which only ions are present in the plasma sheath, with constant density $n_i(x) \equiv n_i = n_s$. In this one-dimensional configuration the Poisson equation deriving from the Gauss theorem (Equation 1.9d) has a simple source term: $d^2\varphi(x)/dx^2 = en_s/\varepsilon_0$, which leads to the analytical solution: $\varphi(x) = -(en_s/2\varepsilon_0)x^2$ from which for the boundary condition $\varphi(s) = -V_0$, (where $-V_0$ is the applied bias voltage) the sheath thickness s may be derived as a function of the plasma characterizing parameters:

$$s = \lambda_{De} \cdot \sqrt{\frac{2 e V_0}{k_B T_e}} \tag{1.46}$$

Equation 1.46 confirms that the sheath scale length is much shorter than the plasma spatial extension. Considering the situation in which the kinetic energy of ions entering the sheath region is small compared with the plasma sheath potential energy the momentum (ion flux), energy conservation and Gauss theorem (Equations 1.9b, c and d) simplify to:

$$\begin{cases} en_i(x)u(x) = J_0 \\ \frac{1}{2}m_i u^2(x) = -e\varphi(x) \\ \frac{d^2\varphi(x)}{dx^2} = -\frac{en_i(x)}{\varepsilon_0} \end{cases} \tag{1.47}$$

where J_0 represents the constant ion current. By simultaneous solution of the three equations in Equation (1.47) it is possible to obtain a new differential equation for the sheath potential, which further multiplied by its first derivative $d\varphi(x)/dx$ and finally integrated by setting the following boundary conditions:

$d\varphi(x)/dx \equiv 0$ and $\varphi(x) \equiv 0$; leads to the solution:

$$\varphi(x) = -\sqrt[3]{\left(\frac{3}{2}\right)^4} \sqrt[3]{\frac{m_i}{2e}\left(\frac{J_0}{\varepsilon_0}\right)^2} \sqrt[3]{(x)^4} \tag{1.48}$$

By imposing the following boundary conditions $\varphi(s) \equiv -V_0$ in Equation (1.48), and solving for J_0 we get the well-known Child's law:

$$J_0 = \frac{4}{9}\frac{\varepsilon_0}{s^2}\sqrt{\frac{2\,e\,V_0^3}{m_i}} \tag{1.49}$$

In particular, Equation (1.49) provides a more refined relation between plasma sheath potential and sheath thickness than Equation (1.46):

$$s = \frac{\sqrt{3}}{2}\lambda_{De} \sqrt[4]{\left(\frac{2\,e\,V_0}{k_B T_e}\right)^3} \tag{1.50}$$

For collisional sheaths and other plasma sheaths regimes not falling into the considered simplified configurations, a dedicated analysis is necessary, as reported in Ref. [18].

1.7
Summary

In this chapter we have provided some fundamental and introductory concepts to develop a more rigorous and comprehensive understanding of the plasma state as well as providing basic tools to better appreciate the technical content of this book. Characterizing parameters and scaling quantities are provided in order to facilitate the parameterization of different plasma states and to allow classification of processing plasmas used for industrial applications.

Statistical and fluid description of plasmas are presented to better understand the fundamentals of plasma reactivity and plasma–surface interactions as the characterizing features of technological plasmas for surface functionalization.

References

1. Tonks, L. and Langmuir, I. (1929) Oscillations in ionised gases. *Phys. Rev.*, **33**, 195.
2. Sweet, P.A. (1958) Electromagnetic phenomena, in *Cosmical Physics*, (ed. B. Lehnert), Cambridge University Press, London.
3. Priest, E.R. (1981) *Solar Flare Magnetohydrodynamics*, Gordon and Breach Science Publishers, UK.
4. Priest, E.R., Foley, C.R., Heyvaerst, J., Arber, T.D., Culhane, J.L. and Aton, L.W. (1998) Nature of the heating Mechanism for the Diffuse Solar Corona. *Nature*, **393**, 545.
5. Petscheck, H.E. and Hess, W.N. (eds) (1966) AAS-NASA Symposium of the Physics of Solar Flares, NASA Spec. Publ. SP-50, 1064, p.425.

6. Lui, A.T.Y. (1987) *Magnetotail Physics*, The John Hopkins University Press, Baltimore and London.

7. Heppner, J.P. Ness, N.F. Scearce, C.S. and SkillmanExplorer, T.L. (1963) 10 Magnetic Field Measurements. *J.Geophys.Res.*, **68**, 1.

8. Gloeckler, G., Hamilton, D., Ipavich, F.M., Studemann, W., Wilken, B., Kremser, G. and Hovestadt, D. (1985) Composition of the thermal and superthermal ions in the Near Earth Plasma Sheet, Proceedings on the Chapman Conference on Magnetotail Physics.

9. Bame, S.J., Aderson, R.C., Asbridge, J.R., Baker, D.N., Feldman, W.C., Gosling, J.T., Hones, E.W., McComas Jr., D.J and Zwickl, R.D. (1983) Plasma regimes in the deep geomagnetic Field Tail: ISEE 3. *Geophys. Res. Lett.*, **10**, 912.

10. Bryant, D.A. (1993) Space plasma physics, in *Plasma Physics*, Chapter 9 (ed. R. Dendy), Cambridge University Press.

11. International Atomic Energy Agency, (1956) Volume 13: Legal, administrative, health and safety aspects of large-scale use of nuclear energy, *Proceedings of the international conference on the peaceful uses of atomic energy; held in Geneva 8 August – 20 August 1955*, United Nations.

12. Butt *et al.* (1958) Proceedings of the 2nd Conference on the Peaceful uses of Atomic Energy, United Nations, Geneva, Vol. 32, p. 48.

13. Bellan, P.M. (2006) *Fundamentals of Plasma Physics*, Cambridge University Press.

14. Denes, F.S. and Manolache, S. (2004) Macromolecular plasma-chemistry: an emerging field of polymer science. *Prog. Polym. Sci.*, **29**, 815–885.

15. Shishoo, R. (2007) *Plasma Technology in Textile Processing*, Woodhead Publishing Ltd. and CRC Press LCC.

16. Chiellini, E. (ed.) (2008) *Environmentally Compatible Food Packaging*, CRC Press and Woodhead Publishing Ltd.

17. Soloshenko, I.A. *et al.* (2000) Sterilization of medical products in low-pressure glow discharges. *Plasma Phys. Rep.*, **26** (9), 792–800.

18. Lieberman, M.A. and Lichtemberg, A.J. (1994) *Principles of Plasma Discharges and Material Processing*, John Wiley & Sons, Inc., New York.

19. Krall, N.A. and Trivelpiece, A.W. (1986) *Principles of Plasma Physics*, San Francisco Press, Inc..

20. Schmidt, George, (1979) *Physics of High Temperature Plasmas*, Academic Press.

21. Fridman, A. (2008) *Plasma Chemistry*, Cambridge University Press.

22. Stix, T.H. (1962) *The Theory of Plasma Waves*, McGraw-Hill, New York.

2
Plasma Systems for Surface Treatment

Guy Buyle, Joachim Schneider, Matthias Walker, Yuri Akishev, Anatoly Napartovich, and Massimo Perucca

2.1
Introduction

Plasma processing has been widely used in various surface engineering applications for several good reasons. First of all, it is crucial that plasma modification influences only the outermost molecular surface layers without changing the bulk properties, that is, plasma treatment is a true surface treatment. This also holds for thermolabile materials provided that the energy input is duly controlled to prevent substrate overheating and the potential resulting change in bulk properties. Further, dry plasma technology offers an environmentally friendly alternative to the conventional wet chemical methods of improving the surface properties of a substrate. Modification of the surface properties of heat-sensitive materials like fabrics and polymer films, has received widespread attention because of their practical industrial importance. Another advantage of plasma treatment is that the resulting surface properties depend on easily adjustable discharge parameters like electric power, treatment time, gas composition, and gas pressure. As a result the reaction process can be controlled well, making nonthermal plasma (NTP) treatment a versatile and multifunctional process that can be widely applied.

A simplified definition of plasma refers to it as a gas which contains a certain fraction of charged particles. Hence, the logical way to characterize plasmas is by looking at the dependence of the particle density versus the particle energy [1]. In this density–energy space, one can define different domains corresponding to different plasma regimes, including, for example, the tokamak plasmas (which are very high temperature) or the plasmas used for light sources. As a plasma is generated by placing a gas inside an electric field, which then ionizes the gas (break down), one also speaks of a 'discharge'.

Most of the industrially used plasma technologies for surface processing are based on the use of so-called nonthermal reactive plasmas, see also Chapter 1. They are characterized by charged particle densities in the range of $10^6 - 10^{20}$ particles/m^3 and average particle energies in the range 0.1–10 eV. They typically consist of a gaseous mixture of charged particles (electrons, ions) and neutral activated species

Plasma Technology for Hyperfunctional Surfaces. Food, Biomedical and Textile Applications.
Edited by Hubert Rauscher, Massimo Perucca, and Guy Buyle
Copyright © 2010 WILEY-VCH Verlag GmbH & Co. KGaA, Weinheim
ISBN: 978-3-527-32654-9

including gas molecules, free radicals, metastables, and UV-photons. In this chapter, we focus on these types of plasmas.

To perform plasma surface treatment using nonthermal, reactive discharges, a wide range of plasma systems exist. Indeed, there is no one-to-one link between the classification of plasmas based on charged particle density and energy and the plasma system used as, for example, certain regions are covered by more than one system.

In general, NTPs can be generated both at reduced and at atmospheric gas pressure. These two pressure regimes will be discussed in more detail here as they are the more common types of discharge. They operate typically at a pressure of 10 and 10^5 Pa, respectively. However, a third group, generated at sub-atmospheric gas pressures, does exist. This category aims at an in-between pressure range, that is, typically around 10^3 Pa. They will not be discussed here. More information about these systems can be found in, for example, De Geyter *et al.* [2].

Here, we do not intend to give an exhaustive overview of the different plasma systems available. Instead, we want to give a description of some of the systems that are used specifically for surface treatment, thereby focussing on systems that are employed for the more industrial application-related results discussed in this book.

2.2
Low Pressure Plasma Systems

Low pressure cold plasma technology, also referred to as *vacuum plasma technology*, found its origin, proliferation, and maturation in the processing of semiconductor materials and printed circuit boards, where this type of processing reduced the need for wet chemical methods.

It is no coincidence that the first industrial applications were at low pressure because this type of discharge is easier to stabilize than the atmospheric ones. At low pressure, it is quite straightforward to produce a homogeneous plasma discharge in relatively large volumes. The drawback is that the generation of large-scale nonthermal reactive plasmas at low pressure requires the use of large-size metallic vacuum chambers and complex vacuum equipment (e.g., vacuum pumps, gas flow controllers, pressure gauges, gaskets). Plasma processing at low pressure is typically implemented as a batch process, that is, a discontinuous process that requires loading and unloading the samples in batches via a load lock in and out of the vacuum chamber. It can also be realized in a continuous fashion by using additional vacuum intermediate chambers for a gradual transition to the required low pressures and by providing sufficient pumping capacity.

In the next sections, we focus on the microwave systems as this type of source was used to obtain the barrier properties discussed in Chapter 8 and part of the work involved equipment optimization. Further, a brief description is given of the capacitively coupled discharge and of the physical vapor deposition (PVD) lateral arc rotating cathodes (LARC) system as they are used for obtaining antifouling and anti-wear coatings, as discussed in Chapters 7 and 9, respectively.

2.2.1
Microwave Systems

2.2.1.1 Introduction

Microwave plasmas have been studied since the early 1950s mainly by introducing dielectric tubes filled with gas at low pressure into various waveguides and/or cavities. Moisan *et al.* [3] developed a first series of single, compact, and efficient surface-wave launchers, based on coaxial and waveguide components, that were particularly suitable for generating long plasma columns by microwaves. They used so-called surfatrons or surfaguides as surface-wave launchers to produce a plasma column in a long, thin tube. The term '*surfatron plasma*' is often used in literature to designate surface-wave sustained plasmas. This plasma source is reproducible and stable, but the plasma is limited in length as well as in volume and is far from being homogeneous.

Komachi and Kobayashi [4] have used this principle of plasma generation and developed a planar source where the plasma is produced below a dielectric plate. In the boundary sheath between this plate and the plasma, surface waves can propagate and sustain the plasma. The microwaves are partially reflected on the surrounding metallic walls of the discharge chamber, resulting in standing waves and in the generation of an inhomogeneous plasma. The maxima and minima of the standing microwave can be observed as bright and dark spots in the pattern of light emission. However, at lower pressure and far from the dielectric plate, this effect is smoothed by diffusion processes, so that this type of source can be used for specific plasma technological applications.

Ganachev *et al.* [5] used a rectangular waveguide with two small coupling slots in the broad side of the waveguide. Microwaves with a frequency of $f = 2.45\,GHz$ and a power of typically 1 kW are fed to the waveguide, and a plasma is generated below a dielectric plate. A plasma radius of 11 cm can be achieved with such a device. This type of plasma source is, for example, used as resist asher in the semiconductor industry.

Surface-wave plasma sources in large diameter dielectric tubes have been successfully developed by Engemann *et al.* [6]. In these so-called slot antenna (SLAN) type of plasma sources, the microwave field is transferred from a rectangular waveguide by a coupling probe into a ring cavity. There the microwaves are radiated through equally spaced coupling slots into a vacuum chamber (quartz or glass tube) where a plasma is formed. Based on this concept, plasma sources of different diameters from 4 to 66 cm have been developed. Additionally, the SLAN sources can be combined with the magnetic field from cobalt–samarium (Co–Sm) permanent magnets producing a magnetic field strength of 0.0875 T inside the quartz tube. In this region the electron cyclotron resonance (ECR) condition is fulfilled and a plasma is generated at low pressure.

2.2.1.2 Standard Microwave System for Textile Treatment

As an example of a standard microwave system, we describe the system from L'Institut Français Textile-Habillement (IFTH) Lyon shown in Figure 2.1a,b. This system was used to perform hydrophilic and hydrophobic textile treatment, as described in Chapter 6 (see especially Sections 6.5.2 and 6.5.3).

The treatment chamber is a vertical cylinder having a diameter of 45 cm and a height of 45 cm having a volume close to 90 l. It is equipped with a system for winding/unwinding the material to be treated. The upper part of the chamber is connected to a quartz tube in which the plasma is generated via a microwave generator operating at 2450 MHz.

Pumping is performed in two steps using a rotary and a diffusion pump and should ensure proper moisture extraction from the material to be treated, necessary for a good plasma quality. Figure 2.1a shows the scheme of the equipment.

The plasma equipment comprises five 'gas' lines and one 'liquid/gas' line. Gas or vaporized liquid precursors can be introduced through the 'gas' lines into the reactor chamber for different processing purposes. Examples are etching/cleaning (e.g., via Ar, He, H_2, . . .), oxidation (e.g., via air, O_2, CO_2, . . .), nitriding (e.g., via N_2, NH_3, . . .), fluorination (e.g., via CF_4, fluorine-containing monomers, . . .), or even thin film deposition (e.g., via acetylene, C_3F_6, propene, . . .). The 'liquid/gas' line operates with liquid products whose vapor pressure is larger than the pressure used during the treatment.

The system is quite versatile as a large number of treatment parameters can be varied:

- type of gas and/or liquid, single, or mixture;
- position of the gas and/or liquid input (introduction into the upper or lower part of the chamber);
- gas and liquid flow (0–$100 \, cm^3 \, min^{-1}$);
- distance between the samples and the plasma source (37–$57 \, cm$) when the treatment is performed in a static way;
- electrical power (700–$1100 \, W$) and frequency (433 or $2450 \, MHz$) of the microwave source;
- static or dynamic treatments, that is, samples can be kept fixed but it is also possible to perform roll-to-roll runs, the winding/unwinding all integrated within the vacuum vessel. Regarding this roll-to-roll treatment, the following throughput is possible: for yarns up to $30 \, m \, min^{-1}$, for woven or knitted fabrics up to $5 \, m \, min^{-1}$ at a width of 20 cm. Thousands of meters of fabric or yarn can be treated per batch.

2.2.1.3 Example: Duo-Plasmaline – a Linearly Extended Plasma Source

The work of the Institut für Plasmaforschung of Universität Stuttgart regarding the design of surface-wave sustained microwave plasmas started in 1994 with the Gigatron [7] and the Duo-Plasmaline [8], which can be considered as a kind of inverse configuration of the above-mentioned surfaguide or surfatron device (cf. Section 2.2.1.1). In the Gigatron, the microwave generator is connected to only

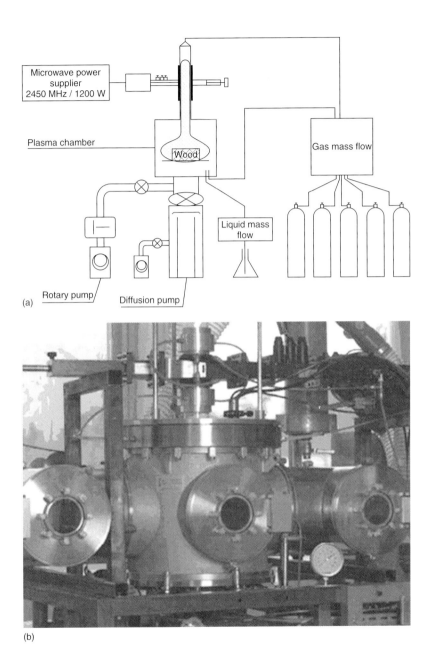

Figure 2.1 Microwave system used for low pressure textile treatment. (a) Schematic representation of the set-up; (b) the main chamber in which the actual treatment takes place.

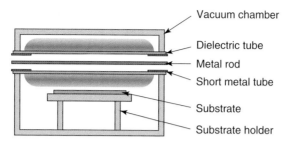

Figure 2.2 Schematic of the Duo-Plasmaline plasma source.

one end, whereas in the Duo-Plasmaline, the microwaves are fed from both ends which considerably improves the homogeneity of the plasma.

In the experimental set-up of the Duo-Plasmaline presented in Figure 2.2, a dielectric tube of glass or quartz with a concentric copper rod and short metal tubes at each end is mounted through a vacuum chamber. Typically, the dielectric tube is 3 cm in diameter and about 80 cm in length. Its inner part is at atmospheric pressure and it is cooled with compressed air. The central copper rod has a diameter of 0.8 cm. A coaxial waveguide is formed by the short metal tubes at the inner surface of the dielectric tube together with the metallic rod, which guides the microwaves into the vacuum chamber.

After having exceeded the threshold of electric discharge, the microwaves at $f = 2.45\,\text{GHz}$ irradiated through the quartz tube into the low-pressure regime inside the vacuum chamber sustain the formation of the plasma around the quartz tube. At low microwave power, the plasma is concentrated only at the two ends of the tube (see Figure 2.3a). With increasing microwave power, the plasma extends from both ends along the tube (see Figure 2.3b), and finally, by further increasing the microwave power, an axially homogeneous plasma is formed (see Figure 2.3c). Different gases can be used in the pressure range of about 1–1000 Pa depending on the plasma technological demands. An example: when using argon as process gas at the pressure of 10 Pa and applying the microwave power of $2 \times 500\,\text{W}$, a linearly extended plasma of about 80 cm in length is formed.

Now, we consider in more detail the electromagnetic wave propagation in a microwave plasma. As derived in Chapter 1 (see Equation 1.17 in Section 1.3.4), for microwaves with a frequency of $f = 2.45\,\text{GHz}$, wave reflection occurs when the critical electron density n_c equals $7.5 \cdot 10^{16}\,\text{m}^{-3}$.

A first consideration yields that the microwave field produces a plasma with the maximum electron density being equal to n_c. In this case, further propagation of the microwaves in the radial direction is not possible. The plasma itself shields the microwaves and forms the outer electrode of the coaxial arrangement. The electromagnetic wave mainly propagates inside the tube along the inner conductor and within the plasma as a radially decaying surface wave [8].

A more detailed view shows that we have to look at the skin depth of the microwaves. Let us now consider a microwave beam propagating in a plasma with an increasing electron density (n_e) profile (see Figure 2.4). As the electron plasma

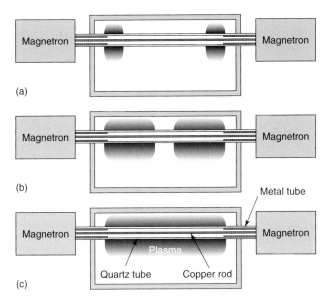

Figure 2.3 Side view of the Duo-Plasmaline for increasing microwave power (from top to bottom). (a) At low microwave power: concentration of the plasma at both ends of the tube only; (b) increase of the microwave power: the plasma extends from both ends along the tube; (c) finally, by further increasing the microwave power, an axially homogeneous plasma is formed.

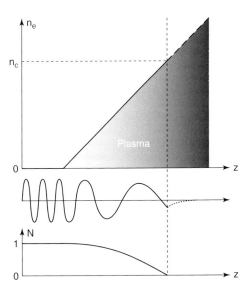

Figure 2.4 Electromagnetic wave propagation and refraction index N in a plasma with increasing electron density.

frequency $\omega_{p,e}$ is given by $\omega_{p,e} = \sqrt{\frac{e^2 \cdot n_e}{\varepsilon_0 \cdot m_e}}$, this will cause a rise of the electron plasma frequency $\omega_{p,e}$, and as a consequence, k will decrease and the wavelength λ, defined as Equation 2.1,

$$\lambda = \frac{2\pi}{k} = \frac{2\pi c}{\sqrt{\omega^2 - \omega_{p,e}^2}} \tag{2.1}$$

will become longer and longer. Finally, the critical electron density n_c will be attained when k is zero and λ is infinite. For higher electron densities ($n_e > n_c$), the dielectric permittivity ε is complex and the wave cannot propagate any further and will be exponentially attenuated within a characteristic length, called the *skin depth* (Equation 2.2) [9]:

$$\delta = \frac{c}{\sqrt{\omega_{p,e}^2 - \omega^2}} \tag{2.2}$$

This means that microwaves with the vacuum wavelength of $\lambda = 0.12\,\mathrm{m}$ approach the cut-off 'point' ($\omega = \omega_{p,e}$) with increasing wavelength.

The region where the wave reflection occurs has a width of only some millimeters. This allows wave tunneling with subsequent resonant absorption which produces an 'overdense' plasma with $n_e > n_c$. This is a typical phenomenon for microwave plasmas where densities of up to $n_e \approx 10\,n_c$ can be achieved.

A Duo-Plasmaline of 3 m length was set up at our institute to demonstrate the potential of this plasma source [10]. When applying the microwave power of $2 \times 1\,\mathrm{kW}$, an axially homogeneous argon plasma is formed along the entire length of 3 m (see Figure 2.5).

The arrangement of several Duo-Plasmalines in parallel can be used to obtain a 2D plasma array. Measurements on such a plasma array with two Duo-Plasmalines, each 80 cm in length and with the distance of 9 cm between them (see Figure 2.6), show that the electron density and the deposition rate are homogeneous on an area of 10 cm × 50 cm at the distance of about 6 cm perpendicular to the Duo-Plasmalines [11]. The effective plasma area can be further increased by applying additional Duo-Plasmalines. A smaller but more flexible equipment called *Plasmodul* has been developed for applications on laboratory level [12, 13]. It consists of several modules made of aluminum, among them gas inlet systems, spectroscopic modules, pumping units, and a plasma array made up of four Duo-Plasmalines [14]. The standard Plasmodul device forms a cylinder with a diameter of 35 cm and a height of 40 cm. The electron density distribution in this plasma array was found to be homogeneous on an area of at least 10 cm × 18 cm [15].

2.2.1.4 Electron Cyclotron Resonance Heated Plasmas

The use of magnetic fields in connection with microwave excitation can serve different purposes. Relative to unmagnetized plasmas, a magnetic field configuration can be used for confinement and shaping of the plasma. Another purpose is microwave power absorption, which is very efficient even at low gas pressures,

Figure 2.5 Duo-Plasmaline with a length of 3 m.

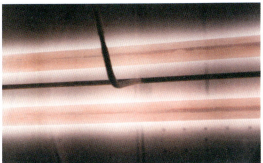

Figure 2.6 A plasma array consisting of two Duo-Plasmalines.

especially when the ECR condition (Equation 2.3) [16]

$$\omega = \omega_{c,e} = \frac{e \cdot B_0}{m_e} \tag{2.3}$$

is fulfilled. Then, for the microwave frequency of 2.45 GHz, the resonant magnetic field strength is given by Equation 2.4:

$$B_0 = \frac{\omega \cdot m_e}{e} = 0.0875 \, T \tag{2.4}$$

The corresponding magnetic fields can be obtained by current driven coils as well as by permanent magnets, such as Co–Sm magnets. An ECR discharge is

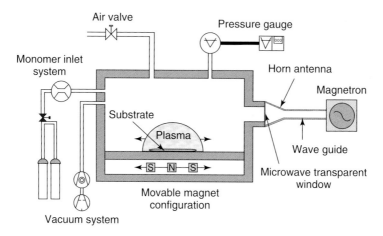

Figure 2.7 Schematic of the experimental set-up of the ECR plasma source.

the only type of microwave discharge that can be operated efficiently at low gas pressures in the range of 1 Pa to approximately 50 Pa. The reduction of the gas pressure leads to an increase of the mean free path of the electrons and to the corresponding decrease of collisions, a fact which directly affects the chemistry in plasma-technological processes.

In our work we used the experimental set-up shown in Figure 2.7. A magnet system is installed underneath a bottom plate inside the vacuum vessel 80 cm in diameter and 50 cm in height. It consists of three rows of Co–Sm permanent magnets arranged on a rectangular iron yoke. The magnets in the middle row are orientated with their north poles facing upwards. The magnets in the left and right row, enclosing the middle row, are orientated in opposite direction, that is, with their south poles facing upwards. This configuration is called '*closed racetrack configuration*' (*magnetron device*) [17, 18]. The entire magnet arrangement has a length of 42 cm and a width of 12 cm and consists of 48 single Co–Sm magnets. The microwaves are radiated via transmission lines and a horn antenna into the vacuum vessel. About 1 cm above the bottom plate of the vacuum vessel, the magnetic field strength B is 0.0875 T. According to Equations 2.3 and 2.4, the resulting electron cyclotron frequency $\omega_{c,e}/(2\pi)$ is equal to the applied microwave frequency of $f = 2.45$ GHz. In this region, the ECR condition is fulfilled and the plasma is preferentially produced and heated. The charged particles undergo intricate motions with specific drift velocities in such an inhomogeneous magnetic field, and often an analytical solution of the equations of motion cannot be found. An overview of the curvature drift and the $\boldsymbol{E} \times \boldsymbol{B}$ drift is given, for example, in Chen [16]. The closed racetrack configuration used here reduces the $\boldsymbol{E} \times \boldsymbol{B}$ drift velocity and improves the plasma confinement, resulting in a homogeneous discharge with high deposition uniformity. Additionally, the magnet arrangement is movable in parallel to the bottom plate inside the vacuum vessel, so that a homogeneous deposition on an area of 50 cm × 40 cm can be obtained.

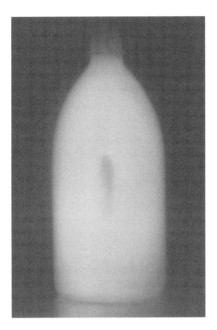

Figure 2.8 An ECR plasma generated in a bottle by a
Co–Sm permanent magnet (dark area in the center of the
bottle).

Furthermore, the ECR technology opens up the possibility to generate a
low-pressure plasma directly inside a hollow body with microwave-transparent
walls, for example, a container or a bottle: when applying a Co–Sm permanent magnet into the inside of the hollow body, the plasma is generated in the ECR region,
that is, only in the internal sphere of the hollow body, without any pressure difference between the internal and the external sphere (see Figure 2.8). This particular
set-up can be used for surface cleaning or for deposition of thin plasma polymerized
barrier films on the inner surface of hollow bodies even complex in shape.

2.2.2
Capacitively Coupled Systems

2.2.2.1 Introduction

In essence, a capacitively coupled system consists of a vacuum chamber with two
flat electrodes. One electrode is at ground potential while the other is connected to a
high frequency power generator. Because of the high frequency of the electric field,
predominantly the electrons are influenced and they become (more) energetic.
Through collisions, the energy is transferred to the heavier particles. But this
transfer of energy from the light electrons to the heavier particles is not so efficient.
Hence, the electron temperature is much larger than the ion and neutral gas
temperatures. Typical values for the electron temperatures are 2–7 eV whereas the
ion and neutral electron temperatures stay roughly below 0.05 eV for gas pressures

below 5 Pa and particle densities below $10^{10}\,\text{cm}^{-3}$. The term *'nonthermal'* stems from this thermodynamic nonequilibrium. Because of the low energy of the heavy particles, this type of discharge is referred to as *'cold' plasma*.

Charged particles tend to get lost at the surfaces that physically limit the discharge area (walls, electrodes). Due to their lighter masses, the electron loss at the electrodes and walls is initially higher than the ion loss. The resulting positive charge in front of the electrodes produces a strong electric field in this region. In these so-called space charge regions or 'sheaths', electrons are repelled from and ions accelerated towards the electrodes. This way, the electron loss is reduced and equalized with the ion loss. Moreover, this process leads to ion bombardment of the walls and electrodes, which can also give rise to damage on the substrate to be treated.

Another drawback is that with the basic configuration the ion density and ion energy cannot be controlled separately. With more advanced systems this drawback can be overcome by tuning the frequency of the applied electric field [19].

2.2.2.2 Capacitive Coupled Plasma for Biomedical Applications

Figure 2.9 gives an example of a capacitive coupled system. This system was used to obtain the antifouling films described in Chapter 7.

This system consists of a home-made stainless-steel reactor (dimensions: $300 \times 300 \times 150\,\text{m}^3$) and comprises two symmetric internal parallel-plate electrodes [20]. The diameter of the electrodes is 140 mm, the inter-electrode distance equals 50 mm. A radio frequency (RF) generator (13.56 MHz) is connected to the upper electrode; the bottom electrode serves as sample holder and remained grounded. Typically, the discharge was run between 10 and 50 Pa using argon, the power was supplied both in continuous and in pulsed mode.

For the gas supply, a shower head configuration as shown in the right part of Figure 2.9 was used to ensure an optimal uniformity. Gas flow rates were monitored

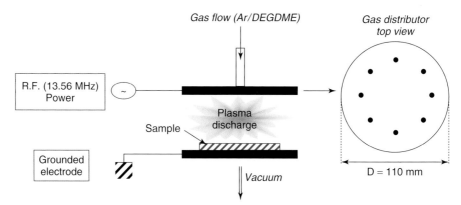

Figure 2.9 Capacitive coupled low pressure system used for the antifouling coatings discussed in Chapter 7. The right part shows the shower head construction used for adding the gas in a uniform way.

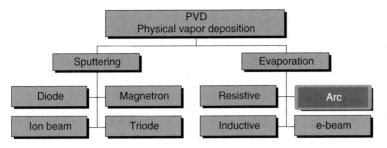

Figure 2.10 General classification of PVD processes and technologies. Highlighted is (Cathodic) Arc PVD deposition which is discussed in Section 2.2.3.2.

via MKS mass-flow controllers and the gas pressure was measured using an MKS baratron.

2.2.3
Physical Vapor Deposition Plasma: LARC®

2.2.3.1 Background
PVD systems include a large variety of technologies used for different applications: cathodic arc PVD, magnetron sputtering, ion beam (i-beam), and electron beam (e-beam) evaporation, as shown in Figure 2.10. The common feature of these systems is that they are run at low pressure and that the main processes involved in the thin film deposition process are physical, for example, the ion bombardment of the target, the resulting removal of material (sputtering), or particle transport to the substrate.

2.2.3.2 Cathodic Arc PVD Systems
In cathodic arc deposition, a high voltage is applied between a cathode and a grounded anode. By choosing the correct configuration of the electrodes and discharge parameters an arc discharge is generated. The cathode arc spot is small ($10^{-8}-10^{-4}$ m in diameter) and concentrates high currents (spot current densities are in the range 10^6-10^{12} Am^{-2}), which via Joule's effect leads to cathode (target) surface local overheating, melting and subsequent evaporation and ionization of the target element(s) (Figure 2.11). This establishes a feedback mechanism close to the cathode surface allowing further electron emission and cathode heating, and therefore emission of ions and droplets. This results in plasma domain expansion from the cathode domain toward the anode, which in turns leads to the filling of the reactor volume. From measurement [21], it is possible to deduce that (i) the plasma contains few neutral species (with energies of 5 eV), (ii) the kinetic energy of the ions is relatively high (20–100 eV) compared with evaporation based technologies (10^{-1} eV) or sputtering (5–10 eV), and (iii) the ion energy is higher than the arc voltage energy. Further, it is reported that emitted electrons have

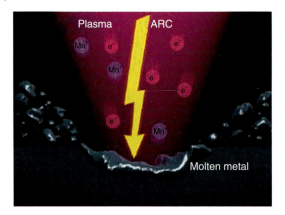

Figure 2.11 Sketch of evaporation process of target material, including generation of droplets of molten metal due to uncontrolled overheating of the surface.

almost near thermal energies, that the electric current carried by ions is around 8% of the total current and that the electric field in the plasma is weak [22].

Since the plasma is generated within the whole volume of the reactor this technique is ideal for coating objects with a complex external surface shape. Indeed, standard flat surfaces are better coated with other techniques, for example, via the sputtering processes.

PVD arc cathodes are provided as planar targets of different shape and dimension, according to the coating unit. The arc spot either follows a random motion path on the target surface or is driven by electromagnetic fields. The arc spot residence time at a given point of the target surface is quite important since it determines the evaporation efficiency of the source material. If the operating conditions are wrong one risks the target melting, which leads to an excess of metal droplets ejected from the target into the reactor chamber. When depositing onto the substrate surface these cause defects in the coating, resulting in inhomogeneity and local concentration of stresses, which all too often cause coating delamination. Since droplets are ejected from the target at low normal angles (the peak of the distribution is at $20°$ with respect to the target surface normal), different solutions have been devised to minimize the droplet fraction. Solutions are based on filtering systems such as a shielded source [23], an enhanced source with field coil [24], a plasma duct filter [25], a knee source [26], or a deflection source [27]. The erosion rate of the macro-particles W_{mp} may be estimated from Equation 2.5:

$$W_{mp} = W_{tot} - \frac{f \cdot m_i}{e \cdot Z} \qquad (2.5)$$

where W_{tot} is the target total erosion rate, f the total ion current fraction (typically about 0.1), m_i is the ion mass, and eZ is the ion charge. Ejection of macro-particles increases with decreasing metal melting point. In case of the deposition of titanium nitride (reactive arc deposition) it was found that the size of the emitted

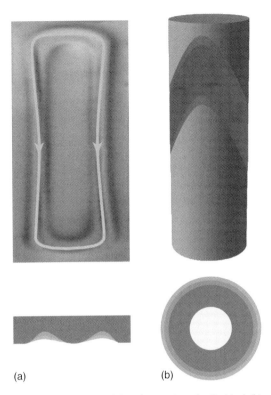

(a)   (b)

Figure 2.12 Sketch of the planar (a) and cylindrical (b) cathode erosion tracks caused by an arc spot current driven motion due to the $\textbf{\textit{E}} \times \textbf{\textit{B}}$ drift.

macro-particles depends on the nitrogen partial pressure: lower sizes are found with increasing nitrogen pressure, while emission speeds range from 0.1 to 800 m s^{-1}, depending on the droplet mass.

One of the most effective approaches to minimize the generation of droplets is to carefully control the evaporation or target material. A common method to drive the arc spot on the surface cathodes is to impose a dipolar magnetic field perpendicular to the electric field to exploit the $\textbf{\textit{E}} \times \textbf{\textit{B}}$ drift of charged particles (including electrons). This drives the arc spot currents into an elliptical path (retrograde motion), providing controlled cathode erosion (Figure 2.12).

So far the process described has been limited to the deposition of metals. In order to synthesize binary, ternary, or quaternary compounds (nitrides, carbo-nitrides) a precursor gas can be injected. A typical example of a binary compound that has been in use for more than 20 years is TiN, deposited as hard coat on machining tools (e.g., milling and cutting tools, inserts, saws, drills, and broaches). Here a titanium cathode is used and nitrogen gas is injected into the plasma reaction chamber, leading to the deposition of TiN onto the substrate surface. By means of adjusting the gas pressure, the arc currents and the nitrogen gas mass flow rate, it is possible to grow sub-stoichiometric, stoichiometric, or super-stoichiometric Ti_xN_y coatings ($x < y$, $x = y$, or $x > y$, respectively). To enhance the film growth rate and to control the residual film stresses, a DC bias potential between 0 and 1000 V is applied to the substrate.

2.2.3.3 Example: Treatment of Food Processing Tools by LARC® PVD System

Since the target material needs to be of high purity, the cost of the process can be significantly lowered by increasing the efficiency of the thermal target erosion. Since 2002, a new solution has been presented with lateral rotating cathodes, the so-called LARC® PVD technique [28]. The rotating cathodes are cylindrical and rotate along their longitudinal axes. As a result the arc spot is guided into an elliptical path along the lateral external cylinder surface. The combined effect of the arc spot motion and the cylindrical cathode rotation results in the arc spot sweeping the target surface continuously along a sinusoidal path whose wavelength ratio with respect to the cylinder circumference is an irrational number.

Cathodic arc evaporation is operated as a batch process. The samples are mounted onto a sample holder, which very often can also be rotated (on a carousel) to obtain highly uniform coatings on complex shapes (see Figure 2.13).

In the following we would like to discuss the use of cathodic arc PVD for the coating of food processing components within the ACTECO project as described in Chapter 9. Prior to the actual PVD treatment the samples are washed. The

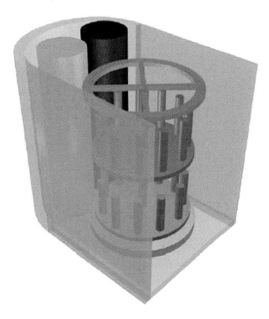

Figure 2.13 Sketch of the Cathodic LARC® reaction chamber in which two cylindrical cathodes are positioned behind the rotating carousel onto which the samples to be coated are mounted.

samples are then loaded into the reactor, which is pumped to the required gas pressure. Then a second cleaning process follows, consisting of heating, purging and etching by energetic argon ion bombardment. The latter is required to remove contaminants and native oxides on (metal) surfaces. In this phase the cathodes are inactive. During the next step a nucleation layer for the coating is prepared by ion bombardment (e.g., by titanium ions) using either one or both cathodes. Finally, the actual growth of the thin film is carried out by reactive ionized gas such as nitrogen (to develop nitrides) or acetylene (to grow carbonitride coatings) in conjunction with evaporation of material from the cathodes. Through careful control of the process parameters, especially the evaporation rate, it is possible to grow monolayers or gradient and multilayered structures. Also the crystalline structure of the thin films can be controlled, for example, nano-crystals embedded in an amorphous matrix. This allows deposition of a new class of super hard nano-composite ceramic coatings [29, 30].

Typical process parameters are arc currents in the order of 50–120 A, a bias voltage in the etching phase of 800 V, which is reduced to 100–150 V in coating phase, resulting in a substrate current typically below 10 A. After the purging cycle at the absolute base pressure of 10^{-3} Pa, the operating pressure is set in the order of 10^{-2} Pa. The potential drop between the cathodes and the chamber walls is some tenths of volts. Electrical heaters inside the chamber provide a process temperature between 350 and 700 K. The typical process cycle duration is about 2.5–4 hours during which a growth rate of about $4-5\,\mu m\,h^{-1}$ is maintained. For more information about the application and the realized hardness improvement of the samples we refer to Chapter 9, its economical aspects are considered in Chapter 11 and a life cycle assessment (LCA) study of the process is presented in Chapter 12.

2.3
Atmospheric Pressure Plasma Systems

More recently much attention has been paid to the development of atmospheric pressure NTP sources. As a result, progress in this field has been impressive in the last years. A number of overviews about NTP production at atmospheric pressures and some applications appeared recently [31–37].

Approaches to develop atmospheric pressure discharges operating at strongly nonequilibrium conditions vary, and a large spectrum of discharge devices already exists. Starting from the dielectric barrier discharge (DBD or so-called 'silent discharge') and corona discharges, which have been known for more than a 100 years but nevertheless have only more recently been used for surface cleaning and activation, a number of different discharge techniques have been successfully developed. This gave rise to the development of sources for NTPs at atmospheric pressure. The interest in this topic stems from the large potential economic benefit expected from numerous NTP technologies. Indeed, it is anticipated that many of these potential applications (e.g., protective coating deposition, toxic and harmful gas decomposition, electromagnetic wave shielding, polymer and fabric surface modifications, and biomedical applications) may see rapid progress. Due to this wide variety of potential applications, requirements for the relevant plasma parameters are also rather diverse.

Along with progress in the discharge techniques, the theory of gas discharge formation and of plasma properties has been developed. The specific features of a NTP produced by a DBD, corona or glow discharge are due to the high sensitivity of their characteristics to the type of gas being used for generating the discharge. In contrast to thermal plasma, which can be completely characterized by a few parameters (gas pressure, temperature, dissipated power density), the number of parameters required for a full description of a NTP is rather large. Typically, electron energy distribution is strongly nonequilibrium, while the ions have nearly the same average energy as the atoms and molecules.

Most of the atmospheric pressure discharges generate nonequilibrium plasmas in the inter-electrode gap based on large numbers of nonstationary micro-discharges, that is, current filaments (so-called *streamers*) of micro- or even nano-second duration. As a result, highly reactive species are generated in the plasma, but due to the high pressure (atmospheric) their lifetime in the plasma is, in general, very short. Therefore, it is necessary either to produce these active species immediately at the surface to be treated or to provide a fast transfer from the plasma region to the surface. Another possibility is to generate the discharge with a gas in which the life time of the active species is longer, which also results in a larger active plasma zone. In practice, this can be realized by using nitrogen instead of standard air. In all atmospheric plasma surface treatment systems one or more of these methods are implemented.

The use of atmospheric pressure allows expensive vacuum equipment to be eliminated and eases the incorporation of a plasma treatment into industrial production lines. Nevertheless, it is only more recently that atmospheric plasmas are gaining interest from industry. The reason is that it is far more difficult to generate a stable, uniform and repeatable plasma discharge with a sufficiently large surface area for surface processing at atmospheric pressure than at low pressure. Indeed, several stabilization approaches are needed, for example, using a high frequency electric field, applying a high gas flow, keeping the inter-electrode distance small, or adding small amounts of noble gases [38].

Another aspect hampering the use of atmospheric pressure plasma systems is the considerable consumption of process gas. Indeed, the main advantage of atmospheric plasma units is their low depreciation costs because of the limited initial investment, but they are characterized by relevant variable costs due to the gas consumption, especially if noble gases are required.

Safety issues may also become important when performing processes with precursors that may impact on the environment at global or local scale. Solutions to these challenges are being devised on a case to case basis, as exemplified in Chapter 12.

In the next sections, we will discuss some atmospheric systems in more detail. The main distinction is made on the solution(s) applied to deal with the fact that the species are only short living. In the corona-type systems, a small distance between the active particles and the substrate is provided. Other types of sources provide a remote treatment by ensuring fast transfer of the active species to the substrate.

2.3.1
Corona-type Surface Treatment

As mentioned before, for so-called 'corona treaters' the fast transfer of active plasma species is ensured by providing a small distance between the substrate and the electrodes. These systems are based on DBDs. The frequency of the applied electric field is typically in the range of 10–40 kHz; and the applied electric field is quite high, as the applied potential is of the order of several kilovolts.

The main drawback of the system is that the substrate to be treated has to pass between the electrodes and the counter electrode. This means the substrates pass through the electric field and, thus, the electrical discharge current passes through the surface while it is being treated. In some cases, this may lead to damage or destruction of the substrate. Especially porous substrates with varying density are prone to this effect.

2.3.1.1 Standard Corona Treatment
The standard corona system consists of naked pure metal blade electrodes above the grounded counter electrode, which is typically the roller over which the substrate travels. However, standard corona equipment usually comprises shielded metal electrodes so that one obtains a narrow strip of a hybrid DBD-corona plasma. A corona discharge is generated between the shielded electrode and the roller. The substrate is treated as it passes through this corona discharge. Typically, standard corona discharge treatment (CDT) operates in ambient air, resulting in the following chemical and physical processes:

- cross-linking
- chain scission
- incorporation of oxygen containing functional groups.

Often this results in an increase of the surface energy of the substrate surface, implicating (i) a better wettability and capillarity and (ii) the availability of chemical bonds for adhesion with, for example, a coating, an ink or an adhesive. Standard CDT can operate at line speeds of a couple of meters per minute up to 1000 m per minute and even higher. The maximum line speed achievable for a given application is determined by the electrical power input, the substrate material, the surface structure and the magnitude of the induced effect that is required.

CDT is often applied to polymer films (PP, PE, PET, PVC, PA), foils (metal), papers, foams, woven fabrics, and non-wovens via roll-to-roll systems. The corona discharge treated substrates are then used in a variety of applications including printing, painting, gluing, laminating, metalizing, coating, and other converting processes. When CDT is used as pre-treatment, as described here, the subsequent process needs to be done within a certain fixed time period after the corona pre-treatment to avoid aging (see also Chapter 6, Section 6.2.2.3). As a result, these corona systems are practically always systems that are integrated in a production line.

2.3.1.2 Controlled Atmosphere Corona Treatment–Aldyne Treatment

Standard corona systems are normally equipped with a hood or enclosure around the treatment zone. In the first place, this is required for safety reasons as it allows the removal of the ozone generated. However, this set-up can also be adapted to control the atmosphere in the treatment zone. As a result, corona discharges can be generated in atmospheres other than ambient air, for example, pure N_2, He, or Ar, with the remaining concentration of air being less than 100 ppm. Such a controlled atmosphere enables the type of functional groups that are introduced onto the surface to be tuned. An example is the use of nitrogen, so that functional groups such as amine, amide, and imide can be imparted. It is also claimed that controlled atmosphere surface activation is less prone to aging compared with standard corona activation. One such system is the Aldyne system from Air Liquide and Softal. A schematic diagram of this Aldyne system and how it is implemented in an industrial unit is shown in Figure 2.14a,b.

Further optimization of the surface treatment is possible based on a N_2 discharge with the addition of specific dopants (e.g., CO_2, H_2, and/or N_2O). The surface chemistry can be tailored to a variety of applications, like increasing the surface energy, wettability, hydrophilicity, adhesion, water-based ink printability, and lamination or metalization. The processed materials can be polymer films, textiles, non-wovens, papers, or metallic foils. Large scale treatment is possible. Depending on the specific treatment conditions and required functionalities, line speeds up to 300 m min^{-1} can be achieved.

2.3.1.3 Liquid Deposition

A further development of the corona or controlled atmosphere corona is the introduction of liquid precursors into the corona discharge zone. Here a liquid precursor dosing system is added to the standard corona device, as sketched in Figure 2.15. This technology has been developed by, for example, Ahlbrandt System GmbH who commercialized it as the 'AS Corona Star' system. Typical plasma operation parameters are an electrical voltage of 20 kV, a frequency of 22 kHz, electrical power of 1000 W, and air as discharge gas.

Water-based solutions of the precursor are sprayed into the corona discharge, which results in coating of the substrate, typically some nanometers thick. These coatings can have different properties, for example, antistatic, antifogging, release liners, primer layers, or disinfectants. An example of an interesting application of this technology is the deposition of antimicrobial layers [39].

The Aldyne system, mentioned in the previous section, can also be adapted in this way. Indeed, by adding liquid precursor molecules (such as DMSO) to the gas mixture in the discharge, thin-film coatings can be achieved for targeted applications like barrier coatings, anti-fog coatings, or anti-wear/anti-scratch. A similar approach, but using tetraethoxy silane (TEOS) as precursor, was used

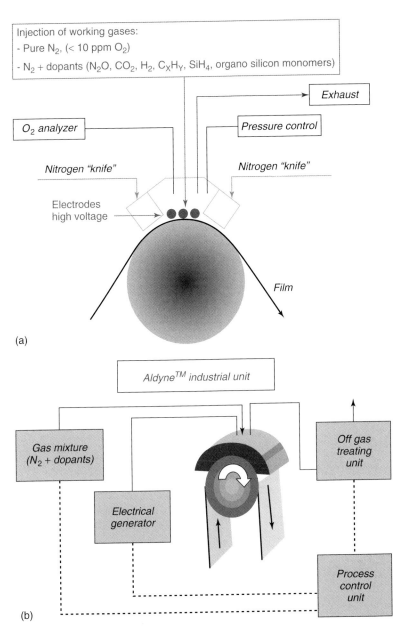

(a)

(b)

Figure 2.14 The Aldyne system. (a) Schematic representation of the Aldyne process, of particular importance are the gas knifes which control the gas mixture atmosphere within the reaction volume; (b) schematic of an industrial unit.

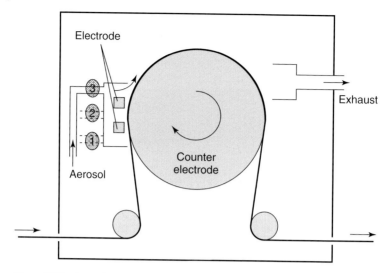

Figure 2.15 General principle of adding a liquid precursor in aerosol form to a corona discharge. The sketch is based on the design of the 'AS Coating Star' from Ahlbrandt Systems GmbH.

to treat PET fabrics, see Chapter 5, Section 5.3.2.5. As can be seen from the time-of-flight secondary ion mass spectrometry (ToF-SIMS) images there, this method allows very uniform thin coatings to be obtained.

2.3.2
Remote Surface Treatment

Corona systems and large area atmospheric pressure plasma processing is often limited to substrates that are flexible and have a predominantly 2D profile. In order to treat more rigid substrates or devices that have complex geometries a variety of nonthermal equilibrium 'remote surface treatment' systems have been developed, in the form of 'plasma torches', 'plasma pencils', or 'plasma jets'. Several novel types of plasma sources providing remote treatment at atmospheric pressure are described by Akishev *et al.* [40].

These sources basically implement the second strategy to deal with the short life-time of the active species in atmospheric pressure plasmas, that is, to provide a fast transfer of active species from the plasma region to the surface. This is achieved by blowing the active species out of the plasma source and towards the surface to be treated: a gas stream is directed between two electrodes, to which an electrical potential has been applied, resulting in the formation of a plasma in jet or torch form.

Several systems are available, their main difference being the electrode design. A first solution is based on a RF powered metal needle acting as a cathode, surrounded by an outer cylindrical anode. An alternative to this design is the use of two co-axial tubes as electrodes. Designs vary as to whether dielectric material is used in conjunction with one or both of the electrodes. Other devices used a needle electrode to form the plasma and the substrate is used as a grounded electrode. Also designs based on parallel plate technology exist. In this case, a plasma is generated between the parallel plate electrodes and exits the device via the process gas flow.

Various nozzles have been designed to further direct the plasma and to control its focusing/spreading, which is important for controlling activation, cleaning, or etching of various substrate surfaces, including biomedical applications [41].

In general, instability problems, like, for example, arcing, can be prevented by using gas mixtures containing He (which limits ionization), high flow velocities, and/or by properly spacing the inter-electrode distance.

Remote treatment works well for many applications, but it does not provide a 100% use of the active species that are produced in the plasma source, due to inevitable losses during their transport by gas flow. However, in contrast to corona devices, the remote plasma sources can be applied to materials of any thickness or rigidity. Moreover, the substrate does not pass through the electric field, thus mitigating the risk of damaging it.

In the next sections, we will highlight three different types of remote surface treatment systems: two were used in the experiments related to the modeling (see Chapter 10) and one is already a commercially available plasma jet system.

2.3.2.1 Plasma Sources Used for Modeling

Chapter 10 of this book deals with the understanding of the physics and chemistry of atmospheric NTPs. For this work, two different remote plasma sources have been used. These sources are referred to as *Source No1* and *Source No2*. We will now discuss the operation of these plasma sources in more detail.

A side view of the plasma emitted from sources No1 and No2 is presented in Figures 2.16a and 2.17a, respectively. Plasma source No1 (details reported by Akishev *et al.* [42]) generates a plasma jet in ambient air with the air stream having an average velocity about of $30-50$ m s^{-1}. This atmospheric pressure plasma source can work in two modes, namely (i) diffusive and (ii) filamentary or streamer mode. Figure 2.16a shows the discharge in streamer mode air flow, clearly showing the effect of blowing off the streamers with the gas flow (bright lines) from the experimental observations the radius of the streamers was determined to be 0.5 mm.

Typical waveforms of discharge voltage and current are shown in Figure 2.16b. For the current pulses strong fluctuations in the amplitudes and repetition frequencies of around 20 kHz were recorded.

Experiments performed with plasma source No1 also revealed that a homogeneous discharge in air was about six times less effective for treating a polymer surface than a discharge with same average energy loading showing numerous nonstationary streamers. For this reason, our main concern for this type of plasma source was to realize a reproducible nonstationary streamer mode.

Plasma source No2 (Figure 2.17a,b) generates a plasma jet in pure (99.999% of purity) nitrogen having an average gas velocity of about $30-50$ m s^{-1}. In the

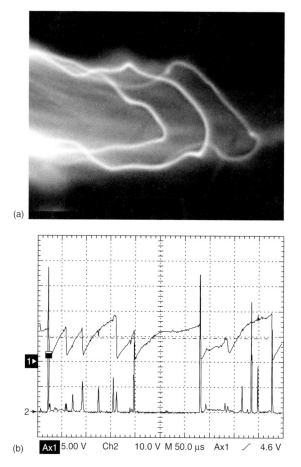

(a)

(b)   Ax1 5.00 V   Ch2   10.0 V M 50.0 µs   Ax1   ∕   4.6 V

Figure 2.16 Atmospheric pressure plasma source No1 used for remote surface treatment outside of the discharge zone. (a) Side view of plasma source No1 in the streamer mode; ambient air, gas flow velocity is 40 m s^{-1}; average electric power is $P = 35$–70 W; the maximum length of the streamer plasma jet in air is about 3 cm; (b) The waveforms of discharge voltage (upper curve) and current (lower curve) in the nonstationary streamer mode. Scales: voltage is 5 kV per unit, current is 200 mA per unit, time is 50 µs per unit.

experiments devoted to remote treatment this steady-state atmospheric pressure plasma source was used only in diffusive mode. The length of the plasma jet depends on the type of ambient gas. In the case of nitrogen it can reach 15 cm, making the plasma jet an effective instrument for surface treatment. However, our measurements revealed that the composition of active species in nitrogen plasma jet included not only excited nitrogen molecules, but also other species generated in the plasma due to the existence of traces of, for example, oxygen and hydrocarbons.

(a)

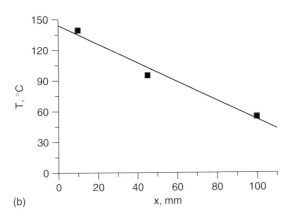

(b)

Figure 2.17 Steady-state atmospheric pressure plasma source No2. (a) Side view of diffusive plasma jet generated in pure (99.999%) nitrogen; (b) Longitudinal distribution of gas temperature along the plasma jet ($x = 0$ is the nozzle of plasma source). Average electric power is $P = 70$ W, N_2 flow velocity is 30 m s^{-1}, diameter of plasma jet at outlet is about 8 mm. The maximum length of diffusive plasma jet in pure nitrogen is 10–15 cm.

2.3.2.2 Example: AcXys Plasma Jet

A special type of plasma jet system, developed by AcXys Technologies, is also referred to as '*plasma knife*'. Here the jet is extended as to form a 'curtain' of plasma, as depicted in Figure 2.18a,b. The heart of the plasma source (Figure 2.18a) consists of two concentric cylindrical electrodes, between which the plasma discharge is maintained (excitation region). The outer electrode has an inlet for the introduction of the gas to be excited and an outlet, in the form of a slit, through which the excited and/or unstable gas can leave the excitation region.

Such plasma jets come in different lengths varying from 60 to 400 mm. For a 60 mm version, the power can be up to 1000 W, which requires cooling and gas flows of some tens of litres per minute. The discharge can be run with ambient air or with nitrogen, but also addition of other gasses (oxygen or hydrogen) is possible for obtaining specific surface chemistries. By combining this plasma jet system

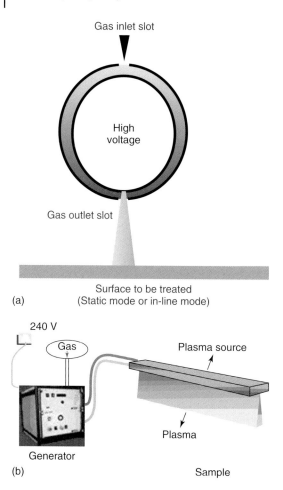

Gas inlet slot

High
voltage

Gas outlet slot

Surface to be treated
(a)   (Static mode or in-line mode)

240 V

Gas

Plasma source

Plasma

Generator
(b)   Sample

Figure 2.18   The AcXys Technologies plasma jet source (a)
sketch of the electrode design and gas flow; (b) sketch of
the general principle of the plasma jet commercialized by
AcXys Technologies.

with a liquid precursors dispensing system it is possible to deposit, for example,
SiO_x coatings [43].

2.4
Summary

To benefit from the potential of a plasma surface treatment, a large variety of
different systems have been developed and implemented, both at low pressure
and at atmospheric pressure. These pressure regimes have their own specific
advantages and disadvantages.

Since they operate in a closed environment, low pressure plasma systems offer, in general, as their main advantage a very large process flexibility that is very well controllable. In this chapter we have discussed the microwave systems in more detail, since this type of source yields a very uniform plasma over a large area.

When discussing the atmospheric pressure systems, distinction was made between systems where the active species are generated in the immediate proximity of the surface, and remote surface treatment systems where the active particles are transferred to the surface via a gas flow. Compared with the low pressure systems, atmospheric plasma sources have the main advantage of not requiring complex vacuum equipment. On the one hand, these discharges are more difficult to stabilize so that realizing a uniform discharge at large scale, which is suitable for industrial applications, poses several technological challenges. Applicability and sustainability of atmospheric pressure applications should also be carefully evaluated with respect to relevant process gas consumption and with reference to environmental and safety issues. However, recent developments have yielded successful solutions for these obstacles.

Acknowledgment

The authors thank Dr. M. Grushin and Dr. N. Trushkin for their helpful contribution that was supported in part by RFBR (No 08-02-00601a).

References

1. Fridman, A. and Kennedy, L.A. (2004) *Plasma Physics and Engineering*, Taylor & Francis Inc., New York.

2. De Geyter, N., Morent, R., and Leys, C. (2006) Penetration of a dielectric barrier discharge plasma into textile structures at medium pressure. *Plasma Sources Sci. Technol.*, **15** (1), 78–84.

3. Moisan, M. and Pelletier, J. (eds) (1992) *Microwave Excited Plasmas*, Plasma Technology Series, **vol. 4**, Elsevier, Amsterdam.

4. Komachi, K. and Kobayashi, S. (1989) Generation-induced plasma source mass spectrometry for elemental analysis. *J. Microw. Power Electromagn. Energy*, **24** (3), 140–149.

5. Ganachev, I.V. and Sugai, H. (2002) Production and control of planar microwave plasmas for materials processing. *Plasma Sources Sci. Technol.*, **11** (3A), A178–A190.

6. Werner, F., Korzec, D., and Engemann, J. (1994) Slot antenna 2.45 GHz microwave plasma source. *Plasma Sources Sci. Technol.*, **3** (4), 473–481.

7. Petasch, W., Räuchle, E., Weichart, J., and Bickmann, H. (1995) Gigatron® – a new source for low-pressure plasmas. *Surf. Coat. Technol.*, **74-75** (Pt 1), 200–205.

8. Räuchle, E. (1998) Duo-PLasmaline, a surface wave sustained linearly extended discharge. *J. Phys. IV Fr.*, **8** (Pr7), 99–108.

9. Lieberman, M.A. and Lichtenberg, A.J. (1994) *Principles of Plasma Discharges and Materials Processing*, John Wiley & Sons, Inc., New York.

10. Schulz, A., Walker, M., Feichtinger, J., Räuchle, E., and Schumacher, U. (2003) Investigations and applications of plasmas generated by the Duo-Plasmaline. Proceedings of the

Vth International Workshop on Microwave Discharges: Fundamentals and Applications, July 08–12, 2003, Greifswald, Germany, Institut für Niedertemperatur-Plasmaphysik e.V., Greifswald, pp. 231–241.

11. Kaiser, M., Baumgärtner, K.-M., Schulz, A., Walker, M., and Räuchle, E. (1999) Linearly extended plasma source for large-scale applications. *Surf. Coat. Technol.*, **116-119**, 552–557.

12. Walker, M., Baumgärtner, K.-M., Schulz, A. and Räuchle, E. (1999) Silicon nitride films from the PLASMODUL® – a new microwave plasma device. Proceedings of the 14th International Symposium on Plasma Chemistry ISPC-14, August 02–06, 1999, Prague, Czechoslovakia, VOL. III, pp. 1427–1432.

13. Schulz, A., Baumgärtner, K.-M., Feichtinger, J., Walker, M., Schumacher, U., Eicke, A., Herz, K., and Kessler, F. (2001) Surface passivation of silicon with the Plasmodul®. *Surf. Coat. Technol.*, **142-144**, 771–775.

14. Walker, M., Quell, S., Feichtinger, J., Hirsch, K., Schneider, J., Schulz, A., and Schumacher, U. (2001) Spectroscopic investigations of a SiH_4 – NH_3 microwave plasma. Proceedings of the 15th International Symposium on Plasma Chemistry ISPC-15, July 09–13, 2001, Orléans, France, VOL. V, pp. 1853–1858.

15. Krüger, J., Kubach, T., Feichtinger, J., Hirsch, K., Lindner, P., Quell, S., Schulz, A., Stirn, R., Walker, M., and Schumacher, U. (2003) Spectroscopic investigations of pulsed microwave generated Ar, N_2 and silane plasmas. *Surf. Coat. Technol.*, **174-175**, 933–937.

16. Chen, F.F. (1985) *Introduction to Plasma Physics and Controlled Fusion. Volume 1: Plasma Physics*, 2nd edn (revised edition of: *Introduction to Plasma Physics.* 1974), Plenum Press, New York.

17. Tuda, M., Ono, K., Ootera, H., Tsuchihashi, M., Hanazaki, M., and Komemura, T. (2000) Large-diameter microwave plasma source excited by azimuthally symmetric surface waves. *J. Vac. Sci. Technol. A*, **18** (3), 840–848.

18. Geisler, M., Kieser, J., Räuchle, E., and Wilhelm, R. (1990) Elongated microwave electron cyclotron resonance heating plasma source. *J. Vac. Sci. Technol. A*, **8** (2), 908–915.

19. Gans, T., Schulze, J., O'Connell, D., Czarnetzki, U., Faulkner, R., Ellingboe, A.R., and Turner, M.M. (2006) Frequency coupling in dual frequency capacitively coupled radio-frequency plasmas. *Appl. Phys. Lett.*, **89**, 261502.

20. Brétagnol, F., Lejeune, M., Papadopoulou-Bouraoui, A., Hasiwa, M., Rauscher, H., Ceccone, G., Colpo, P., and Rossi, F. (2006) Fouling and non-fouling surfaces produced by plasma polymerization of ethylene oxide monomer. *Acta Biomater.*, **2**, 165–172.

21. Sanders, D.M., Boercker, D. B., Falabella, S., Coating technology based on vacuum arc, A review. (1990) *IEEE Trans. Plasma Sci.*, **9** (6), 883–894.

22. Martin, P.J. (1995) Cathodic arc deposition, in *Handbook of thin film Process Technology*, VOL. 1 (eds D.A.Glocker and S.I. Shah), IoP Publishing, Bristol.

23. Brandolf, H.E. (1985) Vapor deposition apparatus and method, US 4511593.

24. Sanders, D.M. Pyle, E. A., (1987) Magnetic enhancement of cathodic arc deposition. *J. Vac. Sci. Technol. A*, **5** (4), 2728–2731.

25. Akesov, I.I., Bolokhvotikov, A.N., Padalka, V.G., Repalov, N.S., and Khoroshikh, V.M. (1986) Plasma flux motion in a toroidal plasma guide, *Plasma Phys. Control. Fusion*, **28**, 761–770.

26. Fallabella, S. and Sanders, D.M. (1992) Comparison of two filtered cathodic arc sources, *J. Vac. Sci. Technol.*, **10**, 394–397.

27. Osipov, V.V., Padalka, V.G., Sablev, L.P., and Supak, R.I. (1978) Ustanovka Dly Nanesenia Pokryty Osazhdeniem Ionov; Izvelkayemykh iz Plazmy Vakuumnoi Dugi. *Prib. Tekh. Eksp*, **6**, 173–175.

28. *http://platit.com/concept* (accessed on 26 July 2009).

29. Veprek, S. and Reiprich, S. (1995) A concept for the design of novel superhard coatings. *Thin Solid Films*, **268**, 64–71.

30. Veprek, S. (1999) The search for novel, superhard materials. *J. Vac. Sci. Technol. A*, **17** (5), 2401–2420.

31. Kunhardt, E.E. (2000) Generation of large-volume, atmospheric-pressure, nonequilibrium plasmas. *IEEE Trans. Plasma Sci.*, **28** (1), 189–200.

32. Chang, J.-S., Lawless, P.A., and Yamamoto, T. (1991) Corona discharge processes. *IEEE Trans. Plasma Sci.*, **19** (6), 1152–1166.

33. Napartovich, A. (2001) Overview of atmospheric pressure discharges producing non-thermal plasma. *Plasmas Polym.*, **6** (1), 1–14.

34. Becker, K., Barker, R., Kogelschatz, U., and Schoenbach, K. (2004) *Non-Equilibrium Air Plasmas at Atmospheric Pressure*, IoP Publishing, Bristol.

35. Conrads, H. and Schmidt, M. (2000) Plasma generation and plasma sources. *Plasma Sources Sci. Technol.*, **9** (4), 441–454.

36. Kogelschatz, U. (2002) Industrial innovation based on fundamental physics. *Plasma Sources Sci. Technol.*, **11** (3A), A1–A6.

37. Fridman, A., Chirokov, A., and Gutsol, A. (2005) Non-thermal atmospheric pressure discharges. *J. Phys. D: Appl. Phys.*, **38** (2), R1–R24.

38. Bárdoš, L., Baránková, H. (2006) Atmospheric Plasma – Yes or No. Publication of the Vacuum Coating Industry – SVC Bulletin, Summer 2006, 42–48.

39. Buyle, G., Rogister, Y., Lefranc, O., and Paul, A. (2008) Antimicrobial coatings on textile fabrics via continuous atmospheric plasma deposition. Proceedings of the 8th AUTEX Conference, June 24–26, 2008, Biella, Italy.

40. Akishev, Yu.S., Grushin, M.E., Napartovich, A.P., and Trushkin, N.I. (2002) Novel AC and DC non-thermal plasma sources for cold surface treatment of polymer films and fabrics at atmospheric pressure. *Plasmas Polym.*, **7** (3), 261–289.

41. Stoffels, E., Flikweert, A., Stoffels, W., and Kroesen, G. (2002) Plasma needle: a non-destructive atmospheric plasma source for fine surface treatment of (bio)materials. *Plasma Sources Sci. Technol.*, **11** (4), 383–388.

42. Akishev, Yu.S., Aponin, G.I., Grushin, M.E., Karal'nik, V.B., Pan'kin, M.V., Petryakov, A.V., and Trushkin, N.I. (2008) Alternating nonsteady gas-discharge modes in an atmospheric-pressure air flow blown through a point-plane gap. *Plasma Phys. Rep.*, **34** (4), 312–324.

43. Benhalima, F. (2006) Surface engineering through atmospheric plasma technology, Presentation at 1st ACTECO Training Day – Lyon (France), October 18th, 2006.

3
Plasma-surface Interaction

Domenico D'Angelo

3.1
Introduction

In plasma chemistry several different elementary reactions can occur simultaneously, as explained in Chapter 1. Native chemicals are transformed to final products in a succession of several steps where ionization is the key process. This is the reason why ionization is often regarded as the first of the elementary plasma chemistry processes to be considered in a plasma process.

Ionization processes can be generally catalogued as (i) direct ionization of non-excited neutral species by electronic impact; (ii) direct ionization of preliminarily excited neutral species by electronic impact; (iii) ionization by collision with heavy particles; (iv) photo ionization; and (v) ionization of the surface (electron emission).

Other parameters that have to be taken into account are the reaction rate, the electron energy distribution function and the cross-section (which is an essential characteristic of each individual elementary process), the mean free path, and the rate constants [1].

Here we will briefly discuss the particularities of plasma chemistry and the relation between the processes in the gas phase and those on the surface to be treated.

In agreement with the literature [2] we subdivide plasma processes into five subsequent steps, which either can take place in the same volume, or can be spatially separated as in remote source processing.

- Creation of primary plasma (i.e., electrons and ions in the production of ionizing plasma).
- Transfer of primary to secondary chemistry (like dissociation of the injected monomer).
- Plasma surface interactions (radicals arrive at the surface and lead to the formation of a chemically bounded surface layer).
- Recirculation and production of new molecules (in the same volume or mixing with injected monomers).
- Clustering of recirculating species (nucleation and formation of dust).

Plasma Technology for Hyperfunctional Surfaces. Food, Biomedical and Textile Applications.
Edited by Hubert Rauscher, Massimo Perucca, and Guy Buyle
Copyright © 2010 WILEY-VCH Verlag GmbH & Co. KGaA, Weinheim
ISBN: 978-3-527-32654-9

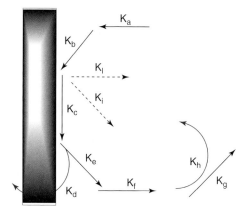

Figure 3.1 Schematic of the processes that can occur when a reactive species interacts with a surface. The K_x represent the kinetic rate constants of the different processes considered, see also text.

These steps can be combined by taking into account their different kinetic constants. The relation between the kinetics of surface processes and the kinetics of processes taking place within the plasma near the surface can be seen in the (simplified) scheme of Figure 3.1.

1) The molecules entering the plasma are converted into activated species with a kinetic rate constant K_a, following particular reaction channels. It is important to underline the specificity of each system, and it is assumed to have a single constant K_a for the activation (i.e., the reaction system evolves following specific kinetic conditions which determine the final dominant dynamics). If this is not the case there will be problems with the reproducibility and uniformity, thus undermining the industrial technological relevance of the process. The process can be represented formally as: $A \xrightarrow{K_a} A^*$.

2) The activated species arrive at the surface and can adsorb there (K_b), they can make a chemical reaction with the surface (K_c), or can spread into the same surface (K_d).

3) Desorption leads to the removal of species from the surface, for example, plain desorption of a surface species A(s) into the gas phase (A) with a remaining surface vacancy (s) can be written as $A(s) \xrightarrow{K_e} A + s$. Another possibility is associative desorption: $A(s) + B(s) \xrightarrow{K_e} AB + 2s$, with a different reaction constant K_e, and migration of the reaction product AB into the gas phase (K_f).

4) Reactive species can couple in the plasma to form larger nuclei of material and dust particles (K_g).

5) Finally, the product resulting from the recombination between desorbed species and activated species in the plasma can return to the surface (K_h).

This shows that in spite of complexity of the system, it is possible to reasonably describe its behavior in (dynamic) equilibrium in terms of dominant dynamics (the concepts presented here are in progress and will be published soon). This was confirmed by characterizing several plasma systems by plasma optical emission spectroscopy (OES); plasma mass spectrometry (MS); surface X-ray photo electron spectroscopy (XPS). These considerations provide the basis for the following discussions. Most of the phenomena mentioned take place in the discharge volume.

In the following we consider the ionization processes which are relevant for surface modifications of the materials that undergo plasma treatment. For this purpose, the interaction between plasma and materials will be considered by dividing them into surface etching, plasma grafting, polymer deposition, and polymer production. These processes will be discussed in more detail in the next sections.

3.2
Polymer Etching

When a polymer is exposed to a plasma for a certain time, a loss of weight occurs, caused by the ablation of the first polymeric layer close to the surface. The extent of this loss mainly depends on the polymer type and the plasma energy; this process also depends on the breaking of polymeric chains, which produces volatile low molecular weight fragments. Generally, the most sensitive polymers include functional groups containing oxygen (ethers, esters, ketones, carboxylic acids), while polyolefins are less sensitive.

The type of gas employed to etch the plasma is also relevant. Noble gas-based plasmas (i.e., Ar and He) do not induce prominent alterations of the chemical composition, while halogenated or oxidant gases, in addition to the weight loss, also produce surface chemical modifications. This feature is often used for surface decontamination, to remove organic micro-pollutants from the surface, and to increase the wettability and adhesion properties of subsequent finishing treatments.

Generally speaking, the etching affects the surface energy of the base material and the surface tension of a liquid in contact with the surface. Indeed, all solids have a defined surface energy, which is a function of surface chemical composition (or rather to the polarity induced by the species which are present on the surface) and of the surface area (including morphological aspects).

Liquids, such as water, have a high surface tension ($72 \, \mathrm{mN \, m^{-1}}$ at $20 \, ^\circ\mathrm{C}$) due to strong interactions between the polar water molecules. On the contrary, oils (Nujol oil $31.2 \, \mathrm{mN \, m^{-1}}$ at $20 \, ^\circ\mathrm{C}$) have low surface tensions as there are only very weak interactions between their apolar molecules.

It is important to understand how the surface roughness affects the interaction between the liquid and the material contacting as, in reality, few material surfaces are truly flat [3]. The surface roughness will enhance the repellence effects seen on a flat surface of the same chemical composition. If a droplet of water is in contact with a flat material and has a contact angle greater than 90°, on the same material with a rough surface the contact angle normally increases and shows

Table 3.1 Water contact angle on polyethylene (PE) films treated with different plasmas.

Plasma	θ (degrees)
CO_2	8
CO	16
NO	25
O_2	35
NO_2	37
Untreated	102

super hydrophobicity (lotus effect). On the other hand, if the droplet of water is normally less than $90°$ on a surface, that is, inherently hydrophilic, roughening the surface (i.e., noble gas-based plasmas) will result in an even lower contact angle.

Table 3.1 shows some water contact angles, measured on polyethylene (PE) samples [4] after treatment in different plasma atmospheres. It illustrates how powerful plasma treatment can be for tuning the surface energy.

3.3
Plasma Grafting

By choosing suitable precursors to be activated in the plasma, it is possible to modify the surface chemical characteristics of natural and synthetic polymers. It is interesting to observe that these surface treatments occur at a temperature close to room temperature and, therefore, they cannot change the physical properties of the polymer. Moreover, their action is restricted to the contact zone with the material. Depending on the choice of precursors, it is possible to select and promote different chemical transformations via the insertion of specific functional groups like oxydrilic, carboxylic acid, amine, fluorine, chlorine, and many others.

The result of this process is a new product with the same mechanical and physical properties as the bulk material, but with a different surface chemical composition, so that the surface has different chemical properties than the initial polymer.

Plasma treatments that employ oxidant precursors (e.g., air, oxygen, nitrogen) on synthetic polymers (e.g., polypropylene (PP), polycarbonate, poly (ethylene terephthalate) (PET)) or natural polymers (e.g., wool, cotton, silk) strongly increase their wettability, which is linked to the formation of polar groups on the surface. The reference method to evaluate the treatment effectiveness is to measure the change of the contact angle of a water drop on the surface. The contact angle (θ) is the angle between the liquid drop, air, and the solid (see Section 5.1.7 in Chapter 5); its value is the result of the interphase surface tensions (solid–liquid, solid–vapor, liquid–vapor) in accordance with Young–Dupré equation (see Equation 5.5).

Figure 3.2 Left: diffusion of a reactive dye on untreated PET; right: uniform diffusion pattern of the dye on PET fibers treated with oxygen plasma.

By increasing the contact angle, the surface wettability will be reduced and vice versa. Considering a PET fabric, it is possible to see different values of surface wettability if the nature of the precursors is changed in the plasma. Their partial pressure and the plasma generator power have an influence as well. These different results are related to the increase of hydrophilic and hydrophobic behavior subsequent to the insertion of functional groups.

On the one hand, the employment of oxygen-based plasmas or organic precursors containing hydroxylic and carboxylic acid functionalities allows the material surface to be modified; it increases the wettability, and strongly decreases the contact angle. This is generally referred to as *unspecific functionalization*. Nevertheless, this is an important industrial process, used mostly to increase the adhesion properties of materials or the dye-uptake (Figure 3.2) and printability of textiles [5]. The introduction of a well-defined functionality by using monomers with retention of their structure can be divided into direct and indirect methods. In the first case the monomer is exposed directly to a low power plasma, in the second only the substrate is activated and subsequently exposed to the monomer. To retain the monomer structure during direct plasma processing, the main parameters involved are low power input, high working pressures (low mean free path), and a short residence time of the molecule in the plasma. Yasuda used an external parameter, the so-called 'Yasuda factor' to express the plasma energy density. The factor is defined as W/FM, where W is the power, F flow rate and M molecular weight. A smaller Yasuda factor corresponds to less fragmentation [6].

On the other hand, the employment of organic precursors containing hydrophobic functional groups allows one to modify the surface properties of hydrophilic materials, like cotton and linen, making them highly hydrophobic.

3.4
Chemical Kinetics

To characterize the complex system of the plasma–surface interactions, it is necessary to study the kinetic aspects.

The chemical kinetics offer a set of irreplaceable methods for the comprehension of chemical reaction mechanisms and for the optimization of their yields. Almost always the concept of chemical kinetics is linked to chemical reaction rates. It is important to remember that kinetics can be subdivided in three types: (i), phenomenological, whose objective is the experimental determination of the reaction speed; (ii) interpretative, to explain experimental data on the basis of reaction mechanisms models; (iii) theoretical, to calculate the parameters of experimental kinetics laws. We remember that surface reaction mechanisms, for most plasma processes, are not yet known or experimentally characterized, although a lot of progress has been made in this field [7, 8].

Looking at phenomenological chemical kinetics, fundamental data to be considered are the concentrations of reactants and products, as well as their temporal evolution.

Different methods are used to follow the concentration, like OES (see Chapter 4) and MS. General experimental information that can be obtained is the dependence of the reaction rate on the composition and the energy of the different reacting species. For an in-depth study of basic questions related to reaction rate, equations of the kinetics and their determination from experimental data, we refer to the classical approach [9]. Here we try to provide some insight into the reaction kinetics of complex systems like plasmas, and focus on the study of plasma polymerization reactions at surfaces.

In order to highlight the similarities and differences between conventional and plasma polymerization, we briefly recall the fundamental processes involved.

3.4.1
Chain Polymerization

In conventional chain polymerization, an initiator (I) produces primary radicals (R•) that attack monomers and produce further, different radicals. These radicals attack and bind more monomers, in a reiterated process that leads, for each activated monomer, to fast growth of an individual polymeric chain. First, we introduce the following symbols: M is the monomer, R• represents the primary radicals, $P_1•$, $P_2•$, $P_3•$, . . . , $P_n•$ are growing radicals with respectively 1, 2, 3, . . . , n monomeric units. Without transfer reactions, the following fundamental kinetics scheme can be set up (Equation 3.1):

$$I \longrightarrow nR•$$
$$R• + M \longrightarrow P_1• \qquad -\frac{d[M]}{dt} = k_i[R•][M] \qquad (3.1)$$

where k_i is the kinetic rate constant of this initiating process. Hence, in this case, a radical (R•) appears in the reaction.

Equation 3.2 describes the subsequent propagation steps with the related reaction constant k_p accounting for the global propagation phase:

$$
\begin{aligned}
P_{1\bullet} + M &\longrightarrow P_{2\bullet} & k_p \\
P_{2\bullet} + M &\longrightarrow P_{3\bullet} & k_p & \quad \ldots \text{and so on, up to:} \\
P_{n-1\bullet} + M &\longrightarrow P_{n\bullet} & k_p
\end{aligned}
\tag{3.2}
$$

The final coupling and dismutation processes are represented in Equation 3.3:

$$
\begin{aligned}
P_{n\bullet} + P_{m\bullet} &\longrightarrow P_{n+m} & k_{t,a} & \quad \text{(coupling)} \\
P_{n\bullet} + P_{m\bullet} &\longrightarrow P_n + P_m & k_{t,d} & \quad \text{(dismutation)}
\end{aligned}
\tag{3.3}
$$

where P_n and P_m are molecules of 'inactive' polymer with respectively n and m monomeric units.

The rate of monomer disappearance (polymerization rate) is given by the amount at the beginning and the propagation rates in Equation 3.4:

$$
-\frac{d[M]}{dt} = v_i + v_p = k_i[R\bullet][M] + k_p[P\bullet][M]
\tag{3.4}
$$

where v_i and v_p represent beginning rate and propagation rate, respectively, and $[P\bullet] = \sum_n [P_{n\bullet}]$ represents the total concentration of polymeric growing radicals, regardless of their length. On average polymeric chains contain a fairly large number of monomeric units, so for each start reaction there are hundreds of propagation reactions. Hence, it is reasonable to neglect the monomer consumption of the start reactions, and Equation 3.4 becomes Equation 3.5:

$$
-\frac{d[M]}{dt} \approx v_p = k_p[P\bullet][M]
\tag{3.5}
$$

Also for the concentration of the radicals, equations are required (Equations 3.6 and 3.7):

$$
-\frac{d[R\bullet]}{dt} = v_R - v_i = k_i[R\bullet][M]
\tag{3.6}
$$

$$
-\frac{d[P\bullet]}{dt} = v_i - v_t = k_i[R\bullet][M] - k_t[P\bullet]^2
\tag{3.7}
$$

where v_R represents the rate of primary radical production (to be defined for each case), and v_t the rate of the termination reaction and $k_t = k_{t,a} + k_{t,d}$ is the global constant of termination. Note that it is necessary to assume the same reactivity for all macroradicals, to obtain the term $k_t[P\bullet]^2$ in the right hand side of the equation.

These equations are not a simply solvable system of differential equations because of experimental difficulties in the measurements of the radical concentrations. Therefore, we introduce the approximation of stationary state, which means that the term $\frac{d[R\bullet]}{dt}$ is negligible with respect to the other terms in Equation 3.6 because the time dependence of the concentration of radicals $R\bullet$ is very low. But the stationary state concerns all species in the system, and thus, also the time dependence of the growing radicals (i.e., the term $\frac{d[P\bullet]}{dt}$) can be neglected.

Hence, the Equations 3.5–3.7 become:

$$v_p = -\frac{d[M]}{dt} = k_p[P\bullet][M]$$
$$0 = v_{R\bullet} - k_i[R\bullet][M]$$
$$0 = k_i[R\bullet][M] - k_t[P\bullet]^2 \tag{3.8}$$

From this set, we can obtain the relations:

$$[R\bullet] = \frac{v_{R\bullet}}{k_i[M]}$$
$$[P\bullet] = \frac{v_{R\bullet}^{1/2}}{k_t^{1/2}} \tag{3.9}$$
$$v_p = -\frac{d[M]}{dt} = k_p[M]\frac{v_{R\bullet}^{1/2}}{k_t^{1/2}}$$

The validity of these equations is experimentally confirmed in several cases, including particular reactions with high efficiency initiators.

3.4.2
Plasma Polymerization

Next we will discuss the case of plasma polymerization on a substrate.

First it has to be realized that the power density of the plasma and the time of monomer residence time in the plasma are the most influential parameters for the composition of the polymer layer deposited on the surface. It is important to consider that at low power the monomer structure is normally preserved, but at high power the chemical structure of the deposited layer can be totally different and corresponds only to the monomeric atomic composition. For example, using acrylic acid (AA) ($CH_2=CHCOOH$) as monomer results in a layer containing carbon (C), hydrogen (H), and oxygen (O) whereas the chemical structure is not well defined. Investigations on plasma polymerization of AA by a pulsed plasma [10] gave the following experimental findings, which are important for a mechanistic approach of the description of AA monomer polymerization in a plasma: (i) the concentration of carboxylic acid groups incorporated into the deposited layer shows a maximum at a certain duty time (defined as the 'on' plasma time of the cycle during pulsed power plasma discharge operation); (ii) the maximum AA groups obtained depends on the monomer residence time in the plasma; (iii) the formation of AA oligomers in the plasma gas phase shows a maximum at the same duty time as where the [COOH] maximum in the deposited layer was observed; (iv) unsaturated monomer structures as well as incorporated carboxylic acid groups are essential for a high concentration of COOH groups in the deposited layer.

According to the referenced study [10] the experimental data can be explained by the scheme in Figure 3.3:

Assuming that the excitation processes remain independent of the residence time and the applied pulse frequency, an *accumulated power input time*, t_{APIT} can be

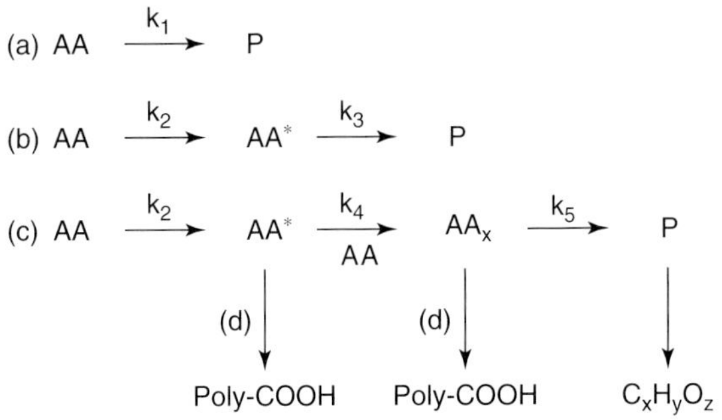

(a) AA $\xrightarrow{\text{k}_1}$ P

(b) AA $\xrightarrow{\text{k}_2}$ AA* $\xrightarrow{\text{k}_3}$ P

(c) AA $\xrightarrow{\text{k}_2}$ AA* $\xrightarrow[\text{AA}]{\text{k}_4}$ AA$_x$ $\xrightarrow{\text{k}_5}$ P

AA* $\downarrow$ (d)   AA$_x$ $\downarrow$ (d)   P $\downarrow$

Poly-COOH   Poly-COOH   C$_x$H$_y$O$_z$

Figure 3.3 The scheme shows (a) fragmentation of AA monomer and recombination into a chemically undefined organic layer, (b) excitation of AA monomer and formation of oligomers (c) the fragmentation of oligomers and deposition in an chemically undefined organic layer, (d) the incorporation of (excited) monomers and oligomers into the deposited poly(acrylic acid) layer.

defined as the product of the mean monomer residence time τ and the duty time t_{on}:

$$t_{\text{APIT}} = \tau \frac{t_{on}}{t_{on} + t_{off}} \qquad (3.10)$$

Assuming first order reaction laws, the following system of liberalized differential equations is derived (Equation 3.11):

$$d[AA]/dt_{\text{APIT}} = (-k_1 - k_4)\,[AA],$$
$$d[AA_x]/dt_{\text{APIT}} = k_4[AA] - k_5[AA_x], \qquad (3.11)$$
$$d[P]/dt_{\text{APIT}} = k_1[AA] + k_5[AA_x].$$

The rate coefficients k_1, k_4, and k_5 are the same as those used in Figure 3.3. The observed experimental maximum occurrence of carboxylic acid groups, which depends on the duty time and the residence time, can be approximately explained by the kinetic model described.

3.5
Example: Plasma Polymerization

Polymer film deposition is the most innovative industrial application of plasma technology, using either atmospheric pressure plasma (APP) equipment with dielectric barrier discharge (DBD) configuration or plasma enhanced chemical vapor deposition (PECVD).

When an organic precursor, which contains structures that are able to make a polymer, is introduced into the plasma (with suitable energetic conditions), polymer products are stratified onto the substrate surface. Precursor activation takes place because of the collision with high-energy free electrons of the plasma, which sometimes leads to precursor fragmentation and to the formation of highly reactive radical and ionic species. The mechanisms that lead to polymerization are still the subject of many studies, but for some of them it was possible to identify main 'reactive channels', although quantification of reaction constants is still an objective. We introduce two representative examples: plasma polymerization of 2-hydroxyethyl methacrylate (HEMA) and plasma polymerization of hexamethyl disiloxane (HMDSO).

3.5.1
Plasma Polymerization of HEMA

3.5.1.1 Theoretical Background

The monomer used as precursor in this process (carried out at Environment Park in the framework of the ACTECO project – see Preface) is HEMA. HEMA is a quite large organic molecule: $CH_2=C(CH_3)-C(=O)-O-CH_2-CH_2-OH$, which can polymerize to poly(HEMA) by opening of the C=C double bond. Usually chain propagation reactions of HEMA polymerization can be represented by HEMA attachment to a radical $R(\bullet)$:

$$R(\bullet) + CH_2=C(CH_3)-C(=O)-O-CH_2-CH_2-OH$$
$$\longrightarrow (RH_2)C-C(\bullet, CH_3)-C(=O)-O-CH_2-CH_2-OH \qquad (3.12)$$

Plasma initiation of the chain polymerization is due to formation of:

- Primary free radicals $R(\bullet)$
- Positive ion radicals $CH_2(\bullet)-C^+(CH_3)-C(=O)-O-CH_2-CH_2-OH$
- Negative ion radicals $CH_2(\bullet)-C^-(CH_3)-C(=O)-O-CH_2-CH_2-OH$.

All these precursors are capable of initiating HEMA polymerization and are formed from the adsorbed monomers by electron/ion bombardment and UV radiation from plasma. Formation of a positive ion radical can be schematically represented by the ionization process reported in Equation 3.13:

$$CH_2=C(CH_3)-C(=O)-O-CH_2-CH_2-OH$$
$$\longrightarrow CH_2(\bullet)-C^+(CH_3)-C(=O)-O-CH_2-CH_2-OH + e^- \qquad (3.13)$$

Subsequently, (Equation 3.14) the positive ion radical initiates the propagation reaction (sequential attachment of additional HEMA molecules):

$$CH_2(\bullet)-C^+(CH_3)-C(=O)-O-CH_2-CH_2-OH$$
$$+ CH_2=C(CH_3)-C(=O)-O-CH_2-CH_2-OH$$
$$\longrightarrow CH_2(\bullet)-C(R_1,R_2)-CH_2-C^+(R_1, R_2) \qquad (3.14)$$

Formation of a negative ion radical (which is also a center of polymer growth) from an adsorbed HEMA molecule on the surface is due to direct electron attachment (Equation 3.15):

$$CH_2=C(CH_3)-C(=O)-O-CH_2-CH_2-OH + e^-$$
$$\longrightarrow CH_2(\bullet)-C^--(CH_3)-C(=O)-O-CH_2-CH_2-OH \qquad (3.15)$$

Similarly to reaction of the positive ion radical, the negative ion radical operates as a nucleus of polymer growth and also initiates a sequence of chain propagation (Equation 3.16):

$$CH_2(\bullet)-C^-(CH_3)-C(=O)-O-CH_2-CH_2-OH$$
$$+ CH_2=C(CH_3)-C(=O)-O-CH_2-CH_2-OH$$
$$\longrightarrow CH_2(\bullet)-C(R_1,R_2)-CH_2-C^-(R_1, R_2) \tag{3.16}$$

Preliminary polymer substrate treatment in an O_2-containing plasma, as well as the partial pressure of HEMA and the discharge power are of uttermost importance to deposit the poly(HEMA) thin film on a substrate of interest.

Treating the polymer substrate surface in an O_2-containing plasma leads to the insertion of specific functional groups, such as oxydrilic, carboxylic, and peroxide groups. As explained by Fridman [1], the formation of the organic peroxide compounds occurs via a chain process, which starts with the insertion of molecular oxygen and the formation of an initiator (Equation 3.17):

$$R(\bullet) + O_2 \longrightarrow R-O-O(\bullet) \tag{3.17}$$

Further propagation of the plasma-initiated chain leads to production of the organic peroxides and restoration of organic radicals (Equation 3.18)

$$R-O-O(\bullet) + RH \longrightarrow ROOH + R(\bullet)$$
$$R-O-O(\bullet) + R_1-R_2 \longrightarrow ROOR_1 + R_2(\bullet) \tag{3.18}$$

At this time the polymer substrate activated in plasma is able to initiate the graft polymerization of the gas phase monomer.

3.5.1.2 Example: Polymerization of HEMA on PET Fabric

As an example we show results of a polymerization process carried out on PET fabric using the HEMA monomer. The chemical surface composition (in atomic percent), measured by XPS, is reported in Table 3.2.

Table 3.2 Measured chemical surface composition in atomic percent (%).

Samples	% C	% O	O/C
Theoretical PET	71.43	28.57	0.40
Reference PET	70.12	29.27	0.42
Theoretical poly(hydroxyethyl methacrylate)	66.67	33.33	0.50
Sample 1	66.63	32.22	0.48
Sample 2	68.25	30.62	0.45
Sample 3	67.33	31.58	0.47
Sample 4	67.57	31.31	0.46
Sample 5	67.29	32.03	0.48
Sample 6	67.73	31.27	0.46

Table 3.3 Carbon peaks and their association with specific chemical forms (in atomic %).

C1s

Sample	C=C	C–C C–C–O	C–O C–O–C=O	O–C=O
Theoretical PET	42.86	–	14.29	14.29
Reference PET	40.64	–	16.66	12.82
Theoretical poly(hydroxyethyl methacrylate)	–	33.34	22.22	11.11
Sample 1	–	36.81	18.24	9.33
Sample 2	–	38.89	18.12	8.82
Sample 3	–	37.75	17.88	8.99
Sample 4	–	38.19	17.78	8.88
Sample 5	–	35.90	20.44	9.64
Sample 6	–	37.62	19.08	8.99

Table 3.4 Oxygen peaks and their association with specific chemical forms (atomic %).

O1s

Sample	O=C	O–C
Theoretical PET	14.29	14.29
PET reference	13.29	15.98
Theoretical poly(hydroxyethyl methacrylate)	11.11	22.22
Sample 1	19.35	12.87
Sample 2	18.71	11.91
Sample 3	19.71	11.87
Sample 4	18.64	12.67
Sample 5	18.74	13.29
Sample 6	19.03	12.24

Different sample numbers refer to different locations on the same specimen to prove treatment uniformity along the electrodes.

In the following we report the deconvolution of the C_{1s} (Table 3.3) and O_{1s} (Table 3.4) peaks characteristic for nontreated PET as well as for PET after plasma polymerization of HEMA.

Three components have been considered to fit the carbon C_{1s} peak for the PET reference and for (HEMA) plasma-treated PET. These components are characteristic for the different chemical environments of carbon and oxygen. For the carbon C_{1s} peak, we consider:

at 284.7 eV: C=C, C–C, and C–H bonds
at 286.3 eV: C–O bonds
at 288.6 eV: O–C=O bonds.

Further, two components have been taken to fit the oxygen O_{1s} peak;

at 531.7 eV: O=C bonds
at 533.3 eV: O–C bonds.

The XPS analysis shows essentially that all samples (1–6) have a similar carbon and oxygen content, close to the theoretical values for poly(HEMA).

Therefore, it seems that a poly(HEMA) layer is deposited on all surface samples. Furthermore the layer thickness is more than 10 nm because a $\pi-\pi^*$ shake-up peak, which would be characteristic of the PET substrate, is not detected.

3.5.2
Plasma Polymerization of HDMSO

Another plasma polymerization process applied to textile fabrics has already been transferred to a pilot plant. This process, aimed at creating hydrophobic and durable coatings, is used to deposit siloxane coatings derived from HMDSO precursor employed in plasma discharges.

In the following we discuss in more detail the chemical aspects of this process. Similar to what was discussed in detail in the previous example, in this case there are also several reaction channels, which can be traced back, respectively, to the ionic and the radical reaction mechanism.

In the case of ionic polymerization developed with PECVD, using HMDSO, we can suppose [11, 12] the following mechanisms (Equation 3.19):

$$e^- + (CH_3)_3\text{–SiOSi}(CH_3)_3 \longrightarrow (CH_3)_3\text{–Si}\bullet + (CH_3)_3\text{–SiO}^+ + 2e^-$$
$$e^- + (CH_3)_3\text{–SiOSi}(CH_3)_3 \longrightarrow (CH_3)_3\text{–Si}^+ + (CH_3)_3\text{–SiO} + 2e^- \quad (3.19)$$
$$e^- + (CH_3)_3\text{–SiOSi}(CH_3)_3 \longrightarrow (CH_3)_3\text{–SiOSi}^+(CH_3)_2 + CH_3 + 2e^-$$

These equations are considered the most probable fragmentation steps, which lead to the generation of three different ions. One of the possible ionic polymerization reactions of these ions is (Equation 3.20):

$$(CH_3)_3\text{–SiOSi}^+(CH_3)_2 + (CH_3)_3\text{–SiOSi}(CH_3)_3$$
$$\longrightarrow (CH_3)_3\text{–SiOSi}(CH_3)_2\text{–O–Si}^+(CH_3)_2 + Si(CH_3)_4 \quad (3.20)$$

The CH_3 group can be eliminated, and after that, the Si–O–Si bond can be established by reaction with the HMDSO molecule. This way, $Si(CH_3)_4$ is eliminated from HMDSO.

Under plasma conditions, ion-radical and radical generation is more likely than the generation of simple ions, and therefore we can also suppose a radical mechanism for the polymerization for HMDSO.

In many cases, the monomer is activated by collision with carrier gas atoms, in this case helium atoms. An example of HMDSO fragmentation by collision with a helium atom is given below, with possible subsequent reaction channels:

$$(CH_3)_3-SiOSi(CH_3)_3 + He \longrightarrow [(CH_3)_3-SiOSi(CH_3)_3]^+ + He + e^-$$

$$[(CH_3)_3-SiOSi(CH_3)_3]^+ + e^- \longrightarrow (CH_3)_3-SiO + Si(CH_3)_3$$

$$\longrightarrow CH_3 + (CH_3)_2-SiO + Si(CH_3)_3$$

$$[(CH_3)_3-SiOSi(CH_3)_3]^+ + e^- \longrightarrow (CH_3)_3-SiOSi-(CH_3)_2CH_2 + H$$

$$[(CH_3)_3-SiOSi(CH_3)_3]^+ + e^- \longrightarrow (CH_3)_3-SiOSi-(CH_3)_2 + CH_3 \qquad (3.21)$$

In parallel to these reactions reticulations occur due to removal of H^+ ions or methylic groups of the main chain and subsequent insertion of bridges $[OSi(CH_3)_2]$ between contiguous chains.

The addition of nonpolymerizable molecules, such as the reactive gas O_2, during plasma polymerization has a strong effect on the final film composition. In PECVD from an HMDSO/O_2 mixture a SiO_x film is deposited.

The O_2/HMDSO ratios are important, in fact ratios larger than 10 : 1 (better 20–50:1 or even higher) allow SiO_x deposition. Several studies investigated the nature of generated intermediate species. In particular D. Theirich et al. [13] presented a study based on MS and IR absorption spectroscopy to identify possible intermediate precursors. Since the debonding energy of the Si-O bond is two times higher than the Si-C bond, the dissociation process also induces a dominant production of the neutral radicals CH_3 and $Si_2O(CH_3)_3$. The deposition of SiO_x is not due to the deposition of species such as SiO radicals but proceeds mainly by removing the carbon from previously deposited $Si_xO_yC_zH_t$ radicals. The carbon-containing species are first deposited and then carbon is removed by oxygen etching, producing CO_2 in parallel.

When transferring these processes to continuous plasma processing units at atmospheric pressure with DBD units, the results obtained are generally poor, because the ratio O_2/HMDSO should be larger than 100 : 1 and the exposition times of the surface to be treated must be long (more than 15 s); these aspects imply low process speed and very high oxygen consumption (resulting in high process costs). From XPS analyzes made on PET samples treated with O_2 : HMDSO = 20 : 1 (and a residence time in the plasma of more than 15 s) in an atmospheric plant, it was shown that the deposited film matrix still contains a large quantity of organic precursor. Only with a permanence time of around 30 s and with an oxygen content of around 75 : 1 it is possible to obtain an almost exclusively inorganic matrix (SiO_x).

3.6
Conclusion

In this section we have presented the basic processes which need to be taken into account in surface plasma processing, and more specifically in plasma polymeriza-tion and the interaction of polymeric surface substrates with a plasma. Although there are many possible reaction channels in the plasma environment, we stressed the relevance of specific dominant dynamics, depending on the process parameters The two examples presented here show what is more generally introduced at the theoretical level to provide an interpretational framework that highlights the complexity of plasma processing for surface functionalization.

References

1. Fridman, A. (2008) *Plasma Chemistry*, Cambridge University Press.
2. Schram, D.C. (2002) Plasma processing and chemistry. *Pure Appl. Chem.*, **74** (3), 369–380.
3. Coulson, S. (2007) *Plasma Technologies for Textile*, Woodhead Publishing Ltd.
4. Canton R., Durante S., Rabezzana F. (2003), Manuale trattamenti e finiture, p. 112, Tecniche Nuove.
5. Wakida, T., Tokino, S., Niu, S., Lee, M., Uchiyama, H., and Kaneko, M. (1993) Dyeing properties of wool treated with low-temperature plasma under atmospheric pressure. *Textile Res. J.*, **63** (8), 438–442.
6. Yasuda, H. (1985) *Plasma Polymerization*, Academic Press, Orlando, FL.
7. d'Agostino, R. (1990) *Plasma Deposition and Etching of Polymers*, Academic Press, Boston.
8. d'Agostino, R., Favia, P., Oehr, C., and Wertheimer, M.R. (2003) *Plasma Processes and Polymers*, Wiley-VCH Verlag GmbH, Weinheim, Germany.
9. Atkins, P. (1986) *Physical Chemistry*, Oxford University Press, Oxford.
10. Behnisch, J., Mehdorn, F., Hollander, A., and Zimmermann, H. (1998) Mechanistic approach to the plasma polymerization of acrylic acid by a pulsed MW (ECR) plasma. *Surf. Coat. Technol.*, **98**, 875–878.
11. Magni, D., Deschenaux, Ch., Hollenstein, Ch., Creatore, A., and Fayet, P. (2001) Oxygen diluted hexamethyldisiloxane plasmas investigated by means of in situ infrared absorption spectroscopy and mass spectrometry. *J. Phys. D: Appl. Phys.*, **34**, 87–94.
12. Raskova, Z. (2006) Plasma diagnostics during thin films deposition, Dissertation thesis.
13. Theirich, D., Soll, Ch., Leu, F., and Engemann, J. (2003) Intermediate gas phase precursors during plasma CVD of HMDSO. *Vacuum*, **71**, 349–359.

4
Process Diagnostics by Optical Emission Spectroscopy

Giacomo Piacenza

4.1
Introduction

The combined use of different plasma diagnostic approaches is fundamental for the analysis, understanding, and development of plasma technological processes.

The study of the spatial and temporal evolution of internal plasma parameters such as particle densities and energies in both plasma bulk and its boundaries, their electrical and magnetic characterization, allows operators to find and design specific and efficient plasma processing conditions tailored to desired functionalities and applications. Moreover diagnostics allow sharper control of surface treatment by tracing possible deviations of plasma parameters from set processing values. Real time and on-line plasma system monitoring diagnostics for industrial applications will be the challenge of next generation of plasma units for assuring due process control and tracing. Plasma diagnostics will also be valuable tools to achieve process eco-efficiency. Indeed, providing due feedback to a process if its parameters and properties deviate from set values serves to avoid generating scrap material and hence improves its environmental sustainability as well as reducing the processing cost to achieve maximum efficiency. Although some plasma diagnostics devices may be quite expensive and are usually employed only for research purposes, other solutions may be devised for industrial applications. In this section, an overview of the most representative optical plasma diagnostics methods is presented as an introduction to this complex discipline. A detailed treatment and in depth examination of the following techniques and theories can be found in the selected books cited in the chapter bibliography [1–5].

4.2
Optical Emission Spectroscopy

Optical emission spectroscopy (OES), which generally exploits a much wider spectral range than just the visible wavelengths region, is commonly used as a remote diagnostic technique based on the analysis of spectra of atoms, molecules,

Plasma Technology for Hyperfunctional Surfaces. Food, Biomedical and Textile Applications.
Edited by Hubert Rauscher, Massimo Perucca, and Guy Buyle
Copyright © 2010 WILEY-VCH Verlag GmbH & Co. KGaA, Weinheim
ISBN: 978-3-527-32654-9

and ions present in the plasma and emitting electromagnetic radiation due to electronic transitions. The radiation wavelengths range from about 100 to 900 nm, that is, from deep ultraviolet, through visible light (from violet at 380 nm to red at 760 nm) to the near infra-red wavelengths.

4.2.1
Theory of Optical Emission

Natural and artificial plasmas (laboratory and processing plasmas) are place of constant excitation of atoms and molecules to an excited state through single or multiple electron impacts. During the relaxation phase to a lower and possibly intermediate energy state a photon is released with an energy corresponding to the energy difference between these two energy states.

Artificial, homogeneous industrial plasmas, are partially ionized, which means that a certain portion of electrons are free rather than being bound to an atom or a molecule. Since the motion of electrons and positive ions can be considered as decoupled at sufficiently large frequencies the plasma responds strongly to electromagnetic fields, and therefore it is electrically conductive. In the plasma, permanent excitation and radiative relaxation or three-body recombination processes of atomic states takes place. In the excitation phase, energetic electrons undergo collisions with the neutral atoms and generate excited electronic states, free radicals, ions, and additional electrons. In the case of an excited neutral, an electronic de-excitation process follows: because of the instability of the excited state the electron returns to its stable ground energy state with the spontaneous emission of a photon, which has a characteristic specific energy which equals the difference between the two energy levels. Its frequency is related to the photon energy by the relation in Equation 4.1

$$E = h\nu \tag{4.1}$$

The de-excitation process can be schematically written as Equation 4.2:

$$A^* \Rightarrow A + h\nu \tag{4.2}$$

The time scale of the de-excitation is very short if the transition is an electric dipole transition, on the order of 10^{-8}–10^{-7} s. On the other hand, such an electronic transition can also be induced by the presence of other photons in the vicinity of the excited atoms. The process is called *stimulated emission*: in most plasma systems, spontaneous emission is the most common de-excitation mechanism while stimulated emission can be neglected.

Another photon emission process is radiative recombination: a free electron is captured on a bound level of an ion with the spontaneous emission of a photon. The emitted photon carries away the momentum and energy excess in the conservation balance of the electron–ion recombination process represented by the Equation 4.3:

$$e^- + A^+ \Rightarrow A + h\nu \tag{4.3}$$

OES is a straightforward diagnostic technique because it is related only to electronic state transitions allowed by quantum mechanics: these atomic spectra are sharp, nearly monoenergetic, and have well-defined peaks corresponding to transitions between various electronic states. Molecules, however, are characterized by more complex spectra as they have a larger number of electronic states, and their vibrational and rotational states are superimposed on their electronic states.

Optical spectra, provide the 'plasma fingerprint' even if their interpretation can be non-trivial. They allow, for instance, control of the reproducibility of processing discharges and are a valuable tool for identifying the presence of specific plasma components and energetic states.

When analyzed in more detail, such spectra reveal that electronic molecular states are coupled to molecular vibrational and rotational states. Since the energy differences between these states are small, the emission peaks are broadened. Furthermore the peaks are Doppler-shifted due to the motion of the emitting molecules, which can lead to overlapping emission peaks, and therefore the spectra consist of multiple bands rather than of sharp atomic emission peaks at specific, well known frequencies.

The energy differences between different atomic and molecular levels (electronic, vibrational, and rotational) are summarized in Table 4.1; a schematic drawing of energy levels in an atomic model with excitation and relaxation of electronic states and the photon emission process is depicted in Figure 4.1.

Table 4.1 Energy differences between electronic, vibrational, and rotational states of molecules and atoms (only electronic states).

Energy level	Energy (eV)	Energy (cm^{-1})
Electronic	0.8–18	6500–14 5000
Vibrational	0.02–0.6	200–5000
Rotational	0.00001–0.0006	0.1–5

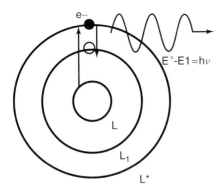

Figure 4.1 Photon emission associated with the relaxation of electronic states.

At higher pressures, collisions take place which broaden the emission energy, but such broadening is usually not observed in low pressure discharges used in plasma processing.

Optical emission also occurs in the following processes: exothermic chemical reactions (chemoluminescence); collisions between ions and neutrals, sputtering products that are released in an excited state, excited products of electron impact dissociation, absorption of photons, and subsequent fluorescence or phosphorescence, collisional relaxation of metastable species, and more.

4.2.2
Spectroscopy

OES, which measures the light emitted from a plasma as a function of wavelength, time, and location, is the most commonly used plasma diagnostic probe, which is used in different surface functionalization processes for both laboratory applied research and industrial optimization: for example, in textile, packaging, and medical application sectors.

The success and diffusion of this spectroscopic technique is based on the fact that it is non-intrusive, inexpensive, and that it can be easily integrated in plasma processing reactors and systems.

As discussed before, the energy of optical photons emitted from a plasma is characteristic of the energy difference between the electronic states of those molecular and atomic species which are present.

OES spectra can be considered as the fingerprint of a particular plasma species excitation situation (e.g., neutral, excited, ions, atoms, and molecules energy transitions) which is related to a specific processing condition. The comparison between spectra from the same processing reactor recorded at different moments can be used to monitor the evolution of the plasma state during a specific process. Furthermore within a specific processing step it can be used to analyze the plasma drift resulting, for example, from the change of plasma parameters.

It is also possible to measure the concentration of atomic species semi-quantitatively, by comparing the intensities of different spectral lines, and furthermore to obtain information related to plasma parameters such as the electron temperature T_e, electron density n_e, and ionization fraction. For example, for low density plasmas (where $n_e \sim 10^{18}$ m^{-3}) in coronal equilibrium, the power P_{ij} radiated per unit volume into a spectral line of the primary ionic species (with an upper state labelled j and a lower state labelled i), can be related to the electron density n_e and electron temperature T_e through Equation 4.4:

$$P_{ij} = K_{ij} n_e^2 \xi_{ex} \ (T_e \chi_j) \tag{4.4}$$

The constant k_{ij} is dependent on the atomic rate coefficient for the considered transition, and the excitation rate coefficient $\xi_{ex} \ (T_e \chi_j)$ is dependent on T_e and on the excitation potential χ_j through the following Equation 4.5:

$$\xi_{ex} \ (T_e \chi_j) = T_e^{1/2} \exp \ (-e\chi_j / K_B T_e) \tag{4.5}$$

Temperature measurements of atomic species can be obtained through the measure of line widths. In particular the line widths due Doppler broadening, are a rather direct measurement of atomic species temperature: that is, the thermal movement of emitting species shifts their apparent frequency [1].

Electron temperature measurements can be reliably determined by laser light plasma scattering from the electrons in plasma (so called Thompson scattering). The electron temperature is specifically related to the Doppler broadening of the scattered laser line [6].

The intensity ratio of a neutral and an ion line can be used to estimate the ionization fraction. Considering hot and highly ionized plasmas, line broadening contains a large amount of information. Doppler broadening is related to the velocity of the emitting ion or atom, while Stark broadening (also called *pressure broadening*) is a density related effect: at high densities, collisions interrupt the emission of radiation, and therefore the lifetime of a specific state will be shorter than without collisions, which results in a broadening of the emission line.

OES is commonly used to monitor plasma deposition and plasma polymerization processes. The analysis of the intensity variations of different lines of specific elements or of molecular emissions as a function of plasma parameters can be correlated to the chemistry of the deposited films and the chemistry of the deposition process. Grinevich *et al.* [7] reported the use of OES as an analytical and monitoring tool for Ti/hydrocarbon plasma polymer film deposition processes using an unbalanced magnetron with a Ti target operated in the DC mode in a working gas mixture of Ar/*n*-hexane. The films deposited combined the biocompatibility and osteogenesis enhancement of titanium with the wear resistance, hydrophobia, and blood compatibility of hydrocarbon plasma films for the adhesion, proliferation and maturation of vascular endothelial cells in orthopedic prostheses. The analysis of the ratio between the OES emission lines of Ti (363.5 nm) and Ar (420.1 nm) served as an indicator of the amount of Ti built in the deposited composite film.

OES had been also used to calibrate the sterilizing efficiency of microwave plasma discharges for both medical and food packaging applications [8]. The efficiency curve for sterilization of E. coli by photons (cell deactivation) has two maxima in the UV region at 220 and 263 nm; the authors utilized absolutely calibrated OES in the UV region in order to increase the intensities of processing plasma emission lines nearer to those of the UV sterilization absorption maxima of *E. coli*.

4.2.3
OES Bench and Set-up

An optical emission measurements bench set-up is shown in Figure 4.2; emission spectra are measured through a quartz window or directly in the discharge zone.

The data are collected at the open end of an optical fiber and analyzed by a monochromator or a multichannel optical analyzer with diffraction gratings designed to work generally in the wavelength range between 200 and 1000 nm. Spectrometers equipped with a monochromator measure different emission wavelengths selected by the rotation of a diffraction grating with high resolution, with

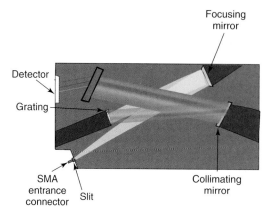

Figure 4.2 OES optical bench set-up.

the drawback of a time of acquisition in the order of a minute for the whole wavelength range scan. With such a configuration it is not possible to possible to analyze the real time electronic plasma states evolution.

In contrast, a multichannel optical emission analyzer, which is equipped with diode array detectors, collects the spectra from a monochromator with a single grating. This technique allows collecting multiple emission spectra in the selected wavelength range within the integration time of 10^{-3} s, and so it allows to monitor different plasma state transitions and to characterize different plasma regimes.

The fundamental components in an OES set-up are: a high transparency optical window in the reactor (for low pressure plasma processing), optical fiber, entrance slit, collimating mirror, grating, focusing mirror, detector collection lens, the detector, and UV filters.

For low pressure plasma processing the optical window in the reactor chamber is made of quartz or sapphire, which have high optical transmission at short wavelengths. Films deposited on the windows during plasma processes can selectively absorb emission and consequently affect the spectra. Such problems can be reduced by purging the window surface with the input gas or by the heating the window itself.

The optical fibers used for spectroscopy can be selected with different core diameters. Solarization resistant assemblies and jacketing fiber protections are needed for application in the deep UV (radiation with wavelengths of less than 300 nm) due to silica transmission degradation.

The most commonly used optical bench design is the symmetrical Czerny–Turner set-up: photons emitted by the plasma exit the optical fiber pathway and pass through a slit which acts as optical aperture. Hence, by decreasing its width it is possible to increase the optical resolution but on the other hand the signal strength will be decreased.

Special longpass filters can be installed after the slit in order to block second and third order effects. The photons are then focused by a collimating mirror toward the gratings. There the plasma emitted light is dispersed by a number of equally spaced grooves with a density of the order of thousand lines/mm: the density of grooves determines the dispersion and the bandwidth. The higher the groove density the better the resolution but at the same time a smaller spectral range can be recorded. The angle of the grooves determines the most efficient wavelength range. Gratings are generated either by diamond ruling or holographic exposure of a photoresist layer. For a particular application a specific grating design must be selected because each grating has a specific optimized wavelength range with best efficiency (>30%).

The focusing mirror reflects the dispersed light toward the detector plane. Photodetectors can be based on different technologies such as photomultiplier tubes, photodiode arrays, and charged coupled devices (CCDs). Nowadays, the second and third technology are the most commonly used in portable compact spectrometers because they can be integrated and are relatively affordable while the first one is a very sensible detection technology mainly used for research applications.

The photomultiplier tube is based on the photoelectric effect: the incident photons transfer their energy and liberate electrons from the surface of a photocathode. The photoelectrons are then accelerated, amplified by a set of dynodes, and collected at the anode. The resulting anode current is proportional to the number of incident photons but amplified by a factor of 10^7 (total number of photoelectrons to the power of the number of dynode stages). The response time is typical of the order of few nanoseconds, while the sensitivity is very high and the dark current is low, permitting the detection of trace elements with concentration. Depending on the nature of the entrance material and on the sensitive layer of the photocathode it is possible to select between different types of photomultiplier where each type has a high sensitivity in a specific and narrow wavelength range (some are best in the far UV while others have their peak sensitivity in the visible range).

Photodiode array detectors are basically multichannel light detectors consisting of a linear array of thousands of pixels. The detectors are closely spaced 'capillaries' with low work function surface materials that emit secondary electrons which generate the cascade.

CCDs are integrated circuits: they are composed of a detection area and an adjacent zone; the absorbed photons transfer the energy to the electrons in the light sensitive semiconductive detection area and those electrons are excited into the conduction band. The electrons are then driven toward an adjacent capacitor by an applied voltage. Charge accumulation occurs until the capacitor is discharged during readout.

Advantages of the CCD detectors are their small dimensions (e.g., up to 3648 pixels with a pixel size in the order of $8 \times 200 \ \mu m^2$) and their quantum efficiency is up to 0.9 (0.9 e charge accumulation per incident photon).

4.3
Optical Absorption Spectroscopy

Optical absorption spectroscopy allows the concentration of specific species in a plasma, for example, the densities of metastable atoms, to be monitored and measured with high sensitivity. Knowledge of the concentration of metastable atoms is of great importance, in particular for glow discharge plasmas where these species are involved in many chemical reaction paths such as Penning ionization and various other energy transfer processes.

Absorption measurements are possible by using optical sources (e.g., intense standard calibrated pulse lamps, tunable diode lasers) emitting convenient spectral lines that can be absorbed by the metastable atoms, that is, the same resonant lines as emitted from the plasma (resonant absorption spectroscopy [9]).

4.3.1
Actinometry

The actinometry technique based on optical emission is a useful approach for the analysis of the kinetics of excitation and relaxation in a plasma. Excited neutrals are created by electron impact in the plasma, thus the excited neutrals concentration can be deducted and quantitatively calculated by the correlation of data on the emission intensities with the knowledge of the electron distribution function.

The optical emission intensity I_λ, related to the photon excitation emissions of the free radical X from the ground state X to the excited state X^* and subsequent relaxation is related to the free radical concentration n_X and to the electron distribution function $f_e(r, v, t)$ by the following Equation 4.6

$$I_\lambda = \alpha_{\lambda X} n_X \tag{4.6}$$

where Equation 4.7

$$\alpha_{\lambda X} = k_D(\lambda) \int_0^\infty 4\pi v^2 dv \, Q_{X^*} \sigma_{\lambda,X}(v) v f_e(v) \tag{4.7}$$

Here the proportional factor $\alpha_{\lambda x}$ is the integral over the line-width of the relation between the electron distribution function $f_e(v)$, the quantum yield of photon emission from the radical excited state Q_{x^*}, and the emission cross-section of a photon (with wavelength λ) due to electron impact excitation of X, $\sigma_{\lambda A}$. The constant term k_D in Equation 4.7 is the photodetector response constant at the selected wavelength.

Typically, even if the $\sigma_{\lambda A}$ is known, the electron distribution function is unknown, that is, $f_e(v)$ generally does not have a single-temperature Maxwell–Boltzmann distribution. Specifically the high-energy tail of the electron distribution function, that is, close to the radical excitation energy E_{X^*}, has a strong shape variability as the plasma discharge parameters (operating pressure, power, frequency) are varied. Hence, also the cross-section changes and the Equation 4.7 fails. Therefore, OES measurements of I_λ can be regarded only as qualitative estimation of the radical density n_x.

In the actinometry technique, developed by Coburn and Chen [10], a small quantity of inert gas (the actinometry gas) of known concentration is introduced into the plasma system. An inert gas, such as Ar, does not react with the species present in the plasma and therefore does not interfere with the plasma system.

The inert trace gas (T) is chosen with a corresponding exciting threshold close to the excited state of the radical X. The ratio of the emitted intensities is therefore a measure for the n_x radical concentration as in the Equation 4.8

$$n_x = C_{XT} n_T \frac{I_\lambda}{I_{\lambda'}} \tag{4.8}$$

I_λ is the emitted intensity of the actinometry gas. The correlation constant C_{XT} is related to the two cross sections at the excitation threshold, and an ideal actinometry gas should have a cross section for excitation that is identical to the that of the species of interest.

The actinometry technique has been used by Favia [1, 11] and co-authors to monitor in a semi-quantitative way the distribution of the plasma emitting species in NH_3/H_2 and $O_2/H_2O/H_2$ as a function of the process parameters (i.e., pressure, power, feed gas composition, and flow rates) in RF plasma glow discharges during PE surface modification for biomedical applications; such plasmas are used to introduce, respectively, surface $-NH_2$ and $-COOH$ functionalities. By correlating the X-ray photo electron spectroscopy (XPS) electron spectroscopy for chemical analysis (ESCA) analysis of the treated surfaces and the trend surface density $-NH_2$ and $-COOH$-species and (measured by UV-Vis colorimetric techniques) with the corresponding actinometer analysis, they were able to study the chemical reaction path of the two reactive plasmas as function of different plasma conditions. This enabled them to identify the best processing conditions and to design the most efficient process in terms of efficient surface species selectivity.

4.4
Laser Induced Fluorescence (LIF)

Laser induced fluorescence (LIF) is a plasma diagnostic technique that can be used to identify atomic species (e.g., atomic oxygen) and combined with OES actinometry it is useful to improve and corroborate the quantification of the concentration of specific plasma species and plasma parameters.

This technique is based on the relaxation fluorescence emission of the plasma induced by excitation with probe laser photons. The most commonly used set-up involves the two phonon excitation processes: an excimer laser or a Nd:YAG laser is used to cause two (or multiple) photon excitation.

The laser beam is focused by an optical set-up (focusing and collecting lenses and optical fibers), while the fluorescence emission is focused to the entrance slit of a spectrometer with a highly sensible image-intensified CCD camera detector or to the entrance slit of a monochromator with photomultiplier detector. LIF and time-resolved LIF have been used.

Gerassimou et $al.$ [12], for example, analyzed via the LIF technique the behavior of different ionic species of an N_2 plasma sheath (i.e., the plasma state that is created in front of a metal surface during the interaction between plasma and metal). The authors studied the N_2^+ ion lifetime evolution (with tens of nanosecond resolution), its concentration and its rotational temperature as a function of the distance between the metal and plasma, and as a function of the plasma parameters. The N_2 ion analysis is fundamental for the study of the physical and chemical interactions between the plasma and the metal surface.

This diagnostic technique is both non-invasive and local because it uses intersecting beam paths. Furthermore, it is the only way to measure Ti without using a large energy analyzer. One laser, tuned to a particular transition, is used to raise ions to an excited state along one path through the plasma. The excited ions fluoresce, emitting light at another frequency, and this light is collected by a lens focused to one part of the path, providing the localization.

Doppler broadening of the line yields the ion velocity spread in a particular direction. The equipment needed for LIF, however, is large, expensive, and difficult to set up. Therefore this technique it is available in relatively few laboratories and, although a powerful method, at the moment is rarely applied to diagnose industrial processing plasmas.

4.5
Conclusion

In this chapter we have introduced an overview of the most used *OES* techniques for the *plasma diagnostics* as process development and monitoring tools. After an introduction of the optical emission theory in plasma, we described the OES concept and which plasma physical quantities can be measured (such as: the electron temperature T_e, electron density n_e, atomic species temperatures, and ionization fraction). Then we have reported a synthetic description of the most used *optical bench set-up* scheme and we considered its fundamental components. Finally we introduced two fundamental optical absorbing spectroscopy techniques: *actinometry* and LIF. While the first is commonly used to monitor the concentration of specific radical species, the LIF technique can monitor their velocity and localization.

References

1. Hutchinson, I.H. (1987) *Principles of Plasma Diagnostics*, Cambridge University Press.
2. Griem, H.R. (2005) *Principles of Plasma Spectroscopy*, Cambridge Monographs on Plasma Physics, Cambridge University Press.
3. Lieberman, M.A. and Lichtenberg, A.J. (1994) *Principles of Plasma Discharges and Materials Processing*, John Wiley & Sons, Inc., New York.
4. Chen, F.F. (2002) *Lecture Notes on Principles of Plasma Processing*, Plenum/Kluwer Publishers.
5. Ochkin, V.N. (2009) *Spectroscopy of Low Temperature Plasma*, Wiley-VCH Verlag GmbH, Weinheim.
6. Shul, R.J. and Pearton, S.J. (eds) (2000) *Handbook of Advanced Plasma Processing Techniques*, Springer-Verlag.
7. Grinevich, A., Koshelyev, H., Biederman, H., Boldyryeva, H.,

Bacakova, L., and Pesicka, J. (2005) Plasma polymer processes in biomedical applications. Proceedings of the 14th Annual Conference of Doctoral Students – WDS 2005, Part III, pp. 545–549.

8. Messerer, P., Halfmann, H., Czichy, M., Schulze, M., and Awakowicz, P. (2005) *Plasma Sterilisation and Surface Modification of Thermolabile Materials Surface Engineering: Science and Technology II*, TMS (The Minerals, Metals & Materials Society), p. 205.

9. Mitchell, A.G.G. and Zemanski, M.W. (1961) *Resonance Radiation and Excited Atoms*, Cambridge University Press.

10. Coburn, W. and Chen, M. (1980) Optical emission spectroscopy of reactive plasmas: A method for correlating emission intensities to reactive particle density. *J. Appl. Phys.*, **51**, 3134.

11. Favia, P., D'Agostino, R., and Palumbo,F. (1997) Grafting of chemical groups onto polymers by means of RF plasma treatments: a technology for biomedical applications. *J. Phys IV Fr.*, 7 (C4), 199–208.

12. Gerassimou, D.E., Cavadias, S., Matras, D., and Rapakoulias, D.E. (1990) Nitrogen ion dynamics in low pressure nitrogen plasma and plasma sheath. *J. Appl. Phys.*, **67** (1), 146–153.

5
Surface Analysis for Plasma Treatment Characterization

Amandine David, Yves de Puydt, Laurent Dupuy, Séverine Descours,
Françoise Sommer, Minh Duc Tran, and Jocelyn Viard

5.1
Introduction to Surface Characterization Techniques

The main applications of plasma treatments in material science are surface functionalization and coating. The properties induced by plasma functionalization and coating are primarily governed by the chemical composition and structure of the outermost atomic layers of the treated surface. In order to understand, to optimize, and to validate the plasma interaction processes on surfaces it is mandatory to have an accurate knowledge of the treated surfaces. Surface-sensitive analytical methods provide an exhaustive characterization of plasma-treated surfaces, including thermodynamics, structure, morphology, and chemical composition. They also encompass all types of materials: polymers, textiles, glasses, biomaterials, films, fibers, and powders. Table 5.1 summarizes important analytical methods used to characterize plasma-treated surfaces:

- XPS (X-ray photo electron spectroscopy) to obtain qualitative and quantitative elemental surface composition, chemical bonding analysis, depth profiling, and mapping;
- ToF-SIMS (time of flight secondary ion mass spectrometry) in static mode to achieve high resolution molecular analysis, chemical depth profiles, and molecular imaging;
- AFM (atomic force microscopy) for high resolution morphology, roughness, adhesion, surface and friction forces, and visco elastic module measurements;
- SEM (scanning electron microscopy) with X-ray analytical facilities (X SEM) (X-ray scanning electron microscopy) and variable pressure scanning electron microscopy (VP SEM);
- TEM (transmission electron microscopy) to determine the structure of deposited layers, thickness and homogeneity of the coatings;
- Contact angle to measure surface energy, wettability, and adhesion;
- Gas and water vapor permeation to measure the efficiency of barrier coatings.

Plasma Technology for Hyperfunctional Surfaces. Food, Biomedical and Textile Applications.
Edited by Hubert Rauscher, Massimo Perucca, and Guy Buyle
Copyright © 2010 WILEY-VCH Verlag GmbH & Co. KGaA, Weinheim
ISBN: 978-3-527-32654-9

Table 5.1 Overview of considered surface micro and nano analysis techniques.

	XPS	ToF-SIMS	AFM	X–SEM	TEM–STEM–STEM–EDX
Elemental analysis					
Qualitative					
Detected elements	$Z >2$ (He)	All	–	$Z >11$ (Na) with Be windows	$Z >6$ (C)
Isotopic analysis	No	Yes	Nonanalytical	No	No
Overlapping, interference	No	Possible	–	Yes	Yes
Relative intensity factor	10	$>10^6$	–	10	10
Detection limit (atom. %)	0.1%	10^{-6}	1 atom	0.1–1%	0.1–1%
Bulk: Cmin (atom %)	1%	10^{-6}–10^{-9}	–	0.1–1%	0.1–1%
Surface (monolayer %)	0.1%	10^{-6}	1 atom	–	–
Depth resolution	2–5 nm	0.3–0.5 nm	0.1 nm (mean roughness)	Few nanometers for SE for backscattered e^- $\sim 1\ \mu m$ for X-rays	Section thickness
Quantitative	Good	Difficult	Nonanalytical	Medium	Medium
Matrix effect, chemical effect	Low	Important	Yes	Yes	Yes
Angular effect	Yes	Yes	–	Yes	Yes
Accuracy	3–10%	<10%	–	2–10%	2–10%
Analysis time	Tens of minutes	Minutes	Seconds	Few seconds to minutes	Few seconds to minutes
Chemical and molecular analysis					
Chemical speciation	Good	Good	No	No	No
Molecular information	Fair	Excellent	Possible	No	No
Structural information					
Crystallography	Possible	Possible	Yes	No	Yes
Conformation	No	Possible	Possible	No	No
Unsaturation, crosslinking	No	Possible	Possible	No	No

Mapping – microanalysis					
Elemental	Yes	Yes	Nonanalytical	Yes (X SEM)	Yes
Chemical bonding, molecular	Yes	Yes	Nonanalytical	No	No
Lateral resolution	3 – 10μm	0.1 – 1μm	Atomic	~1 nm on SEM images	Atomic on TEM-STEM images
Image time	~20 min	~5 min	~5 min	~few seconds for SEM images	~few seconds for TEM image
Depth profile					
Angle distribution (nondestructive)	Yes	–	No	No	No
Ion etching (destructive)	Yes	Yes	No	No	No
Topography, roughness	No	No	Yes	Yes	No
Thermal properties	No	No	Yes	No	No
Mechanical properties	No	No	Yes	No	No
Surface energy, forces	No	No	Yes	No	No
Other features					
Analysis of insulating materials	Yes	Yes	Yes	Yes	Yes
Charging effect	Yes	Important	Possible	Yes	Yes
Neutralization	Efficient	Efficient	Efficient	Yes with metallization or VPSEM	Efficient
Analysis of powders, fibers	Yes	Yes	Yes	Yes	Yes
Polymers, biological samples	Yes	Yes	Yes	Yes	Yes
Irradiation damage	Low	Limited	No	Low with care	Low with care
Perturbation, artifacts	Low	Important	Low in no contact mode	No	No
Analysis conditions					
UHV	Yes	Yes	Possible	Yes	Yes
In atmosphere, in liquids	No	No	Yes	Yes (VPSEM)	No

These analytical methods can be used to obtain a variety of conceptual knowledge of treated surfaces that can be further used to tailor plasma processes for specific applications. In the following a short introduction to the principles of these techniques will be given, and it will be shown how they can be used to characterize plasma-treated surfaces of polymers, textiles and biomaterials.

5.2
X-ray Photoelectron Spectroscopy (XPS) or Electron Spectroscopy for Chemical Analysis (ESCA)

XPS, also known under the acronym ESCA (electron spectroscopy for chemical analysis) [1], is based upon the determination of the kinetic energy of photoelectrons expelled from core levels of sample surface atoms by absorption of monochromatic soft X-ray photons in the so called photoelectric effect.

Core level binding energies measured by XPS are specific to the emitting elements in a reliable way and shift due to changes in their chemical bonding environments. The technique provides quantitative information on the chemical structure, atomic composition, and chemical bonding state. The sampling depth is approximately 3λ (λ being the inelastic mean free path of electrons in the solid), which is usually smaller than 10 nm. Since the irradiation damage by kiloelectronvolt X-rays is very low even in the case of polymers, organic, or biological samples, XPS is practically nondestructive. There are a large number of applications, and any sample compatible with ultra-high vacuum (UHV) can be analyzed: metals, ceramics, semiconductors, composite materials, polymers, and biopolymers. XPS is one major tool for surface analysis [2, 3] and is specifically suitable for characterizing surface treatments by plasma.

- **Spectroscopy** The main information provided by XPS is the determination of the surface elemental composition and stoichiometry, chemical bonding structure, chemical functional groups, and oxidation states grafted by plasma treatment, which allows to determining the involved chemical process and the reaction yields.
- **Concentration Depth Profiles** By varying the detection angle θ relative to the sample normal, sampling depths can be varied as $3\lambda \cos \theta$ and concentration depth profiles can be quantitatively plotted for an overall depth of less than 10 nm in a nondestructive way. This method is often used to demonstrate the surface location of grafted species and superficial segregation. The thickness of homogeneous coatings which are thinner than 10 nm can also be measured by using angle-resolved XPS. For larger depth or a thickness of up to a few micrometers, ion etching can be used for plotting destructive depth profiles and for measuring the thicknesses of coatings.
- **Lateral Surface Imaging and Analysis** New generations of XPS spectrometers allow to obtain a two dimensional surface imaging with spatial resolution $\leq 3\,\mu m$. Quantitative XPS microprobe analysis can be performed with a spot size $\leq 10\,\mu m$. Chemical imaging is used to observe the heterogeneity of sample

surfaces, to characterize contamination areas, uniformity (or heterogeneity) of the functionalization or coatings, or microstructures.

5.2.1
Principles of XPS

The XPS principle is based on the determination of the intensity distribution as a function of kinetic energy of the photoelectrons emitted by a sample under the irradiation of soft X-ray photons with an energy $h\nu$. Such spectra typically show photoelectron peaks from specific atomic core levels. The kinetic energy E_K of the photoelectrons is given to a good approximation by $E_K = h\nu - E_B$ where $h\nu$ is the energy of the incident X-ray photons, and E_B is the binding energy of the photoelectrons. The energy conservation law of the photo electric effect allows to determine the core level binding energy E_B from the measurement of the kinetic energy E_K of the corresponding photoelectrons. The binding energies E_B of the core levels are characteristic of the atomic number Z of the emitting atom and of the quantum numbers n, l, j of the ionized orbital. The XPS peaks are labeled using the Z_{nlj} notation as for instance C_{1s} or Au $4f_{7/2}$.

Figure 5.1 shows a wide scan XPS spectrum obtained for a SiO_2 sample excited by monochromatic Al K_α radiation at the correspondent energy $E = h\nu = 1486.6$ eV. The main features in this spectrum are the appearance of core level XPS peaks assigned to the ionization of the atomic orbitals Si_{2s}, Si_{2p}, and O_{1s} specific of SiO_2. The peaks correspond to the primary photoelectrons emitted without energy loss. Each peak is accompanied at lower kinetic energy by a background of secondary electrons resulting from photo electrons with inelastic energy loss.

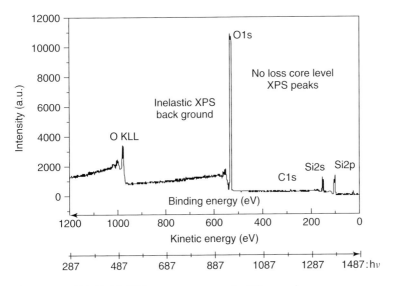

Figure 5.1 1000 eV-wide XPS survey spectrum of a SiO_2 sample.

The peak labeled O_{KLL} is the Auger emission following the de-excitation of the XPS O_{1s} core hole. Note also the C_{1s} XPS signal due to adventitious carbon contamination on the surface of the SiO_2 sample, evidencing the sensitivity of XPS to surface contamination.

5.2.2
XPS Core Level Chemical Shift

The core level binding energy E_B of an element changes when the chemical environment of this element changes. XPS core level chemical shifts, ΔE_B, can be qualitatively understood by the transfer of valence electron charge and change of oxidation state of the element. Withdrawal of valence electrons and increase of oxidation degree induce an increase of binding energy E_B of core levels of this element. Vice versa, addition of valence electron charge in a chemical reduction of the element results in decreasing the XPS core level binding energy of this element.

Examples of XPS ΔE_B chemical shifts are presented in Figure 5.2. When elemental silicon Si^0 is oxidized into SiO_2, (Si^{+4}), the Si_{2p} binding energy increases by 4.2 eV. The three different, chemically non-equivalent C atoms in poly(ethylene terephthalate) (PET) give three components at different binding energies in the C_{1s} spectrum, aromatic C–C at 284.8 eV, C–O at 286.8 eV, and O–C=O at 289.10 eV.

Since the analysis of plasma-treated polymers represents an important XPS application, the chemistry of carbon with air, oxygen, fluorine, and nitrogen plasmas is of great interest. So the C_{1s} binding energy, E_B (eV) and its chemical shift, ΔE_B, with O, F, and N bonding are important values. Table 5.2 summarizes

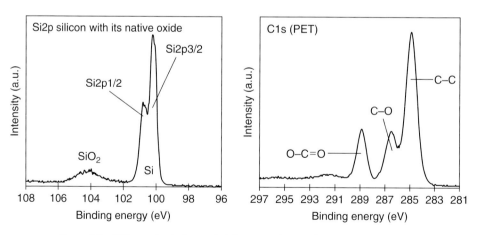

Figure 5.2 XPS core level chemical shifts: Si_{2p} (elemental silicon with native oxide layer) and C_{1s} (PET).

Table 5.2   Examples of C_{1s} chemical shifts.

Functionality group	Binding energy, E_B (eV)	Chemical shift, ΔE_B (eV)
C–C, aliphatic (reference)	285.00 ± 0.00	0.00 ± 0.00
C–C, aromatic	284.73 ± 0.04	-0.27 ± 0.04
C=C	284.66 ± 0.15	-0.34 ± 0.15
C–O	286.60 ± 0.30	1.60 ± 0.30
C=O, O–C–O	287.90 ± 0.30	2.90 ± 0.30
–O–C=O	289.10 ± 0.15	4.10 ± 0.30
C–F	287.90 ± 0.30	2.90 ± 0.30
C–CF	285.40 ± 0.30	0.40 ± 0.30
C–F$_2$	290.90 ± 0.30	5.90 ± 0.30
C–F$_3$	292.70 ± 0.30	7.70 ± 0.30
C–N<	285.90 ± 0.30	0.90 ± 0.40
–CN	286.70 ± 0.30	0.70 ± 0.40
–N–C=O	288.10 ± 0.30	3.10 ± 0.40

some of these values for the carbon chemical functionalities encountered most frequently.

5.2.3
Quantitative Analysis

The intensity I_A of a XPS photoelectron peak for an element A homogeneously distributed in a sample of thickness z is expressed as Equation 5.1:

$$I_A = K \cdot N_A \cdot \sigma_A \cdot \lambda_A \cdot T_A \cdot \left[1 - \exp \left(-\frac{z}{\lambda_A \cos \theta} \right) \right] \tag{5.1}$$

where
  N_A is the atomic concentration of the element A in the sample.
  σ_A is the photoelectron cross section (ionization probability of the core level). The σ_A values have been calculated for most of the electronic levels excited by Al K_α radiation (Scofield cross-section factors).
  λ_A is the inelastic mean free path of the considered photoelectron. In a first approach, for kinetic energy in the range of $100–1000$ eV, we may assume that values for λ_A are in the range from 1 to 3 nm and are varying as the square root of kinetic energy $\lambda \sim (E_K)^{1/2}$
  T_A is the transmission function and detection efficiency of the spectrometer; it depends also on the kinetic energy
  θ is the analysis angle relative to the sample normal.
  K is a constant that includes all other experimental factors – detection geometry, solid angle, analyzed area, photon flux – it is identical and constant for all photoelectron peaks during the same experiment.

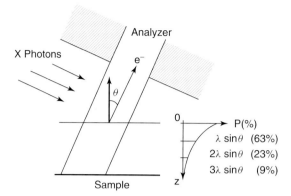

Figure 5.3 Detection geometry and XPS intensity distribution according to depth z.

The expression $\left(1 - \exp\left(-\dfrac{z}{\lambda_A \cos\theta}\right)\right)$ weighs the XPS intensity emitted from different depths z below the sample surface. It is shown in Figure 5.3 that 63% of the XPS signal is emitted from the first $\lambda \cos\theta$ (<3 nm) below the sample surface underlining the surface sensitivity of XPS. λ may therefore be also regarded as the mean XPS analysis depth. The surface sensitivity can be enhanced by increasing the detection angle θ in a grazing angle experiment: at $\theta = 70°$, $\cos\theta = 0.3$ and the analysis depth z is in the range between 0.3 and 1 nm.

5.2.4
Quantitative Analysis of Nitrogen Plasma-Treated Polypropylene

The XPS technique can be used to quantitatively monitor surface plasma treatment of polymers. Figure 5.4 shows XPS spectra (wide scan survey and C_{1s}, O_{1s}, and N_{1s} regions) obtained for untreated and atmospheric pressure nitrogen plasma-treated polypropylene (PP) film, respectively.

The XPS spectra of Figure 5.4 show that before plasma treatment the surface of the film corresponds to pure PP: only a C_{1s} signal is detected (H is not detected by XPS). After nitrogen plasma treatment, O and N atoms are chemically grafted onto the PP surface as C–O, C=O, N–C, N–C–O, and N– $(CO)_2$ assigned according to the XPS chemical shifts. From these XPS data we can quantify the chemical composition and modifications of the surface treated under specific conditions, by using the equations from the previous section.

Chemical characterization of the plasma-treated surface by XPS gives significant information on the chemistry of the plasma–surface interaction and of the chemical properties conferred to the surface by the treatment as evidenced for instance by the correlation of the contact angle, (indicative of the wettability and adhesion) with XPS data, as shown in Figure 5.5. The more PP is grafted with nitrogen and amine chemical functions, the lower is the water contact angle and therefore the more wettable and adhesive is the PP.

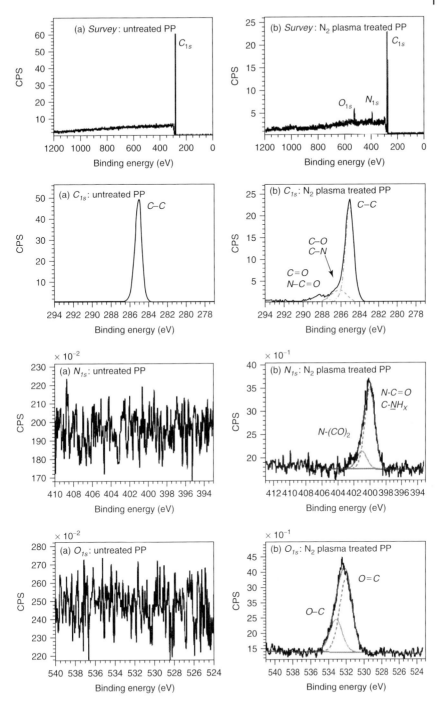

Figure 5.4 Specific spectra of a PP film (a) before and (b) after nitrogen plasma treatment.

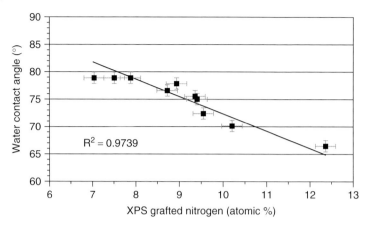

Figure 5.5 Correlation of the grafted amount of nitrogen and amines measured by XPS with water contact angle, which is indicative for wettability and adhesion.

5.2.5

Angle-Resolved XPS Depth Profiling and Surface Sensitivity Enhancement by Grazing Angle XPS Detection

Plasma treatment is a very shallow surface treatment, and the surface techniques used to investigate the modifications performed on the samples should have the lowest possible analysis depth. Regarding XPS, the analysis depth may be reduced by using grazing angle detection at the largest possible angle relative to the sample surface normal direction. This possibility is demonstrated for the case of plasma fluorination of poly(butyl terephthalate) (PBT). XPS spectra of Figure 5.6 recorded in normal direction with respect to the PBT sample surface show actually that after plasma fluorination the C_{1s} and O_{1s} are almost completely attenuated by the F coating which accordingly has a thickness of about 7–10 nm. This layer is therefore well suitable for a case study on depth profiling by angle-resolved XPS and focusing on the outmost surface layers with grazing angle detection.

Figure 5.7 presents the variation of the F/C atomic ratio measured for this sample at different inspection depths according to variable detection angles θ. An enhancement of the fluorine concentration within the first nanometers below the surface is evident. This F enhancement toward the surface of PBT is confirmed by the intensity increase of the CF_X components relative to the C–C one at grazing angles as shown by the C_{1s} spectra in Figure 5.7a. This shows that the CF_X functions are located at the very top of the treated PBT surface.

5.2.6

Determination of Thin Coating Thickness by Angle-Resolved XPS

Angle-resolved XPS offers unique capabilities for measuring in a nondestructive way the thickness of uniformly smooth and very thin (thickness $\leq 7-10$ nm)

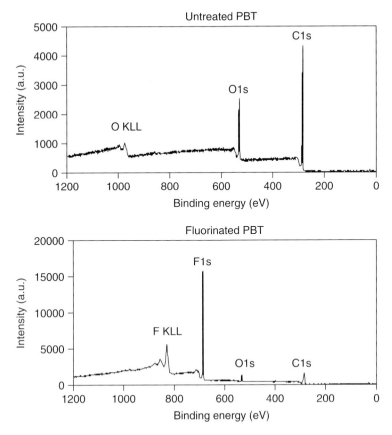

Figure 5.6 Survey spectra of PBT film before and after plasma fluorination with a 7–10 nm thick F coating.

coatings, as shown with the following example of a thin SiO_X coating ($d \le$ 7–10 nm) grown by plasma-enhanced chemical vapor deposition (PECVD) on PET film. In this case, the XPS signals from the SiO_X overlayer (Si_{2p}) and from the substrate itself (C_{1s}), emitted in the direction θ relative to the sample normal are, respectively, given by the relations in Equation 5.2:

$$I_{Si2p}^{d} = I_{Si2p}^{\infty} \left[1 - \exp\left(\frac{d}{\lambda_{Si2p}\cos\theta} \right) \right]$$
$$I_{C1s}^{d} = I_{C1s}^{0} \exp\left(-\frac{d}{\lambda_{C1s}\cos\theta} \right) \tag{5.2}$$

where

I_{Si2p}^{d} and I_{Si2p}^{∞} are the intensities of the Si_{2p} peak from the SiO_X overlayer with thickness d and from an infinitely thick (bulk) layer (i.e., with a thickness $d > 10$ nm), respectively.

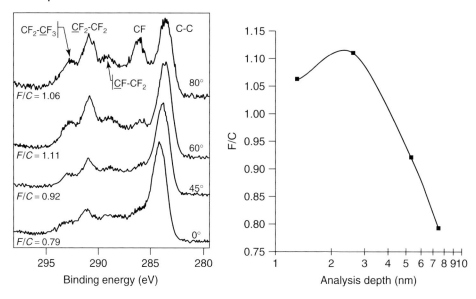

Figure 5.7 Depth profiling of F/C ratio by angle-resolved XPS and surface enhancement of F and CF$_X$ components in the topmost 3 nm of the surface as evidenced by grazing angle detection.

I_{C1s}^{d} and I_{C1s}^{0} are the intensities of the C$_{1s}$ peak from the PET substrate covered by SiO$_X$ with a thickness d and from uncovered PET, respectively.

These expressions can be rearranged to give Equation 5.3:

$$\mathrm{Ln}\left[1 - \left(\frac{I_{Si2p}^{\theta}}{I_{Si2p}^{\infty}}\right)\right] = -\frac{d}{\lambda_{Si2p}\cos\theta}$$

$$\mathrm{Ln}\left(\frac{I_{C1s}^{\theta}}{I_{C1s}^{0}}\right) = -\frac{d}{\lambda_{C1s}\cos\theta} \tag{5.3}$$

Hence, if the growth of the SiO$_X$ overlayer onto the PET substrate follows a layer-by-layer mode with a homogeneous thickness d, the terms at the left hand of the equations above plotted as a function of $1/\cos\theta$ give linear plots with slopes equal to $-\frac{d}{\lambda_{Si2p}}$ and $-\frac{d}{\lambda_{C1s}}$, respectively, where λ_{Si2p} and λ_{C1s} are the electron mean free paths corresponding to Si$_{2p}$ and C$_{1s}$ XPS electrons.

This is shown for results obtained for SiO$_X$ deposited onto 12-μm thick PET film by a 45 kHz and 1500 W discharge in a flow of oxygen-diluted hexamethyldisiloxane (HMDSO) precursor molecules mixed in helium at a pressure of 30 Pa in a roll-to-roll process [4]. As shown in Figure 5.8 we actually obtain linear plots for the values Ln$[1 - (I_{Si2p}^{\theta}/I_{Si2p}^{\infty})]$ and Ln$(I_{C1s}^{\theta}/I_{C1s}^{0})$ as a function of $1/\cos\theta$, which gives clear evidence that the SiO$_X$ overlayer is uniform. Further, the thickness d of the SiO$_X$ layer is determined from the slopes d/λ of these linear plots. This gives

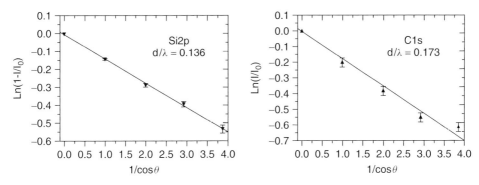

Figure 5.8 Angle-resolved XPS determination of thickness of SiO$_X$ layer uniformly deposited on PET by PECVD.

values of 0.37 or 0.44 nm, depending on whether the Si$_{2p}$ peak or C$_{1s}$ peak, and the corresponding inelastic mean free paths λ for electrons from these orbitals, is used for evaluation. Using λ_{Si2p} (2.72 nm) or λ_{C1s} (2.67 nm) in SiO$_2$ gives the thickness with an accuracy of 10%.

This methodology for determining the thickness of SiO$_X$ is then applied to follow the growth of SiO$_X$ by measuring the SiO$_X$ thickness as a function of the exposure time of the substrate to the plasma expressed as the reciprocal of PET winding speed. The growth of SiO$_X$ with time is found to be linear with a deposition rate in the order of 0.4 nm s^{-1} under these conditions (Figure 5.9).

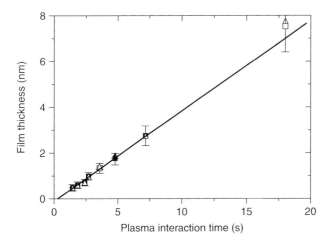

Figure 5.9 SiO$_x$ coating thickness deposited on PET surface evaluated from both XPS C$_{1s}$ substrate and Si$_{2p}$ coating shows a linear dependence in function of the web interaction time in the plasma. SiO$_x$ coating deposition rate is measured equal to 0.4 nm s^{-1}.

5.2.7
Mapping

XPS provides very useful capabilities for mapping non-equivalent chemical sites of the same element with different core level shifts. Figure 5.10 presents examples of XPS chemical images obtained from 10 μm diameter carbon fibers coated by fluorination plasma. The chemical contrast is quite satisfactory. Clearly the C–F image (Figure 5.10b) and the F atoms image (Figure 5.10d) are well overlapping. The C–C and C–F images are partially overlapping and complementary (Figure 5.10c). These images indicate that the fluorination coating of the fibers in this case is not homogeneous. They show that some fibers are not at all fluorinated and others are only partially coated.

The conclusion drawn from these XPS images can be confirmed by results obtained with microanalysis of areas selected from the chemical maps (Figure 5.11).

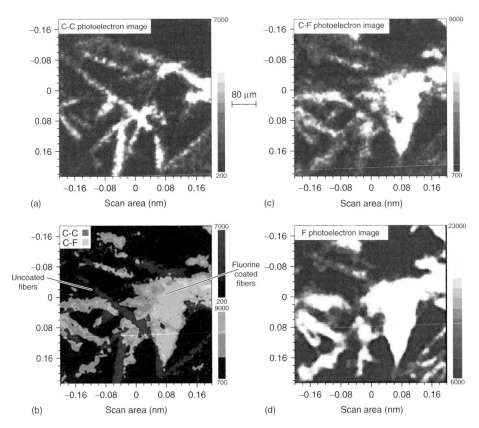

Figure 5.10 Chemical mapping of plasma fluorination of carbon fibers. (a) C–C bonding mapping; (b) C–F mapping; (c) overlay of the previous C–C and C–F mappings; (d) F atoms mapping.

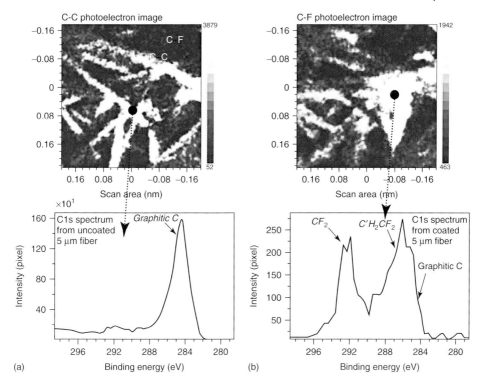

Figure 5.11 XPS microanalysis of two selected areas (●) localized in C–C (a) and C–F$_X$ (b) bonding mappings, respectively.

The C$_{1s}$ spectrum from a 10 μm size spot in the brightest area of the C–C map shows only the C–C component while the C$_{1s}$ spectra from C–F areas show both C–C and C–F components indicating a partial fluorine coating.

5.2.8
Summary of XPS

XPS is well suited for investigating the surface chemistry induced by cold plasma treatment, functionalization, and coating.

XPS is a direct technique without artifacts. All elements except H and He can be detected. The bulk sensitivity is <1% of the atomic concentration and the surface sensitivity is <1% of a surface monolayer (10^{12} to 10^{13} atoms cm^{-2} or less than 1 nmol cm^{-2}). Interference and overlap effects are not very pronounced. The average analysis depth (approximately the same as the photoelectron mean free path λ, with values ranging from 1 to 3 nm) makes it well suited to investigating plasma treatment of surfaces. The maximum inspection depth is around 3λ, that is, between 3 and 9 nm. The surface sensitivity of XPS can be enhanced by grazing angle detection. Surface mapping can be performed with at a lateral resolution of less than 3 µm and microanalysis is possible with a spot size smaller than 10 µm.

Chemical bonding information is provided by the XPS core level chemical shift ΔE_B at a energy resolution below 0.3 eV. Quantitative analysis by determination of the stoichiometry and the relative atomic composition is obtained without a matrix effect, an effect which is common in secondary ion mass spectrometry (SIMS) measurements. The reproducibility is better than 3% and accuracy better than 10%. Depth profiling and thickness measurements of coatings are performed in a nondestructive way by angle-resolved XPS at maximum depth of 7–10 nm. Larger film thicknesses and depth profiles can be determined destructively in combination with ion etching.

Soft X-rays in the kiloelectronvolt range induce little irradiation damage, and organic materials, polymers, biological matter, and any UHV-compatible material can be safely analyzed by XPS. For insulating samples the charging effect can be effectively neutralized by using an electron flood gun.

5.3
Static Secondary Ion Mass Spectrometry by Time of Flight (ToF-SSIMS)

Time of flight static secondary ion mass spectrometry (ToF-SSIMS) provides a mass spectrum of atomic, molecular, and molecular fragment ions from a solid surface. It is based on the detection of positively and negatively charged ions (secondary ions, SIs) which are produced under the bombardment of incident ions (primary ions). The static SIMS technique is characterized by a very low primary ion dose so that minimum damage is induced on the analyzed surface. The nature of the SIs emitted is intimately related to the sample surface chemistry. Moreover, the SIs come from the topmost surface (10 Å), so ToF-SIMS is one of the most surface sensitive techniques. The SI emission phenomenon is sufficiently general to analyze all types of samples (which must be UHV-compatible) and application fields are as large as elemental and molecular surface analysis, depth profiling, and ionic imaging. Only the basic principles of the SIMS technique and application examples are described below. More details on the instrumentation can be found elsewhere, for example, in Benninghoven *et al.* [5] and Vickermann and Briggs [6].

5.3.1
Principles of ToF-SSIMS

The impact of an ion beam with kinetic energy of a few kiloelectronvolts produces the emission of a variety of particles: photons, secondary electrons, radicals, neutral particles (atoms and molecules), and SIs (positive and negative) [5, 6]. This entire collection of emitted particles is characteristic of the sample surface chemistry. In SIMS, SIs are extracted and mass analyzed. They only represent a small part of the emitted particles. The neutral particles can also be studied by post-ionization techniques (secondary neutral mass spectroscopy, SNMS).

5.3.1.1 Secondary Ion Emission

The primary ion energy required for SI emission is in the order of a few kiloelectronvolts (5–25 keV). The incident primary ions produce a collision cascade within a few tens of angstroms below the surface. The implantation depth, that is, the distance traveled by the primary ions within the sample, ranges from 4 nm to more than 15 nm, depending on the substrate or the primary ion nature and energy. The collision cascade transfers energy to atoms and molecules in the surface region and disrupts chemical bonds. When the transferred energy is sufficient, emission of secondary particles from the first atomic or molecular layers of the surface can occur. This makes SIMS one of the most sensitive techniques for analysis of the topmost surface layer (10 Å).

5.3.1.2 Static and Dynamic Modes

The intensity of the primary ion beam defines two types of SIMS analysis:

- 'Static' SIMS uses a very low primary ion current dose ($Ip < 1$ nA cm^{-2}), hence, the sample sputtering yield is very low (1 Å h^{-1}). Only a small fraction of the molecular (or atomic) layer is eroded during the analysis. Molecular ions and fragments from the intact surface are detected together with atomic ions.
- 'Dynamic' SIMS uses a high primary current intensity ($Ip > 1$ mA cm^{-2}), hence, the sputtering rate of the sample is important (> 10 μm h^{-1}) and this mode is destructive. Only atomic ions or small clusters are detected and the technique is more bulk sensitive. This mode of analysis does not provide molecular information but is a very high sensitivity elemental trace analysis technique. Plots of the intensity variation of specific peaks versus sputtering time of the substrate provide a concentration distribution of the detected elements as a function of the depth.

5.3.1.3 Molecular SIMS

The difference in the sputtering rate is not the only effect of the primary ion current intensity. In the 'static' mode, each primary ion impact can be considered as an independent or isolated event. The energy transmitted to the surface is sufficiently weak to desorb intact molecular ions or simple and large mass fragments. Static SIMS is a mild ionization mass spectrometry. Moreover, not enough ions are emitted from the surface for forming a high density plasma. Consequently, no rearrangement or recombination of primary ions occurs. Static SIMS produces molecular peaks and fragment ions from species originally present on the surface, which allows structural and chemical surface analysis.

5.3.2
Applications of ToF-SSIMS

SIMS analysis can be done on any sample compatible with UHV: metals, metal alloys, natural and synthetic polymers and tissues, varnishes, coatings, adhesives, crystals, ceramics, resists, glasses, wood, paper, biological samples (nail, hairs, bones, membranes, vegetal tissues, etc.), thin coatings, mono-molecular layers (Langmuir–Blodgett, automatic blending), additives, and surfactants.

SIMS analysis provides elemental and molecular chemical of only the topmost surface of the sample, that is to say the first atomic or molecular monolayer of the surface. Three modes can be employed: spectrometry mode, SI imaging mode, and depth profiling. All three methods will now be described.

5.3.2.1 Spectrometry Mode

The study of the SIs spectral fingerprint allows chemical and elementary analysis, detection of chemical functions, and molecular identification of the outmost surface, with ppb (parts per billion) sensitivity and a mass resolution better than 10 000. The spectra constitute the fingerprint of the surface, and they are sensitive to the chemical nature, composition, physical and chemical phases, aging, crystallinity, crosslinking, oxidation, and more.

This mode is the most commonly used and allows molecular identification of polymers and copolymers, detection and identification of additives, surfactants, contaminants, treatment or cleaning residues, nature of oxides and determination of metal complexes, metals and metal alloy analysis; characterization of coatings and grafting; evaluation of the crosslinking rate, recovery, or degradation.

5.3.2.2 Secondary Ion Imaging

2D mapping of elemental and molecular species can be obtained by scanning the primary ion beam on a particular area of interest, with a lateral resolution around 100 nm on the more recent instruments. The imaging mode is typically used in order to locally analyze defects or contamination, corrosion points, glass defects (e.g., for contact lenses and molds), pigmentation defects of coatings or varnishes, metalization defects (e.g., for silk screen printing, reflectors), microelectronics, optical fibers, soldering, connectors, to analyze particles or to test the coating homogeneity.

5.3.2.3 Depth Profiling

Alternating sputtering and analysis phases allow in-depth analysis to be performed. The application fields are limited to very thin coatings with a thickness of less than 1 μm (thin layers from 1 nm to several tens of nanometers) but the depth resolution can overtake the monolayer without any loss of sensitivity (depth resolution inferior to 1 nm).

The main applications concern are depth profiles of ultra-thin coatings, grafting by plasma technology, metal oxide layers studies, species segregation, oxidation treatment, or dopant distribution in semiconductors.

5.3.2.4 Data Treatment by Multivariate Methods: Multi-Ion SIMS

ToF-SIMS datasets result from parallel ion detection over a wide mass range with high mass resolution. The information contained in the ToF-SIMS datasets is very rich but also very difficult to interpret. ToF-SIMS spectra, for example, contain up to

1000 mass peaks corresponding to as many variables or dimensional relationships. The human brain can easily handle 2D relationships and 3D relationships but with more difficulty. Therefore, to make full use of all the information buried within the ToF-SIMS datasets requires the reduction of their dimensionality. Multivariate methods are widely used to treat large datasets: the aim is to reduce the dimensionality of the data in order to extract relationships among properties like, for example, additive concentration, surface contamination, or ageing. Principal component analysis (PCA) [7] is a multivariate method aimed at the classification of samples: it is used to extract from the complex spectra, the particular features which correlate most strongly with a known variable, for example, surface functionality, processing conditions, or any quality parameter characteristic of the samples. Principal component regression (PCR) and partial least square (PLS) regression are multivariate methods aimed at quantification: they are used to establish relationships between spectra variables and known properties of samples, such as copolymer composition, surface concentration or coverage, or crosslinking density.

For that purpose, Multi-Ion SIMS was developed and commercialized by Biophy Research. It is a user-friendly standalone Windows application developed under Matlab. It allows direct importing of ToF-SIMS data (spectra, depth profiles, and images) for data matrix building and their reduction by PCA, PLS, or PCR methods. An example of its application for the quantification of additives on polymer surfaces is given below.

5.3.2.5 Examples

The combination of spectrometry with the mapping and depth profiling capabilities of ToF-SIMS has been used to evaluate plasma functionalization of textiles and thin film coatings as illustrated by the following selected examples.

5.3.2.5.1 Poly(ethylene terephthalate) Tissue   Figure 5.12 illustrates the chemical mapping of a reference PET tissue and Figure 5.13, a PET tissue treated with TEOS (tetraethoxy silane) by atmospheric plasma. The corresponding spectra are presented in Figure 5.14. The chemical images have been acquired over a 1.5×1.5 mm^2 area: the reference PET fabric surface is fairly clean and characterized by the PET ($C_{10}H_8O_4$) fragment ions shown in Figure 5.14a together with some contaminants: K, Na$_2$Cl, and an additive with a molecular weight of 480 Da. After plasma treatment, the surface of the PET fabric is covered by a thin silicon oxide layer characterized by $Si_xO_yH_z$ fragments (Figure 5.14b) and masking the PET substrate. Other features are also evidenced: increase of the K, NaCl$_2$, and additive ions intensity, mainly on the 'weft' fibers. Figure 5.15 illustrates the high mass resolution capability of ToF-SIMS: on the reference PET fabrics only COH and C_2H_5 fragment ions from PET are detected, while Si containing species can easily be observed on the TEOS-treated sample.

5.3.2.5.2 Polypropylene Packaging   Another example of imaging is given in Figure 5.16 where the surface of a PP food tray packaging has been analyzed.

Field of view: 1500.0 × 1500.0 μm²

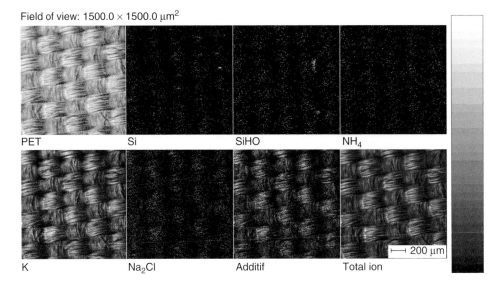

Figure 5.12 Chemical ToF-SIMS mapping of reference PET fabrics. Images are attributed to the indicated element or compound(s).

Field of view: 1500.0 × 1500.0 μm²

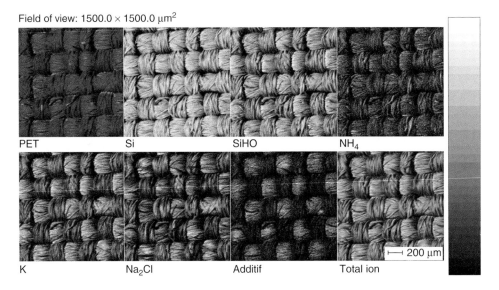

Figure 5.13 Chemical ToF-SIMS mapping of PET fabric treated by a TEOS plasma. Images are attributed to the indicated element or compound(s).

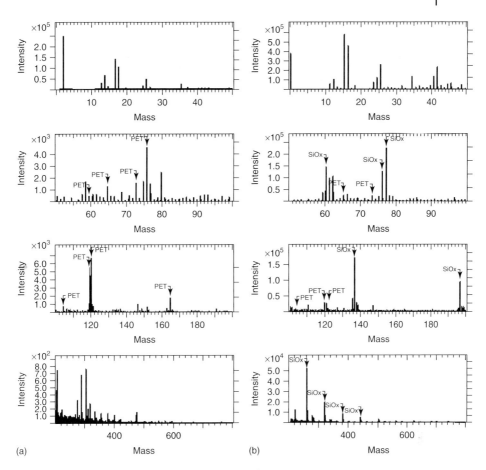

Figure 5.14 (a) Partitioned spectra of reference PET fabrics characterized by the fragment ions of PET ($C_{10}H_8O_4$); (b) partitioned spectra of TEOS-treated PET fabrics characterized by $Si_xO_yH_z$ fragment ions.

Figure 5.16 illustrates the lateral distribution of some fatty acids based processing additive present at the tray surface: the molecular specificity of ToF-SIMS allows palmitic acid ($C_{16}H_{31}O_2$) to be differentiated from stearic acid ($C_{18}H_{35}O_2$) and shows that they are not homogeneously distributed at the tray surface.

5.3.2.5.3 SiO$_x$ Barrier Coating on PET   The depth profile shown in Figure 5.17 corresponds to a SiO$_x$ coating deposited by low pressure plasma on top of a PET substrate in order to improve its barrier properties. Starting from the surface, a contamination layer sits on top of the SiO$_x$ coating characterized by the O and Si ions. At the SiO$_x$/PET interface, a peak of carbon and SiC is observed, which is the signature of a chemical reaction occurring between the SiO$_x$ coating and the PET

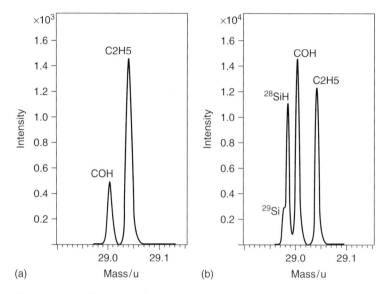

Figure 5.15 High mass resolution spectra at mass 29: (a) reference PET; (b) TEOS treated PET.

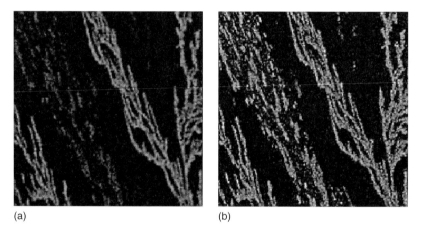

(a)   (b)

Figure 5.16 ToF-SIMS image (101 × 101 μm^2) of contaminants at the surface of a food packaging tray: (a) palmitic acid ($C_{16}H_{31}O_2$); (b) stearic acid ($C_{18}H_{35}O_2$).

substrate. These chemical reactions are very important since they are governing the SiO_x barrier properties and the SiO_x/PET adhesion.

5.3.2.5.4 Anti-UV Additive Qualification on PET Films The last example illustrates the use of multivariate analysis of ToF-SIMS data to quantify an anti-UV additive (Irgafos 168) at the surface of PET films. First, PET films were prepared with addition of defined amounts of Irgafos 168 ranging from 0.016 to 5.74 wt% in

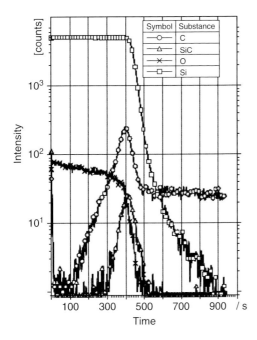

Figure 5.17 Depth profile of SiO$_x$ plasma deposited barrier coating on PET. The erosion time coordinate scales as the depth under the assumption of constant sputtering yield.

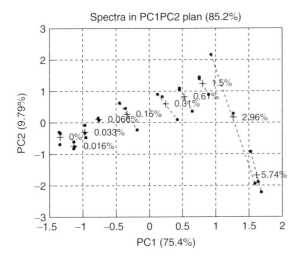

Figure 5.18 Multivariate analysis of ToF-SIMS data for the quantification of an anti-UV additive (Irgafos 168) at the surface of PET films.

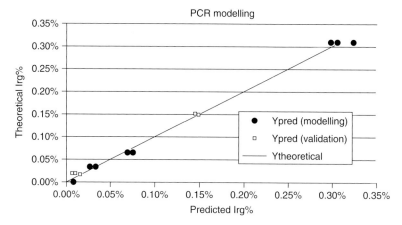

Figure 5.19 PC Regression of ToF-SIMS data for the quantification of an anti-UV additive (Irgafos 168) at the surface of PET films.

order to get a calibration curve for the quantification of Irgafos 168. The PCA reduction of the calibration dataset using Multi-Ion SIMS is illustrated in Figure 5.18, showing that only two principal components PC1 and PC2 are enough to explain more than 85% of the variation between the calibration samples. The points in the figure represent the position in the (PC1, PC2) plan of the individual spectra (intensity of the PET and Irgafos 168 fragment ions) corresponding to the calibration samples. PC1 appears to be a linear function of the Irgafos concentration between 0 and 0.61 wt%, corresponding to the saturation of the uppermost PET surface by the additive. Figure 5.19 shows an example of PC regression using PC1 for the surface quantification of Irgafos 168: predicted values for the calibration dataset (plain circle) and validation samples (squares). This example illustrates the capability of ToF-SIMS to quantify the amount of additives at the surface of unknown polymer samples.

5.4
Atomic Force Microscopy

AFM is based on analysis of the interaction modes between a sample surface and a micrometer scale beam (cantilever) with a nanometric tip. These interactions may be detected by electronic or opto-electronic methods.

5.4.1
Operating Modes in AFM

When the tip is brought into the proximity of a sample surface, forces between the tip and the sample lead to a deflection of the cantilever. Depending on the situation, forces that are measured in AFM include mechanical contact force, van der Waals forces, capillary forces, chemical bonding, electrostatic forces, magnetic forces, and so on. As well as force, additional information may simultaneously be obtained by using specialized types of probe.

5.4.1.1 Contact Mode

The contact mode was historically the first mode used in AFM [8]. In contact mode, the main interacting forces between tip and surface are very short range repulsive forces, so atomic resolution can be achieved (Figure 5.20). Two modes of operation can be applied: constant height mode and constant force mode.

5.4.1.1.1 Constant Force Mode

The constant force mode is most commonly used. The cantilever deflection is maintained constant using the feedback loop to regulate the height position of the sample surface via the displacement of the piezoelectric ceramic. In this case, the image contrast is only due to the z variations of the piezoelectric scanner delivering the topographic information. In constant height mode, the z position of the piezo which carries the cantilever is kept constant, and the topography is sensed by variations of the cantilever bending due to the surface features when the surface is scanned in the x and y directions.

In both cases the deflection of the cantilever is sensed using a laser beam reflected on the rear side of the cantilever. A position sensitive detector monitors the position of the reflected beam.

Friction Mode (Lateral Force Microscopy) In contact mode, scanning the surface with the tip induces friction forces due to the slide and/or adhesion of the tip to the sample. The detection of these forces is possible by measuring the twisting of the cantilever when scanning perpendicular to its axis, that is to say

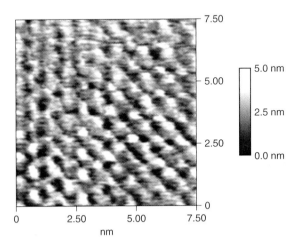

Figure 5.20 Molecular resolution on a phospholipid Langmuir–Blodgett film analyzed in contact mode: the dimensions of the rectangular cell are 0.68 ± 0.02 and 0.93 ± 0.05 nm.

measuring the difference between the output signal on the left and right segments of the photodetectors. The lateral force imaging allows one to differentiate areas with different hardness with one monolayer sensitivity and these forces can be quantified.

Force Curves: Adhesion and Indentation   In contact mode, the detection and measure of the forces interacting between the tip and the surface are obtained using a cantilever with a characteristic spring constant (k). The curve obtained corresponds to the cantilever deflection versus the distance d between the sample and the tip during the cyclic up and down displacements of the piezoelectric toward the cantilever (see Figure 5.21). By calibrating the deflection (z) measured on the photodetector, the force F can be calculated using Hooke's law:

$$F = k \cdot z \tag{5.4}$$

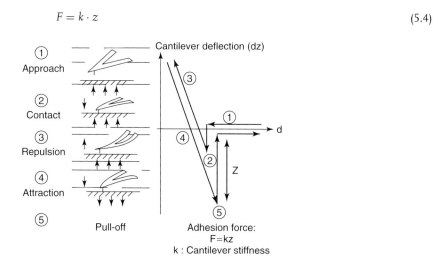

Figure 5.21   Schematic AFM force curve measurement.

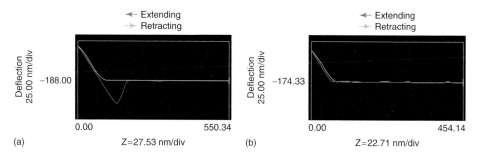

Figure 5.22   Force curves (gray) measured on a biomaterial before (a) and after plasma fluorinated coating (b): On the fluorinated surface the adhesion force is more than 60 times lower than on the untreated polydimethylsiloxane.

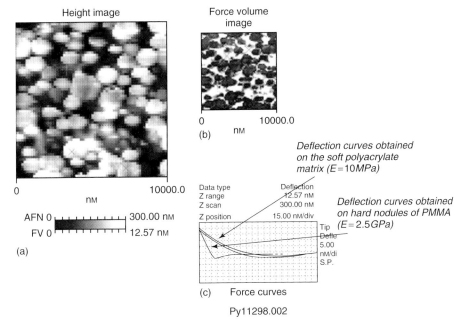

Figure 5.23 AFM force volume mode used for indentation of a biphase polymer [poly(methyl methacrylate) (PMMA), polyacrylate]; the polymer surface was prepared by cryomicrotomy. The height image (a) gives the surface morphology. The contrast is proportional to the distance done by the piezo to reach the imposed maximal deflection. The force volume image (b) gives for each pixel the deflection value for a chosen value (z position) of the piezo A force curve (c) can be obtained on each pixel of the image.

The distance z can be linked to the contact force of the tip on the sample (pull off force) if the spring constant k of the cantilever is known Figure 5.22.

It is possible to obtain a force curve image and to measure adhesion force variations of the tip on the surface. AFM is a useful tool for studying surface interactions by means of adhesion force measurements with a functionalized tip. The tip can be functionalized by chemical methods or by plasma treatment.

It is also possible to indent the sample surface using a diamond tip, to record the indentation curves and to quantify the Young modulus. The principle is the same as with a classical nanoindenter but with the advantage of a precise localization of the indentation point with the force volume imaging mode [9] (Figure 5.23).

5.4.1.2 Resonant Modes

The contact mode allows obtaining the best resolution, but the adhesion forces (particularly capillarity and electrostatic forces) and the friction forces increase the total force and can induce damages of tip and sample. This is the case for soft and fragile materials. For this reason, other operating modes have been developed in order to perform nondestructive analysis.

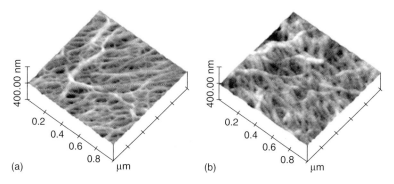

Figure 5.24 3D morphology in contact – no contact mode ('tapping mode'): (a) untreated PP, (b) PP treated by the Aldyne process.

5.4.1.2.1 The Contact – No Contact Mode This mode, developed first by Digital Instruments, is called 'tapping mode' atomic force microscopy (TMAFM). The cantilever oscillates on the surface sample at a frequency close to its resonance frequency. The amplitude is imposed high enough (typically above 20 nm) so that the tip penetrates periodically in and out the contamination layer (or water layer always existing on any surfaces) and the friction forces are removed.

The surface topography is presented by the height image (Figure 5.24). Following the surface morphology, constant oscillation amplitude is used as the feedback signal via the z displacement of the piezoelectric. In tapping mode the lateral resolution can be better than one nanometer.

The morphological modification induced by the plasma treatment (Figure 5.24) is only an increase of the thickness of the thin PP lamellas. The same mean roughness $Ra = 4.3$ nm is measured on the two samples on the 1 μm scan TMAFM images.

5.4.1.2.2 Phase Contrast Mode The phase contrast mode is used with the tapping mode: the phase lag of the cantilever oscillation, relative to the piezoelectric displacement, is monitored and recorded. This phase lag is very sensitive to variations in material properties such as adhesion and/or elasticity at very high resolution.

The following images (Figure 5.25) illustrate the interest of the phase contrast mode. First, a PP tray surface was analyzed as received (a): the phase contrast mode is sensitive to the viscoelastic properties, we observe dark domains due to the presence of additives. The thickness of the domains measured on a height image (not presented) is only 2 nm. The color contrast obtained on these domains indicates that they are softer and/or more adhesive than the substrate. In a second step (b), the PP tray surface was washed with isopropanol, and we observe the disappearance of the dark domains. The additives domains were analyzed by ToF-SIMS and identified as a mixture of palmitic and stearic acids, see also Figure 5.16 in Section 5.3.2.5.2.

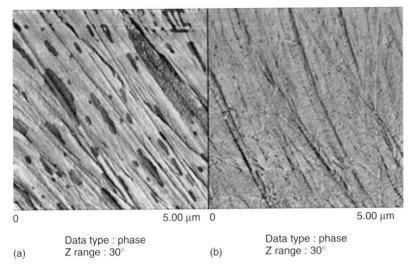

0   5.00 µm 0   5.00 µm

     Data type : phase   Data type : phase
(a)   Z range : 30°   (b)   Z range : 30°

Figure 5.25 PP tray surface analyzed as received (a) in TMAFM, phase contrast mode, the dark domains of additives are removed by the washing with isopropanol (b).

5.4.1.3 Other Modes

Several operating modes have been developed to measure electrical and magnetic properties particularly for microelectronics applications, for example, tunneling atomic force microscopy (TUNA), scanning capacitance microscopy (SCM), and scanning spreading resistance microscopy (SSRM) [10].

Specific modules can be added to an atomic force microscope to obtain physical or chemical information:

- **Thermal microscopy** to study phase transitions in polymers (melting, crystallization, glass, and sub-glass transitions) can analyze the sample between -35 and $+250\,^{\circ}$C.
- **NanoTA**: in nanothermal microscopy (Anasys Instruments), a tip is used first to image the sample. Then this special tip is heated locally on a chosen point and the cantilever deflection is recorded until the melting of the investigated point. Calibration with samples of known melting point is used to measure the local melting point with a resolution of a few nanometers.
- A **microtensile stage** has recently been developed for the atomic force microscope, allowing imaging, in TMAFM, of the formation of cracks during elongation of plasma-coated polymer films [11] (Figure 5.26).

5.4.2
Summary and Outlook

AFM is a recent microscopy technique and numerous operating modes are used today to obtain various analytical information summarized in Table 5.3.

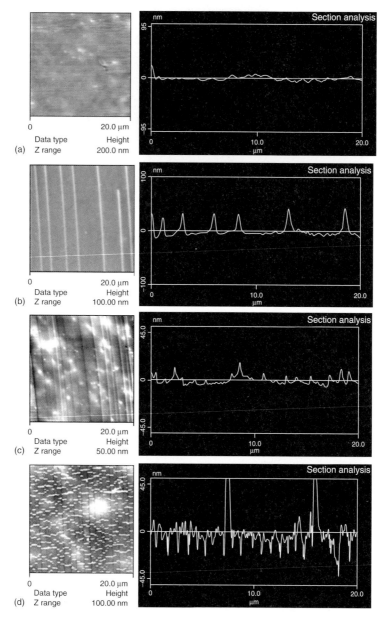

Figure 5.26 Evolution of coating surface pattern investigated by TMAFM with increasing applied strain for a PET substrate coated with a 10 nm thick SiO_x coating (left column) with a corresponding height scan throughout the width (right column). (a) 0% strain: initial nondamaged coating surface. (b) 3% strain: appearance of first wrinkles on the coating surface, without cracking the coating. (c) 5% strain: appearance of first cracks in addition to surface wrinkles. (d) 15% strain: appearance of transverse buckling failures in the coating when cracking process comes close to saturation.

Table 5.3 Analytical information obtained with the major operating modes in AFM.

AFM modes		Analytical information
Contact	Constant force or constant height mode	Morphology from molecular resolution to 100 μm scan, roughness measurements (sensitivity : fraction of nanometers in z)
	Force volume image	Quantification of adhesion forces Indentation (quantification of Young modulus) From few MPa to several GPa
	Lateral force microscopy (friction)	Imaging of hardness differences and quantification of friction coefficient
	Electrical modules (TUNA), (SCM), (SSRM)	Measurement of very low currents (pA), variations of capacitance and resistance
Resonant mode (contact–noncontact)	Height mode	Morphology, roughness measurements
	Phase contrast imaging	Viscoelastic properties
	Magnetic force microscopy (MFM)	Magnetic forces
	Electric force microscopy (EF), surface potential imaging (SP)	Electrostatic forces
	Harmonix mode	Morphology, adhesion, stiffness

New operating modes continue to appear, for example, the Harmonix mode (VEECO) allows adhesion and stiffness images to be obtained simultaneously with morphological and phase contrast images in TMAFM [12]. Future developments on the atomic force microscope will combine this microscopy with chemical analysis; for example, coupled RAMAN–AFM is just beginning to be used.

5.5
Scanning Electron Microscopy (SEM)

5.5.1
Principles of SEM

The principle of SEM is based on the interaction of an electron beam with a surface layer of a specimen and on the detection of the emitted species. The interaction of electrons with matter leads to the emission of electrons in several fields of interest:

- primary backscattered electrons without loss of energy
- primary electrons with loss of energy
- secondary electrons and Auger electrons
- electrons of the continuum.

Photons and X-rays are also emitted in subsequent processes after electron emission.

5.5.2
Imaging in SEM

The detected signals used to form the image when the primary electron beam sweeps the surface sample are backscattered electrons and secondary electrons.

Backscattered electrons have a high energy and can come from depths of several hundreds of nanometers within the specimen at high voltage. They have been elastically scattered by nuclei in the specimen and escape from the surface. Their proportion varies from 0.06 for carbon to 0.5 for gold at 20 keV, so the contrast of the backscattered image is linked to the specimen composition.

Direct secondary electrons are produced by the interaction of the primary electrons with the surface and emitted with low energy (<50 eV), coming from the first nanometers of the surface. As they come from an area defined by the beam size, they can be used for high resolution topographic images.

The standard *source of primary electrons* consists of a tungsten filament with a low beam current. It can be replaced by a lanthanum hexaboride (LaB_6) filament providing a brighter beam. Excellent beam brightness is now obtained with Schottky emission guns such as W/ZrO field emission gun (FEG) and cold or hot FEGs. The brightness improves the signal to noise ratio and allows better resolution. Particularly for polymers, low beam currents, and low accelerating voltages are used to reduce sample damage.

The *detectors* for backscattered electrons are silicon diodes of large area or scintillators. Mixing signals from more than one detector may produce good images. For secondary electrons, a Thornley–Everhart scintillator/photomultiplicator is generally used, which also detects a small fraction of backscattered electrons.

5.5.3
New Generation of SEM

In field emission scanning electron microscopy (FESEM), a short focal final condenser is used to give the highest possible resolution and the specimen is just below the lens. A combination of electromagnetic and electrostatic lens is used in the LEO VP SEM allowing a reduction of lens aberration at low energy. A FEG has lower energy spread and very high brightness so that the resolution at low voltage in a FESEM is as good as the resolution given by a SEM at high voltage. High resolution can be obtained in high vacuum mode: 2.5 nm at 1 kV and 1 nm at 20 kV.

In the HPSEM (high pressure scanning electron microscope) several pressure regions from the FEG (10^{-10} mbar) to the specimen chamber allow working at high pressures of up to 1 mbar with a high resolution. The advantage is that there is no need to coat the sample surface with gold or carbon: at this high pressure level, there is no charging effect for voltages between 10 and 30 kV because the positive ions from the gas neutralize surface charges on the specimen.

Several instruments, for example the LEO 1530VP SEM, combine FESEM at low voltages and variable pressures (from 2×10^{-2} to 1 mbar, adjustable in steps of 0.01 mbar) to low vacuum (2.0×10^{-6} mbar).

5.5.4
Chemical Analysis

When the electron beam interacts with the inner shell electrons, atoms return to their ground state emitting X-rays with well-defined energy characteristics of the emitting atom.

The X-ray energy spectrum is composed of sharp peaks corresponding to K, L, or M emissions and of the continuum forming the background. The continuum is produced when the high primary energy electron beam is slowed by scattering near the nucleus.

Two different types of detector are used in SEM to measure the X-ray intensity: either as a function of the wavelength (wavelength-dispersive X-ray spectroscopy, WDS) or of the energy (energy-dispersive X-ray spectroscopy, EDS).

In EDS, the detector is made of Si(Li) and all elements of $z > 11$ are detected simultaneously. If the detector has an ultra-thin window, elements with atomic number $z > 5$ can be detected.

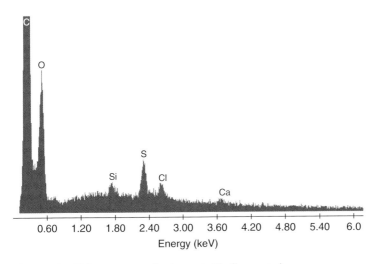

Figure 5.27 EDS spectrum realized on a textile fiber coated with silicon; the silicon signal is detected around 1.76 keV.

In WDS, the X-rays impinge on a bent crystal and are reflected if they satisfy Bragg's law. Several crystals with different lattice spacing are generally used to cover all the wavelengths of interest. Elements of $z > 3$ are detected in WDS.

The two techniques can be quantified; WDS is more sensitive than EDS, but the time of analysis can be very long (5 minutes to several hours) in WDS and this can lead to the degradation of the sample, especially for polymers. The preparation of uniform samples is necessary because the quantification is only possible with flat specimens.

The example in Figure 5.27 presents an EDS spectrum of a textile fiber coated with silicon.

5.5.5
Sample Preparation and Applications

Specimen preparation is quite simple since it is generally not necessary to coat the sample with conductive material. Usually, polymer, fibers, or powders are just stuck on an adhesive carbon substrate.

With the new generation of instruments any type of sample can be analyzed (see Figure 5.28):

- nonconductive materials such as most polymers by means of X-rays analysis in an energy range of 20–30 keV;
- moist samples in combination with a cooling stage;
- the use of low kV imaging in high vacuum mode allows obtaining 5 nm resolution at 200 eV.

5.6
Transmission Electron Microscopy (TEM)

5.6.1
Principles of TEM

TEM is based on the detection of electrons transmitted through the sample and provides structural and morphological information at atomic scale. The use of electrons involves the existence of an UHV in the microscope column and the use of ultra-thin specimens (around 100 nm thick).

The electrons going through the specimen are detected, and can be classified into three types:

- transmitted electrons, which pass through the specimen with little change of direction or wavelength;
- elastically scattered electrons (diffracted electrons), which undergo a directional change due to atomic collision;
- inelastically scattered electrons, which undergo both a directional change and a partial energy loss due to electronic collision.

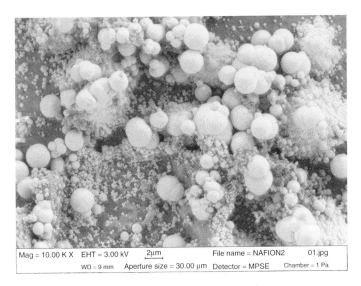

Mag = 10.00 K X   EHT = 3.00 kV   2µm   File name = NAFION2   01.jpg
WD = 9 mm   Aperture size = 30.00 µm   Detector = MPSE   Chamber = 1 Pa

Figure 5.28   SEM image obtained at P = 1 Pa and 3 kV.
The sample is a Nafion membrane coated with palladium
by electroless metalization: small particles 30 nm in diameter and aggregates between 1 and 3 µm in diameter are
observed.

The distinction between transmitted and scattered electrons is related to the
contrast of images in TEM.

Electrons are thermo-ionically emitted by a hot filament (tungsten or
lanthanum hexaboride) or produced by a FEG. Compared with hot cathodes,
the FEG is designed to minimize the diameter of the crossover, to obtain
high brightness and low energy dispersion. The accelerating voltage between
cathode and anode is commonly in the range of 50–120 kV for conventional
TEM, and up to 300 kV for high resolution transmission electron microscopes
(HRTEMs). After the electron gun, several electromagnetic lenses and di-
aphragms are required to focus the electrons and illuminate the sample as
desired.

The sections to be analyzed by TEM are deposited on a grid (copper, nickel,
gold, ...) mounted on a goniometer, within the polar pieces of the objective
lens. It is possible to observe frozen specimens using a cooled stage. In standard
equipment, four or five axes are motorized for specimen displacement: X-axis,
Y-axis, Z-axis, α and β angles for specimen tilt.

The imaging system located after the specimen is composed of an objective di-
aphragm and of several projector lenses to bring the sample image on a fluorescent
screen, for different magnifications. Electron microscope images are recorded by a
charged coupled devices (CCD) camera.

5.6.2
Resolution

The resolution is typically around 0.2 nm but recent progress has been made, resulting in sub-angstrom resolution. The resolution is mainly limited by spherical aberration of the objective lens, characterized by the spherical aberration coefficient C_s. There are only two ways to improve resolution: to shorten the wavelength associated to electrons by increasing the accelerating voltage or to reduce the C_s of the objective lens. Recent HRTEMs are equipped with a C_s corrector, which provides the complete compensation of the spherical aberration in the objective lens.

5.6.3
Image Contrast

Contrast on TEM images has the different following Physical Origins:
- **Amplitude or Diffusion Contrast** This contrast is present in amorphous materials where electrons are transmitted or inelastically scattered. In this case, the image brightness depends on the local electronic density of the specimen (and on specimen thickness): darker regions in bright field image are regions of the specimen of higher scattering.
- **Diffraction Contrast** If the sample is crystalline, incident electrons are elastically scattered (or diffracted) in a specific direction according to Bragg's law. In this case, the diffraction contrast depends strongly on the orientation of the crystal and is associated to imperfections in the sample such as dislocations, inclusions, and multiphases. The diffraction pattern of the crystal can be observed in diffraction mode.
- **Phase Contrast** The passage of the electrons through the specimen introduces a phase lag in the wavefront, depending on the atomic mass of the specimen and on its thickness. The retarded phase will interfere with another wave, giving a phase contrast. But this contrast is also induced in the TEM by imperfect focus. Indeed, defocus causes waves from neighboring regions to overlap in the image and interfere. Phase contrast is more visible for thin and not strongly scattering samples, and must be interpreted with care.

5.6.4
Chemical Analysis

A chemical analysis of a localized area of the sample is possible using a fine and bright electron beam (FEG) with a scanning system scanning transmission electron microscopy (STEM). Two techniques are concerned: electron energy loss spectroscopy (EELS) and EDS.

- **Electron Energy Loss Spectroscopy (EELS)** The energy loss of inelastically scattered electrons, which is characteristic of the atomic and electronic structure of the sample, can be analyzed by EELS. A spectrometer measures the intensity of inelastically scattered electrons at a given energy loss, and a detector associated to the spectrometer registers the spectrum $I(\Delta F)$. The most important part of the spectrum, giving information on the specimen elemental composition, is the region of high energy loss, where the fine structure corresponds to characteristic losses due to deep levels (K, L, ...) excitation. The selection of electrons with a particular energy loss can provide a filtered image, where the contrast indicates the area of the sample responsible for this energy loss. This technique is most useful for light elements ($Z < 20$), and it is also used in polymer systems.

• **Energy Dispersive X-ray Spectroscopy (EDS)** The EDS technique involves the detection of X-rays, having well-defined energies characteristic of the atoms in the specimen. In an energy dispersive X-ray spectrometer, X-rays enter a silicon lithium detector, creating electron hole pairs that trigger a pulse of current through the detector circuit. The number of pairs produced by each X-ray photon is proportional to its energy. The pulses produced are then displayed as an energy spectrum.EDS is most useful for heavier elements ($Z > 11$), therefore it is often applied to find out the nature of fillers and contaminants. For polymers as in SEM, the use of detectors without window or with a thin Mylar window allows lighter elements such as carbon to be analyzed.

5.6.5
Typical Applications of TEM

Typical applications of TEM analysis are:

• layer thickness and interface quality of thin films;
• characterization of microstructures and chemical composition on various materials;
• crystallographic studies (phase identification and separation, defect analysis...);
• evaluation of beam sensitive materials.

The first example corresponds to SiO_x coating on a PP substrate (Figure 5.29), the second example is a multilayer ($SiN_y/SiO_xC_z/SiN_y$) deposited on PP (Figure 5.30). The third example shows a PET fiber treated with HMDSO (Figure 5.31).

5.6.6
Sample Requirements

Since electrons are easily stopped or deflected by matter, specimens have to be very thin to be imaged by TEM. Note that the higher the speed of the electrons (i.e., the accelerating voltage in the gun), the thicker the specimen that can be studied. The choice of the section thickness depends on the accelerating voltage of the gun but also on the nature of the specimen and the goal of the analysis. Section thickness is typically around 100 nm at 100 kV.

Preparation techniques for TEM observations depend on the specimen nature and on the goal of TEM analysis. Thin samples can be obtained by grinding, ultramicrotomy, mechanical polishing followed by chemical dissolution or thinning by FIB (focused ion beam), cleavage, replicas, and other methods.

Ultramicrotomy is one of the most widely used methods in the preparation of polymers for electron microscopy. The steps involved in sample preparation for ultramicrotomy include: (i) specimen staining, (ii) specimen embedding and (iii) sectioning. Staining, involving the addition of heavy atoms (such as osmium tetroxide or ruthenium tetroxide) to specific structures of the polymer, is often required to enhance contrast. In the second step, the specimen is embedded in a resin. The resin must be chemically inert toward the specimen, easy to cut and stable in UHV and toward the electron beam. Transverse sections are then realized by ultramicrotomy, using a diamond knife supplied with a trough filled with water to float the sections that are finally picked onto TEM grids. Ultramicrotomy performed at low temperature and called *cryo-ultramicrotomy* is a sectioning method used for samples too soft at room temperature.

FIB is generally applied to conductive and tough samples such as semiconductors, and is often associated to a SEM to follow sample thinning. Ions (typically gallium) are accelerated with a tension between a few kilovolts and 30 kV and a current from a few pico amperes to a few tens of nano amperes. The ion beam is focused on the surface and is used to make the sample thinner to finally get a section thickness compatible with TEM observations.

Nevertheless, each preparation technique generates its own artifacts. For example, ultramicrotomy can create undulations and compression, while FIB induces selective abrasion and amorphization with gallium implantation.

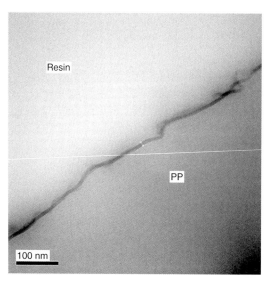

Figure 5.29 PP/SiO$_x$ sample: SiO$_x$ layer thickness: 8 nm.

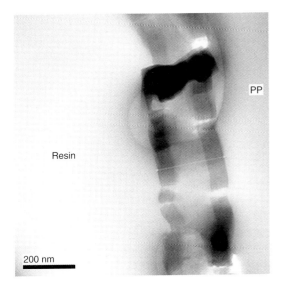

Figure 5.30 PP/SiN$_y$/SiO$_x$C$_z$/SiN: coating thickness: SiN$_y$ inner layer: 70 nm/SiO$_x$C$_z$ intermediate layer: 125 nm/SiN$_y$ outer layer: 63 nm.

Figure 5.31 Treated PET fiber: coating thickness: 11 nm.

5.7
Contact Angle Measurement

The *contact angle* is defined as the angle between the outline tangent of the liquid drop at the contact location and the solid surface (θ) (Figure 5.32).

The contact angle can be measured by various methods (droplet deposition method, Wilhelmy method, and Washburn method). The droplet deposition method is the most commonly used technique: the drop of liquid is deposited on the solid surface, a camera takes a picture just after the deposition, and the

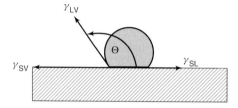

Figure 5.32 Contact angle definition. $\gamma_{SL}, \gamma_{LVL}$, and γ_{SV} are the interfacial tensions between solid–liquid, solid–gas, and liquid-gas, respectively.

droplet profile is analyzed by the software to calculate the static contact angle. In dynamic measurement mode the droplet volume is increased with constant speed and then decreased. Several pictures are acquired with time. The angle calculated during the increase is called the *advancing contact angle*, and during the decrease the receding angle. Several cycles can be performed. The hysteresis between the two angles indicates the evolution of the surface before and after wetting.

5.7.1
Surface Energy Calculation

The contact angle depends on the interfacial tensions liquid-gas (γ_{LV} or γ_L), solid-liquid (γ_{SL}), and solid-gas (γ_{SV} or γ_S).

The Young equation defines the relation between contact angle and interfacial tensions:

$$\gamma_s = \gamma_{sl} + \gamma_l \cdot \cos \theta \tag{5.5}$$

The interfacial tension solid–gas γ_S is called the *surface energy of the solid*. The surface energy is a measure of the potential evolution of the surface to lower its energy state by establishing interactions with another phase. It is generally agreed on to divide the interactions in two types:

- **Dispersive interactions**: Lifshitz–van der Waals interactions always present in materials;
- **'Polar' interactions**: Lewis acid–base interactions due to a difference in electronegativity which can be absent in some materials.

5.7.1.1 Owens and Wendt Model for Surface Energy Calculation
Owens and Wendt [13] supposed that the surface tension consists of the sum of a polar fraction and a dispersive component and that each component of phase 1 can only react with the same component in phase 2. By combining the interfacial tension from the Owens and Wendt model and the Young equation the relation between the contact angle and the surface energy becomes:

$$\gamma_L(1 + \cos \theta) = 2 \left(\sqrt{\gamma_S{}^d \cdot \gamma_L{}^d} + \sqrt{\gamma_S{}^p \cdot \gamma_L{}^p} \right) \tag{5.6}$$

Table 5.4 Contact angle of polyester fabrics after various atmospheric Plasma Treatments.

Sample	N_2 pretreatment	Treatment	θ_0 (°)
0	No	No	93.7 ± 2
1	No	Fluorination	132.8 ± 1.8
2	Yes	No	Absorption
3	Yes	Fluorination	130.5 ± 1.0

$\gamma_L, \gamma_L^{\,d}, \gamma_L^{\,P}$ are known, thus contact angle measurements with at least two different liquids are necessary to determine the surface energy of the solid γ_S. In this approach one of the liquids must be completely apolar.

5.7.1.2 Good and Van Oss Model for Surface Energy Calculation

Good and Van Oss [14, 15] consider the polar component in terms of Lewis acid–base interactions. This means that the polar component is divided into positive and negative fractions. The positive fraction of phase 1 can only react with the negative fraction of phase 2 and vice versa. With the interfacial tension as considered by Good and Van Oss, the Young equation is rewritten as:

$$(1 + \cos\theta)\sigma_l = 2\left(\sqrt{\gamma_s^D \cdot \gamma_l^D} + \sqrt{\gamma_s^+ \cdot \gamma_l^-} + \sqrt{\gamma_s^- \cdot \gamma_l^+}\right)$$ (5.7)

$\gamma_L, \gamma_L^{\,d}, \gamma_L^{\,+}, \gamma_L^{\,-}$ are known, thus to solve this equation and to calculate γ_S, it is necessary to measure the contact angles of at least three different liquids with the following requirements:

- two of them must have $\gamma_L^{\,+}, \gamma_L^{\,-} > 0$;
- one of the polar liquids must have $\gamma_L^{\,+} = \gamma_L^{\,-}$. Water is often used because it serves as neutral point in the Lewis acid–base scale.

The surface energy unit is J m^{-2} but more frequently used are mJ m^{-2}, mN m^{-1}, or even dyn cm^{-1} (1 mJ m^{-2} = 1 mN m^{-1} = 1 dyn cm^{-1}).

Within the ACTECO project (see Preface), the water contact angle of polyester fabrics was measured to evaluate the efficiency of atmospheric plasma treatments (Aldyne) to give hydrophobic properties to the surface (a fluorinated monomer was sprayed and grafted by atmospheric plasma treatment). The results are shown in Table 5.4. The fabric is initially hydrophobic (contact angle of 93.7°). A plasma pretreatment was done to render the fabric hydrophilic (after atmospheric nitrogen plasma pretreatment the water is absorbed before the measurement could be done). After the fluorination treatment, the measurements show that with or without pretreatment, the contact angle is above 130°. These high contact angles are due to the high roughness of the fabric driving the surface toward super-hydrophobicity.

5.8
Conclusions

Quantitative analysis and mapping of surfaces at atomic, molecular, and nanometric resolutions is available from a comprehensive set of techniques. Surface science offers a valuable approach of the nature, composition, and reactivity of the components of plasmas, and is an indispensable tool for process development and quality control of treated surfaces as well as for material surfaces analysis prior to design and surface treatment development.

References

1. Siegbahn, K., Nordling, C., Fahlman, A., Nordberg, R., Hamrin, K., Hedman, J., Johanson, G., Bergmark, T., Karlsson, S.E., Lindgren, I., and Lindberg, B. (1967) *ESCA: Atomic, Molecular and Solid State Structure by Means of Electron Spectroscopy*, Almquist and Wiksells, Uppsala.
2. Briggs, D. and Seah, M.P. (1990) *Practical Surface Analysis, Volume 1, Auger and Photoelectron Spectroscopy*, John Wiley & Sons, Inc., New York, USA
3. Briggs, D. and Grant, J. (2003) *Surface Analysis by Auger and X-ray Photoelectron Spectroscopy*, IM Publications and Surface Spectra Ltd.
4. Magni, D., Deschaneaux, Ch., Hollenstein, Ch., Creatore, A., and Fayet, P. (2001) Oxygen diluted hexamethyldisiloxane plasmas investigated by means of in situ infrared absorption spectroscopy and mass spectrometry. *J. Phys. D: Appl. Phys*, **34**, 87–94.
5. Benninghoven, A., Rudenauer, F.G., and Werner, H.W. (1987) *Secondary Ion Mass Spectrometry: Basic concepts, Instrumental Aspects, Applications and trends*, John Wiley & Sons, Ltd, Chichester, UK.
6. Vickerman, J.C. and Briggs, D. (eds) (2001) *ToF SIMS: Surface Analysis by Mass Spectrometry*, IMPublications and SurfaceSpectra, Ltd, Chichester, UK.
7. Jolliffe, I.T. (2002) *Principal Component Analysis*, Springer Series in Statistics, 2nd edn, Springer, Berlin.
8. Binnig, G., Quate, C.F., and Gerber, Ch. (1986) Atomic force microscope. *Phys. Lett*, **56**, 930–933.
9. Reynaud, C., Sommer, F., Quet, C., El Bounia, N., and Tran Minh, D. (2000) Quantitative determination of Young modulus on a biphase system using atomic force microscopy. *Surf. Interface Anal*, **30**, 185–189.
10. De Wolf, P., Vandervost, W., Smith, H., and Khalil, N. (2000) Comparison of two-dimensional carrier profiles in MOSFET structures with scanning spreading resistance and inverse modelling. *J. Vac. Sci. Technol. B*, **18** (1), 540–544.
11. Rochat, G., David, A., Sommer, F., and Fayet, P. (2008) Mechanical properties of SiOx coated PET films characterized by AFM equipped with microtensile stage. *Vac. Technol. Coat*, June 2008, 72–77.
12. Sahin, O., Magonov, S., Su, C., Quate, C.F., and Solgaard, O. (2007) An atomic force microscope tip designed to measure time-varying nanomechanical forces. *Nat. Nanotechnol*, **2**, 507–514.
13. Owens, D.K. and Wendt, R.C. (1969) Estimation of surface free energy of polymers. *J. Appl. Polym. Sci*, **13**, 1741–1747.
14. Good, R.J. (1992) Contact angle, wetting and adhesion: a critical review. *J. Adhes. Sci. Technol*, **6** (12), 1269–1302.
15. Van Oss, C.J. (1994) *Interfacial Forces in Aqueous Media*, Marcel Dekker Inc.

Part II
Hyperfunctional Surfaces for Textiles, Food and Biomedical Applications

Plasma Technology for Hyperfunctional Surfaces. Food, Biomedical and Textile Applications.
Edited by Hubert Rauscher, Massimo Perucca, and Guy Buyle
Copyright © 2010 WILEY-VCH Verlag GmbH & Co. KGaA, Weinheim
ISBN: 978-3-527-32654-9

6
Tuning the Surface Properties of Textile Materials

Guy Buyle, Pieter Heyse, and Isabelle Ferreira

6.1
Introduction

Textile materials have valuable intrinsic properties, such as flexibility, low specific weight, strength, a large surface to volume ratio, a good 'hand' or touch, and softness. Because of this, they easily accept additional functionalities, for example hydrophobic, oleophobic, or antibacterial properties. It is well known that traditional wet methods of applying these finishes require the use of large amounts of chemicals, water, and energy. Plasma is a dry processing technique and permits reduction of the use of all three resources mentioned. In this chapter, we will discuss where plasma can be used in the textile industry, what plasma can achieve on textile materials and the current state of integration in textile processing.

In this introduction, we will give a brief background about plasmas and their fundamental advantage for materials processing. But first, we will explain why plasma has the potential to have a tremendous impact on the textile finishing industry.

6.1.1
Potential Impact of Plasma on the Textile Industry

Properties of textile materials are typically reached by the introduction of conventional additives by impregnation, mixing with melt polymer before spinning, or impregnation with resin and thermal fixation. All these traditional processes induce pollution, consume a lot of energy, and generally modify the physical properties of the materials.

Many properties of textiles are more related to the outer surface rather than to the bulk, and plasma processing is an attractive alternative to conventional aqueous processes, especially given the commercialization of industrial plasma equipment.

Chemical treatments used in textile finishing need large quantities of water. For example, typically 330 l of water are used to produce 1 kg of wool and 380 l for 1 kg of cotton. Limitation or elimination of water consumption during finishing is economically justified by the continuously increasing costs of supplying water and

Plasma Technology for Hyperfunctional Surfaces. Food, Biomedical and Textile Applications.
Edited by Hubert Rauscher, Massimo Perucca, and Guy Buyle
Copyright © 2010 WILEY-VCH Verlag GmbH & Co. KGaA, Weinheim
ISBN: 978-3-527-32654-9

purifying effluents, and by the high energy costs induced by the after-treatment drying. Environmental protection and safer working conditions also explain the interest of the textile industry in novel dry processes. From a legal viewpoint, the pollution generated in textile finishing is an important problem to be solved in the near future with legislation becoming more and more strict. To give an example, the amount of textile treated in France is about 200 kt per year and this activity generates about 250 kt of oxidative materials (data source: French Ministry of Economy – Sessi 2005). On top of this, one should still add the toxic chemicals being used. As example, Table 6.1 gives the pollution related to typical cotton processing, Table 6.2 for wool.

Plasma technology has the potential to improve or substitute most of the finishing treatments and brings a competitive solution to the environmental problems by

Table 6.1   Overview of pollution induced by cotton finishing.

Process	Water		Biological oxygen demand (BOD)		Solid effluents
	$(l\ kg^{-1})$	(%)	$(mg\ l^{-1})$	(%)	(%)
Steeping	4	1	11 000	54	>50
Washing	20	5	4 500	22	10–25
Bleaching	180	46	1 000	5	3
Mercerizing	7	2	30	–	<4
Dyeing	30	8	1 000	5	10–20
Impregnation	25	7	1 200	6	10–20
Rinsing	110	30	200	1	5
Finishing	5	1	1 500	7	15
Total	381	100	–	100	–

Data source: IFTH.

Table 6.2   Pollution induced by chlorination of wool (AOX = adsorbable organic halogens).

Chlorine/Hercosett	AOX load (g ton^{-1} of wool)	AOX concentration $(mg\ l^{-1})$
Chlorination	80	20
Antichlorination	10	80
Neutralization	140	110
Rinsing	60	15
Resin	30	340
Softener	25	210
Total	350	40

Data source: IFTH.

Table 6.3 Comparison of plasma treatment vs. traditional processes for a range of different parameters.

Parameter	Plasma	Traditional
Solvent	None (gas phase)	Water
Energy	Electricity	Heat
Type of reaction	Complex	Simple
Deepness of the treatment	Very thin layer	Bulk of the fibers
New treatments	High potentialities	Very rare
Equipment	Totally new	Slow evolution
Water and energy consumption	Low	High
Pollution	Very low	High

saving water and energy and by decreasing the effluents to be treated. Plasma technology can substitute, or even improve, most of the finishing treatments as indicated in the Table 6.3. Moreover, it brings a competitive solution to the environmental and economical problems related to water and energy usage and it also decreases the effluents to be treated. The ecological aspects of plasma treatments are covered in more detail in Chapter 12.

6.1.2
Plasma Basics

A *plasma* is referred to as the *fourth state of matter*, next to solids, liquids, and gases. A gas consists of freely moving atoms and/or molecules. All these particles are electrically neutral. A plasma is a gas of which a fraction of its constituents are no longer electrically neutral. Instead, the atoms/ molecules are ionized, that is, they have lost (or gained) one or more electrons. Free electrons are also present in the plasma. Note that the definition of a plasma is not dependent on the equipment needed to generate it (e.g., corona discharge, dielectric barrier discharge (DBD), glow discharge). Consequently, the term *plasma*, as used in this chapter, represents all these types of discharges. More information on the basics of a plasma discharge can be found in Chapter 1.

In practice, one generates the plasma by applying an electric field between two electrodes with a gas in between them. This can be done at atmospheric pressure or in a closed vessel under reduced pressure. In both cases, the properties of the plasma will be determined by the gases used to generate the plasma, as well as by the applied electric power and the electrodes (e.g., material, geometry, size). More information about plasma systems can be found in Chapter 2.

Note also that plasmas can be generated in water, see, for example, Malik *et al.* [1]. This form of discharges is normally not considered when dealing with plasma treatment of textiles. Indeed, the fact that plasma is a dry technology is usually considered one of its main advantages, as mentioned in the previous sections. Nevertheless, plasma discharges in water do exist, and their use on textiles has been reported by Temmerman *et al.* [2] for de-sizing yarns (see also Section 6.3.3.1). However, apart from this, we will not further consider this type of discharge here.

6.1.3
Fundamental Advantage of Plasma Processing

A plasma is basically a gas of which a fraction of the particles is electrically charged. This fraction of charged particles is typically very small in the plasmas used for materials processing (order of 1% or below), but these particles are crucial as they can interact with electric and magnetic fields.

For materials processing, the goal is to initiate physico-chemical reactions. These reactions will only take place if a certain energy barrier (the activation energy) can be overcome (see Figure 6.1). Traditionally, this can be done by heating the material, that is, adding thermal energy. This is a very inefficient process because all particles become energized, whereas only a fraction of them is needed for the reaction. In a plasma, energizing only a limited group of the particles to enable physico-chemical reactions is possible, due to the interaction of the charged particles with the applied electric field, resulting in a much higher efficiency.

To illustrate this, we refer to a life cycle assessment (LCA) study about imparting oleophobic properties on a poly(ethylene terephthalate) (PET) substrate (Chapter 12). That study shows that only about one-third of the energy is needed when a plasma process is used, compared with traditional wet processing. Moreover, the LCA study also shows that the environmental impact of the plasma process is considerably smaller (at least by a factor of 2) with respect to the global warming potential (i.e., contribution to CO_2 emission), acidification, photochemical ozone creation potential, and eutrophication (Table 6.4).

6.1.4
Classification of Plasmas from the Textile Viewpoint

As mentioned, a plasma distinguishes itself from the other states of matter because it contains charged particles. From a fundamental point of view, plasmas can be classified based on the energy and density of these charged particles. This leads to the identification of different types of plasmas, for example, the *Aurora Borealis* or the hot plasmas used for nuclear fusion. This classification is of little use from a practical materials processing viewpoint. Therefore, we only consider plasmas here that are suitable for materials processing.

For this application one uses a nonthermal, low temperature plasma. The expression 'nonthermal' refers to the fact that the ion and electron temperature in the plasma are not at equilibrium. The expression 'low temperature' refers to the limited energy of the ionized species. This is explained in detail in Chapter 2 of this book.

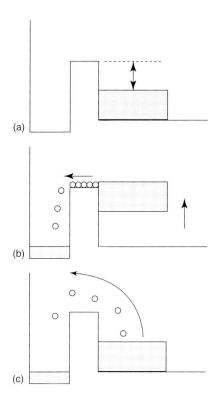

(a)

(b)

(c)

Figure 6.1 Sketch of how the activation energy is over
come in a typical process. (a) Representation of the activation energy to be overcome; (b) the traditional way is
adding thermal energy to all particles; (c) using plasma it
is possible to specifically energize only a small fraction of
the particles.

Table 6.4 Comparison of a traditional and plasma-based
process for rendering PET oleophobic.

Treatment	GWP (kg CO_2)	AP (g eq SO_2)	POPC (g C_2H_4)	EP (g PO_4^{3-})
Plasma	0.4	1.2	0.19	0.26
Traditional	1.67	2.36	2.81	0.76

Different LCA parameters (GWP, AP, POPC, and EP) are compared, the meaning of
which can be found in Section 12.3.3. Clearly, the plasma-based process shows a
smaller impact for all the parameters.

Table 6.5 Overview of the typical pressure ranges encountered for plasmas and their (dis)advantages.

Gas pressure	Strengths	Weaknesses
Low (~0.01 kPa)	Uniformity Flexibility	Batch process Expensive equipment
Subatmospheric (~1 kPa)	Uniformity Flexibility	Batch process or expensive equipment
Atmospheric (~100 kPa)	In-line Speed	Influenced by the environment Less flexible

We will categorize the most common nonthermal low temperature plasmas used for textile processing in two ways: the gas pressure at which they are generated and the geometry of the substrates that can be treated.

6.1.4.1 Pressure-based

As mentioned, one needs a gas to generate a plasma. The pressure of this gas will not only have a large influence on the plasma properties but also on the type of equipment needed to generate the plasma. Indeed, some types of plasma can only be generated at reduced pressure. We distinguish three pressure ranges: low pressure, subatmospheric, and atmospheric plasmas, as indicated in Table 6.5.

Low pressure plasmas are typically in the pressure range of 0.01 kPa. A vacuum chamber and the necessary vacuum pumps are required, which means that the investment cost for such a piece of equipment can be (very) high. These plasmas are characterized by their good uniformity over a large volume. They can be generated using noble gases but also with other gases (e.g., O_2, N_2, or air). The plasma process can be tuned to a very wide range of needs because a large variety of gases can be introduced into the closed vacuum chamber. Roll-to-roll systems are possible for textile treatment. Since it is impractical to perform the treatment in-line with the other process steps, a batch process is usually used for the plasma treatment, implying that the roll has to be placed into the vacuum chamber.

Atmospheric plasmas operate at standard atmospheric pressure (about 100 kPa). Open systems exist which use the surrounding air. Also, systems in a conditioned reactor are possible. In that case, the gas typically used is nitrogen, but sometimes (addition of) a noble gas is needed. Because of the interaction with the atmosphere, one has to be careful about which substances are injected into the plasma. Hence, the range of possible processes is not as large as for low pressure plasmas. On the other hand, these systems are easily integrated in existing finishing lines, a major advantage from an industrial point of view. Of course, for an in-line process to

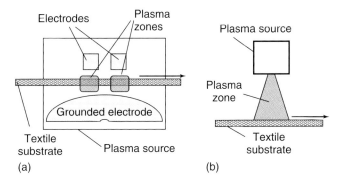

Figure 6.2 Atmospheric plasma sources can be split into a group (a) where the textile substrate has to pass in between the electrodes and (b) where this is not the case.

be feasible, the plasma treatment has to be done at sufficiently high line speeds, which is not evident for textile materials.

In between these pressure regimes, one has the subatmospheric plasmas (typically around 1 kPa). The characteristics of systems operating in this pressure range are a mixture of those of the low and the atmospheric pressure range. Subatmospheric plasmas aim at providing the process flexibility of the low pressure without their more complex and expensive equipment. However, one does need a closed reactor, which means that the process is not immediately compatible with in-line processing.

6.1.4.2 Substrate-based

From a practical point of view, it is important to know what kind of textile substrates and which sizes can be treated. For low pressure systems, the basic limitation is given by the size of the vacuum vessel. If this is made large enough, one can treat all kinds of thicknesses, widths and lengths. For more voluminous textiles the vacuum compatibility (primarily outgassing) has to be kept in mind, but solutions can be provided for this.

Two groups of atmospheric plasma sources can be identified (Figure 6.2): configurations where the textile substrate has to pass in between the electrodes and configurations where this is not the case. For the former there is a limitation on the maximum thickness that can be treated, for the latter not.

Table 6.6 gives an overview of the upper limits for the substrate thicknesses that can be treated, together with the uniformity and the ease of scaling-up to large widths. The overview is made for four common types of atmospheric plasmas: corona, DBD, glow discharge, and plasma jet. Only the latter belongs to the group depicted in Figure 6.2b.

Table 6.6 Overview of the properties of some typical types of atmospheric pressure plasma equipment.

Discharge type	Geometry	Substrate thickness	Uniformity	Scale-up width?
Corona	Line	<10 mm	− (μ discharges)	+
DBD	Plane	<20 mm	−/+ (μ discharges)	−/+
Glow discharge	Plane	<15 mm	+	−
Plasma jet	Line	No limit	+	−/+

The numbers given in the column 'Substrate thickness' are only meant to be indicative. For the last two columns, the following code is used: '+': easy to achieve; '−': (very) difficult to achieve; 'μ discharges': microdischarges [3].

6.2
Plasma Treatment of Textile Materials

6.2.1
Overview of Functionalizations

In general, a very large range of possible plasma treatments on textile can be devised. A lot of research has been done on this and has been reported in the literature. The most common treatments include:

- imparting hydrophilic properties
- increasing adhesion
- enhancing printability, dyeability
- changing the electrical conductance
- imparting hydrophobic and oleophobic properties
- application of antibacterial agents
- application of fire retardant agents
- anti-shrink treatment of wool
- sterilization
- de-sizing of cotton.

Recently an extensive review article was published which summarizes a large number of (academic) research efforts on functionalizing textiles via plasma, and the reader is referred to this paper for a full discussion [4].

Apart from the actual desired functionalization, very often other specifications and requirements are posed for the plasma treatment of the textiles. Indeed, the treatment applied:

- must confer the required property;
- should not disturb possible operations of subsequent finishes;

- should not modify the intrinsic properties of fibers (touch, behavior, texture . . .) in general;
- should resist maintenance operations (stain removal, washing, drying, ironing, drying machine, dry cleaning . . .).

The above mentioned functionalities are based on different physical/chemical interactions, which will be further discussed in next section.

6.2.2
Effect of Plasma Treatment on Textile Substrates

Depending on the nature of the excited gas and the treatment conditions, plasmas can induce different modifications on the fiber surface: etching, fixing functional groups, cross-linking, and polymer deposition.

First, we will give some background on the active species within the plasma that are actually responsible for how a plasma interacts with a (textile) substrate. For this topic, we also refer to Chapter 3 of this book. Then, we discuss the different categories of treatment (activation, functionalization, coating). At the end, the typical effect of aging on the treated surface is dealt with.

6.2.2.1   Interaction of Active Plasma Species with a Surface
In order to understand how plasma processing can lead to a wide variety of different functionalities (see Section 6.2.1), we first look at the ways a plasma can interact with a substrate.

As mentioned, charged particles (ions and electrons) are present in the plasma. Next to these, there are also atoms/molecules, metastables, and radicals present in the active plasma zone, as well as photons (because of the UV light generated). All these particles will interact in their own way with the substrate, leading to a large variety of different surface processes. The depth of the substrate that is influenced by photons can be up to several tens of nanometers, but for all other particles, the interaction depth is limited to about 10 nm or less. This means that the plasma will only affect the outermost (very thin) layer of the substrate, that is, it is a real surface modification technique. This has a positive side (the bulk properties are not influenced) but also a negative one (surface contamination can be detrimental to the plasma process).

6.2.2.2   Basic Plasma Effect on Substrate
The interaction of the active species in the plasma with the substrate can either lead to the addition of particles to the substrate or removal of particles from the substrate.

In the latter case, the plasma treatment can lead to cleaning, to etching, and/ or to sterilization. Applications that can be thought of are a homogenization of fabrics by removing a previously required finish, surface cleaning prior to finishing operations, or increasing the color shade by increasing the micro-roughness. Examples discussed in literature are the removal of sizings (e.g., Cai *et al.* [5]), sterilization (e.g., Roth *et al.* [6]), or antishrink treatment of wool (e.g., by Canal *et al.* [7]). The latter is based on the smoothing of the scales of the wool fibers, thus eliminating the mechanism that leads to shrinking.

In the case of adding particles to the substrate, one speaks typically of activation, functionalization, or finishing/coating. As these processes are more common for textile applications, they will be discussed in more detail. For more specific literature the reader is referred to, for example, a book dedicated to the plasma treatment of textiles [8] or a recent review article [4].

Surface activation refers to the temporal increase of the surface energy. Such a treatment enhances the affinity of the substrate for other substances and is especially needed for synthetic materials which typically have a low intrinsic surface energy (e.g., polypropylene (PP) or polyethylene (PE)). The process is based on the implantation of oxygen, leading to the formation of chemical groups like $-OH$, $=O$, $-COOH$. The groups formed tend to reorient themselves with time (because of the thermal energy present). As a result, the treatment is not permanent and has to be done in-line, as close as possible to the subsequent process step which it is intended to promote. This process is typically achieved using a standard corona in open air.

Functionalization refers to the permanent grafting of chemical groups onto the surface. A typical example is the incorporation of nitrogen based groups (e.g., amines and amides). This way a permanent primer layer can be obtained. The process can be realized, for example, by using a DBD or plasma jet, with nitrogen as process gas.

Plasma finishing/coating refers to the deposition of a very thin coating (in the order of some nanometers) onto the substrates. This is achieved using a plasma device (corona, DBD, plasma jet), in combination with a unit to vaporize a liquid precursor. The precursor can be chosen according to the targeted functionality. Applications include, for example, oleophobic properties, fire proofing, or antibacterial properties. The main advantage is that the functionality can be realized with a very limited add-on, e.g., on the order of 0.2 g m^{-2} for obtaining antimicrobial properties [9]. This way, the typical textile properties (hand, softness, flexibility, etc.) are not influenced (sometimes referred to as '*invisible*' finishing).

Plasma coating with polymerizable gases is a technique which has very large possibilities as it bears the potential of substituting nearly all the traditional wet treatments. For each application, different kinds of chemical products are used for traditional wet finishing. Some examples are given in Table 6.7.

In principle, it should be possible to substitute all these treatments by a plasma deposition process. Here an important efficiency improvement can be obtained, as no extra heating of the substrate is required using plasma polymerization, as indicated in Figure 6.3. Also coatings of metals or alloys can be obtained by a plasma-assisted method, for example, silver coating for antibacterial properties [10].

6.2.2.3 Aging

It is very important to realize that certain changes induced on the surface by a plasma treatment are not permanent. Instead there is aging, that is, the effect of the

Table 6.7 Overview of typical current wet textile finishing processes.

Chemicals	Textiles	Properties
Fluorinated derivatives	Cellulosic, polyamide, polyester	Water repellency
Silicones	Cellulosic	Touch
Phosphorous derivatives	Cellulosic, wool	Flame retardancy

The first column gives the chemicals used, the second the textile material and the third the properties aimed at.

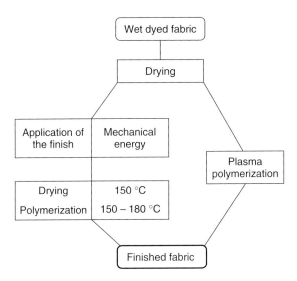

Figure 6.3 Sketch showing the required processes for finishing a fabric, by the traditional route (left branch) and by plasma polymerization (right branch). By plasma polymerization, one can eliminate the drying step required for the traditional method, which represents an important advantage.

plasma treatment decreases with time. Most often, the effect will remain partially and a steady state will be reached, which can take some weeks.

One has to distinguish between plasma treatments that are specifically prone to aging and those that are not. Therefore, we look back at Section 6.2.2.2 where the different effects of a plasma are described. Aging is not really an issue for treatments like etching or coating. In the first case, material is permanently removed, in the latter material is permanently added. For surface functionalization, aging can be an issue. Plasma treatment is a surface effect: only the outermost layers of the surface are treated. For functionalization, the grafted chemical groups can reorient themselves or additives or smaller molecules from within the bulk can migrate to the surface. Both effects typically occur under the influence of thermal energy. They also lead to the same effect: less of the grafted chemical substance is available at the surface, thus the impact of the functionalization will be reduced.

The above reasoning holds especially for surface activation where oxygen-based groups are incorporated, that can relatively easily reorient themselves. Hence, the typical activation process by corona is performed in-line, immediately before the finishing step for which it is intended.

Dye exhaustion has been found to reduce upon aging, indicating a reduced hydrophilicity of the coating [11]. In contrast, recovery of wettability by water immersion was found to be best when water vapor was added to the discharge [12].

Aging is usually regarded as an unwanted side-effect in plasma chemistry. In contrast to this, aging may be taken as an advantage, for example, in $SiCl_x$ coatings. Water vapor from the air will induce hydrolysis resulting in a hydrophilic $Si(OH)_x$ coating [13]. In addition to hydrolysis and oxygen bond formation on long lasting radicals entrapped in the coating, also reorientation of the surface groups results.

The effect of aging will be discussed in one of the case studies in Section 6.5.2.1.

6.3
Integration of Plasma Processes into the Textile Manufacturing Chain

In this section we will probe into the textile manufacturing process and will give references to the use of plasma throughout this production chain, that is, during natural fiber preparation, extrusion, spinning, treatment of fibers, fabrics, non-wovens, and up to the treatment of garments.

Plasma treatment can also be used in the textile industry for effluent treatment. Since this application is not considered a direct textile treatment, it will not be considered here.

Figure 6.4 is a simplified representation of the typical flow from raw materials to finished textile products. From the raw material, natural fibers can be extracted or filaments can be extruded. Natural fibers are processed into a sliver or top and subsequently transformed into a yarn by the spinning process. Another way to obtain yarns is by combining extruded filaments. A single filament can also be used as such, it is then referred to as a *monofilament*, and it can possibly undergo a further finishing process.

Fibers can also directly be processed into a non-woven structure. However, most often yarns are produced which are further processed by traditional textile methods like weaving, knitting, or braiding, to be transformed into, for example, fabrics, ropes, or plaits. Most often, these products undergo further finishing and tailoring steps before a finished textile product is obtained.

In principle, one can envision a plasma treatment in between any two successive steps in the production chain. However, in practice some positions in the manufacturing chain will be much more convenient than others. Here, we basically consider five different positions where plasma treatment is more common:

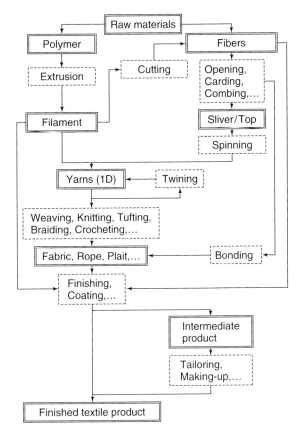

Figure 6.4 General flow of the textile manufacturing process with the positions where processing takes place. Boxes with a double full line frame represent actual products; boxes with a dotted frame line represent processes.

- fiber level
- filament level
- yarn level
- fabric level
- intermediate/finished textile product level.

In the following, we will point out some interesting results obtained at these different steps.

6.3.1
Fiber Level

The initial processes in textile manufacturing, starting from raw materials, are focused on the alignment of natural fibers and compounding of synthetic polymer batches. The use of plasma assisted surface modification at this stage in the production workflow is rather rare. In addition, fibers treated at this stage will quickly divert from the regular workflow and be implemented as such, for example, ultra-thin wool fibers where activated in an argon/radio frequency (RF) low pressure plasma [14]. As a result, hydrogen atoms branch off from the peptide chains and subsequent oxidation in air induces reactive hydroxyl and peroxide groups. In a following step, nano-sized silver particles can be attached providing an antimicrobial activity to the fibers. Besides wool, natural cellulose fibers can be treated as well.

Wu *et al.* reported how PE fibers are treated to improve their affinity with concrete, as such fibers can be mixed in concrete as a reinforcement means [15]. The atmospheric plasma treatment of PP fibers for the same application is also reported [16].

Some work has also been done at sliver level, that is, during the preparations for making fibers. One example is given in Thorsen *et al.* [17], where a pilot-scale corona reactor was adapted to process cotton slivers. Corona treatment improved the cotton spinnability and increased the subsequent yarn and fabric strength and also the abrasion resistance. Furthermore it was reported that the dyeability remained unaltered and that the fiber-hand was unaffected by air corona treatment.

6.3.2
Filament Level

Specific treatment of filaments at this level is limited to specialty products. We consider here the example of monofilaments used as sutures, which are used in medicine to stitch the body after surgery. This type of monofilament is available in both persistent and absorbable form. Plasma treatment of persistent sutures with acrylics effectively reduced bacterial contamination and colonization [18], while plasma coating of absorbable monofilaments showed an increased tensile strength after 120 days of hydrolysis [19].

Treatment of extruded filaments can be done both independently from and integrated into the extrusion process. Part of the extrusion process is the drawing of the filaments, that is, they are elongated to give them their strength. Therefore, it is important whether the plasma treatment takes place before or after the drawing step. Research on this topic has been reported by Warren *et al.* [20], who investigated the influence of the drawing on the plasma treatment effects. They concluded that the level of draw applied to PP tape not only affects the surface structure of the untreated tape but also influences the sensitivity of the tape surface to low pressure oxygen plasma treatment.

A continuous in-line process for low pressure plasma treatment of porous hollow fibers has been disclosed by Oehr *et al.* [21]. Such porous hollow fibers are, for example, used in blood filtration. Using a plasma functionalization, the fibers can be given a region-selective affinity for compounds in the blood that need to be removed, which is beneficial for the filtration capacity.

6.3.3
Yarn Level

After alignment, the fibers or filaments are spun into yarns. This 1D structure is ideally suited for treatments that need to cover the entire surface. This is in clear contrast to a fabric where shadowing effects may reduce effective treatment (see further).

6.3.3.1 Natural Materials

6.3.3.1.1 Cotton   A typical cotton yarn treatment is de-sizing (i.e., the removal of products that were added to the yarn to promote its processing) or the removal of small fibers that are not fully in line with the yarn. Plasma treatment in both helium and air have proven effective in de-sizing of cotton yarns although a cold wash after treatment significantly improved the effect [5]. While de-sizing improves the weaving process, a more hydrophilic surface will improve dyeability of the fiber or the resulting fabric after weaving. Temmerman *et al.* [2] showed that a DC glow discharge, in air or under water, can provide hydrophilicity to the cotton yarn. The effect is induced by local heating, and shock waves in the under-water discharge may enhance the treatment [22].

6.3.3.1.2 Wool   In contrast to cotton, plasma treatments of wool aimed at improved hydrophylicity and dyeing were more successful in an oxygen-rich environment [23, 24]. The impact of the plasma treatment however is not directly on the peptide structure of wool but due to oxidation of the thin fatty-acid-rich lipid layer, referred to as *F-layer*, omnipresent on the fibers. Short plasma treatments (>10 seconds) will increase the oxygen content of the lipid layer, while longer treatments progressively remove the F-layer [25–27]. An advantageous side effect of hydrophilic wool surface is its reduced susceptibility to shrinking and felting [28]. SO–S and SO_2–S groups are formed on the fiber surface by increasing water uptake and subsequent H bonding increases cohesive forces and reduces flexibility. As a result, felting and shrinking is reduced [29]. Still, current plasma treatments for felting and shrinking reduction cannot compete with wet chemical methods where shrinking can be reduced to 1% compared with, for example, 21% after plasma treatment and 69% shrinkage without treatment [30]. An additional side effect often observed in plasma treatment of wool is a reduced whiteness. This is, for example, reported by Danish *et al.* [31] for treatment of Angora rabbit fibers whereas fibrosity and dye uptake typically improve.

6.3.3.1.3 Other Natural Fibers O_2 plasma treatment for 10 minutes at a pressure of 50 Pa of degummed silk (*Bombyx Mori*) yarns resulted in an increased nano roughness, β-sheet molecular ordering, and weight loss, while the crystallinity was reduced [32].

Yuan *et al.* observed an optimal interfacial bonding between sisal and PP in an air plasma, while an argon plasma at atmospheric pressure improved dyeability and surface roughness of bamboo after 4 minutes [33].

6.3.3.2 Non-natural Materials

The previous section showed that plasma treatment of natural fibers and yarns is mainly focused on dye improvement. Also for nylon the dyeability with natural dyes, like cochineal and madder, can be improved using an oxygen plasma [24]. However, for technical synthetic yarns the goal is usually an improved adhesion, for example, for their use in composites. Low pressure plasma treatment of ultra-high strength PE in NH_3 resulted in a fourfold increase in pull-out strength when embedded in epoxy [34]. A sevenfold increase in pull-out strength was achieved by using an oxygen plasma instead [35, 36]. Also, plasma treatment appeared to be better than chemical cleaning of PE fibers for concrete reinforcements with improved flexural and toughness properties [15, 37].

Adhesive strength in composites with PET [38] yarns could be improved by plasma assisted removal of spin finish and oxidation. The need to remove spin finish and moisture is also an issue in high strength aramid or polyamide yarns such as Kevlar [39, 40] and plasma treatment in reactive gases can help to achieve this. In addition, by using gases that have a known reactivity in a plasma, such as O_2, air, or NH_3, functional groups can be created on the surface of the fiber. Results obtained with a T-Peel test showed that an improvement in adhesion to the composite matrix of up to 2486% (!) could be achieved [41].

Coating of inorganic glass fibers with tetravinylsilane in an oxygen plasma improved adhesion to a polyester (PES) matrix by 6.2%, resulting in an increased short beam strength with 32% [42].

In a study by Höcker, acid resistance could be achieved on Nomex with a fluorethane coating [30]. In the same study the hydrophilic character of PET yarns was increased by plasma coating with ethylene to improve the incorporation into a PE matrix.

However, care must be taken in optimizing plasma conditions for the envisioned application and materials. For example on PP yarns, a competition between grafting and etching of newly added functional groups was observed [16]. As a result, longer treatments do not necessarily improve activation. PP coating and grafting with maleic anhydride and acrylic acid by an oxygen plasma resulted in a stable reduction of the water contact angle [30, 43, 44] as also reported in Section 6.5.2.1 of this chapter.

6.3.4
Fabric Level

In this section, we look at the treatment of 'two-dimensional' textile objects. The term two-dimensional is put between quotation marks, because a flat textile structure like a knitted or woven fabric or a non-woven, is not really a 2D object. Indeed, when looking at the scale of the active particles within the plasma, such a structure has to be considered a 3D object, as will be discussed further in Section 6.4.3. Here, we give some references of work done on woven and knitted fabrics and on non-wovens.

6.3.4.1 Woven Textiles

6.3.4.1.1 Natural Materials

Cotton When treating cotton, a cautious balance between plasma reactivity and treatment time is needed to achieve an optimal result. Moreover, repeated, short-term vacuum treatments can be better than one long pass [45]. For example, dyeability improved as more carboxylic groups were grafted on the cotton surface, reaching an optimal surface loading after which dyeability reduced [11]. The same holds for corona treatment where an improved tenacity and wettability was achieved more easily in successive, short passes [46]. Increased dyeability and scouring could also be achieved by pure O_2 plasma treatment [47, 48]. In contrast to this, water and dirt repellency can be provided by either hexamethyldisiloxane, CF, C_2F_6 coating, or metal sputtering [30, 48, 49]. A more fashion-related use of plasma is the decoloration of denim textiles [50].

Wool As described for wool yarns, the fatty acid F-layer can be removed by plasma treatments, meanwhile improving the wettability. The same effect can be reached for wool fibers within a fabric [12, 28]. In addition, dye uptake was shown to be more efficient [51] while color brightness [24] and color depth [28] could be increased in an oxygen discharge or corona treatment, respectively.

Improved hydrophilicity can be advantageous when low shrinking is desired. The altered surface charges additionally influence mechanical, thermal, and permeability properties of the wool fabric [52].

Linen Wrinkles in flax and derived linen fabrics are very persistent and are caused by small fibrils that are not aligned with the yarn. Wong *et al.* [53, 54] report how plasma treatment can help to remove them and longer plasma treatments generally improve the resistance against persistent wrinkles while no further improvement on water uptake is observed after 10 minutes treatment. Plasma etching and shrinking of the fibers was found to be more pronounced in an argon plasma compared with an O_2 plasma. Unfortunately, yellowing of the fabric as a result of the treatment follows the same trend [54]. Dyeing of the fabric could be further improved by acrylic acid binding after plasma activation in air [55].

Silk Bhat and Nadiger [56] investigated the use of low pressure (13 Pa) nitrogen plasma on tasar silk fabrics. They found that the wettability increased while the drying rate and the crease recovery angle decreased. In a different study, low pressure (50 Pa) nitrogen plasma treatment of silk yarns with prolonged exposure times (30–90 minutes), resulted in an increased crystallinity [32]. Using an argon microwave-induced atmospheric plasma, Park *et al.* [57] found that plasma treatment for sterilizing silk showed no adverse effect on the tensile strength or on the surface morphology. As such, this technique was deemed useful for protecting silk artifacts.

6.3.4.1.2 Non-natural Materials   Adjusting the plasma treatment parameters on synthetic woven fabrics allows one to tune the surface energy of the fabric toward the desired application, relatively independently of the intrinsic properties of the synthetic material itself. By selecting the proper plasma settings PES fabrics can be tailored for improved dyeability, air permeability, or water repellency. An increase in surface energy could be achieved on a woven PES fabric by using a commercially available atmospheric device (Ahlbrand Coating Star) at 8 m min^{-1} [58] while improved dyeability was found for fabrics treated in an Ar–O$_2$ atmosphere at atmospheric pressure [59]. The opposite effect was achieved by applying an air-stabilized DC discharge in which a faster response was found using a streamer regime instead of a glow [60]. Improved water repellency could also be achieved by fluorine grafting in a low pressure discharge within 1 minute [61]. However, repeated washing induced polymer rearrangements at the surface due to oxidation and the water repellency capacity decreased significantly.

Similar to the PES fabric treatments, wettability can be improved by plasma grafting of functional groups to the surface. Such grafting can be achieved either directly using well-chosen gases or gas mixtures [6] or in a two-step procedure. An example of the latter process is the following. In a first step SiCl$_x$ is attached to the textile surface in a vacuum discharge. In a subsequent step, that is, when the treated fabric is taken out of the reactor, autohydrolysis by water vapor in the air results in the formation of hydrophilic Si–OH$_x$ groups [13]. For improved wetting the increased O/C ratio is indicative of a proper treatment [62]. By sputtering Cu or Al onto the fabric surface, the opposite, that is, a water repellent effect can be achieved, see, for example, Shahidi *et al.* [49].

In addition to a tailor-made surface hydrophobicity, antimicrobial, or anti-fungal active sites can be introduced on PES woven surfaces by a quaternary ammonium coating [63]. Moreover, the activity of the coating could be induced or recharged by washing in an acidic environment. Antimicrobial properties of metal-sputtered glass fiber fabrics showed a different behavior of the metal applied toward micro-organisms. Copper was found most effective against bacteria and fungi, while silver was found highly effective only against bacteria [64].

Plasma treatment may well improve a specific property of the selected material, but in some cases, other desired properties are negatively affected during the treatment. Plasma treatment of aramid fabrics, for example, was found to improve the interlaminar shear strength but meanwhile decreased the ballistic resistance due the generation of amine groups [65].

6.3.4.2 Knitted Textiles

Treatment of knitted fabrics is much less common than woven fabrics. As a result, only a few references could be found. Except for the first, all the examples regard the treatment of knitted wool fabrics.

Cuong *et al.* [66] report how a glow discharge at atmospheric pressure was generated using a DBD and a mixture of He–Ar inert gases. A knitted fabric based on PET filaments was treated and subsequently grafted with a hydrophilic acrylic acid. They reported an increased hydrophilicity due to the plasma treatment, which was attributed to oxygen containing functional groups incorporated onto the grafted fiber surface and increased surface roughness.

Radetic *et al.* [67] reported the use of low pressure RF plasma for the treatment of knitted wool fabrics, which led to a remarkable increase in dyeing rate: the equilibrium exhaustion was established much faster compared with an untreated or a conventionally chlorinated sample.

Canal *et al.* [7] managed to gradually tailor the surface hydrophilicity of a knitted botany merino wool fabric. First, the knitted fabrics were subjected to a low temperature RF plasma treatment at low pressure. Hydrophobic properties were imparted by a post treatment with hydrocarbon chains of different chain length. They also report the close relationship between the hydrophobic/hydrophilic properties of the wool surface and its shrinkage properties.

Karahan *et al.* [68] treated knitted wool fabrics with argon and air atmospheric plasma. The results showed that there was an increase in thermal resistance, but water vapor permeability, friction properties, pilling tendency, bursting strength, thermal conductivity, and air permeability values decreased.

6.3.4.3 Non-wovens

Although effective for textile plasma treatment, vacuum systems require elaborate pumping equipment. Therefore, the emergence of atmospheric pressure plasma systems provides an important progress for direct in-line industrial implementation. Especially for non-woven substrates the absence of vacuum pumping devices is advantageous because a significant amount of air is trapped inside their interior structure. By using plasma at atmospheric pressure, elaborate pumping is omitted. In addition, the swiftness of the technique makes continuous treatment at an industrial relevant speed possible.

The effect of air permeability was nicely illustrated by Leroux *et al.* Surface treatment was found optimal at 8 m min^{-1} for woven PES, while a 4 m min^{-1} treatment was needed for a non-woven using the Ahlbrandt Coating Star [58]. Other examples of relatively fast surface treatments include for example melt-blown polyurethane and electro-spun nylon, on which a surface energy of more than 70 mN m^{-1} could be maintained for over one year after a 5 second treatment using a glow discharge at atmospheric pressure (also referred to as one atmosphere uniform glow discharge plasma OAUGDP) [6]. On PET and PP, a 0.1 second treatment was found to increase liquid uptake significantly [2]. For sterilization however, a somewhat longer treatment was needed for *E. coli* and virus destruction (30 seconds to 9 minutes respectively) [6].

An example of plasma pre-treatment is the atmospheric nitrogen treatment on PET non-woven which lead to hydrophilization. This proved beneficial for subsequent electroless Ni plating [69]. Polypropylene non-woven fabrics treated by a He atmospheric pressure glow discharge plasma showed increased fiber-to-fiber friction, which was reported to enhance the tensile strength, low-stress mechanical properties, and air permeability [70]. Further, the plasma-treated fabrics gained weight during storage and lost it again after drying, indicating that the plasma-induced surface oxidation leads to moisture absorption from air. This is in agreement with the observed increased surface wettability with prolonged exposure times.

An example for a post-treatment with plasma is given by Virk *et al.* [71]. They showed that an antimicrobial wet finish in combination with a fluorocarbon atmospheric plasma finish can lead to excellent antimicrobial and blood barrier properties on non-wovens used for surgical gowns. Further, it was reported that the plasma treatment did not alter the weight, thickness, stiffness, air permeability, breaking strength, or elongation of the treated substrates.

Use of low pressure plasma (an argon-based RF plasma at 250 mTorr) as pre-treatment of non-woven PET is described [72]. After the plasma pre-treatment, subsequent UV-induced grafting of acrylamide and itaconic acid was performed to obtain a hydrophilic non-woven suitable for further application of antimicrobial agents.

6.3.5
Intermediate/Finished Textile Material

In the above sections, examples for treating textiles at different basic textile levels were given: fibers, filaments, yarns, fabrics (woven, knitted, non-woven). In the following we consider examples for more complex textile structures, that is, intermediate or finished textile products.

A first example regards (tufted) carpets. During the production of such carpets, one needs to coat and/or laminate the back of the intermediate carpet, to ensure a proper binding of the carpet pole and/or to add a backing fabric to the carpet. The adhesion of the coating and the penetration of the coated agent can be improved by, for example, corona treatment [73].

Here, we would also like to mention the possibility of treating medical implants. One example is textile-based stents, which are tubular medical devices that are used to keep arteries open to maintain vital blood flow. To improve their performance, for example, with respect to biocompatibility or antifouling, these devices can be plasma treated (e.g., coated) [74]. Another example is a scaffold for tissue engineering. Tuzlakoglu [75] described how scaffolds from wet spun starch-based fibers are treated with plasma under argon atmosphere. Both treated and untreated scaffolds exhibited a similar ability for osteoblast-like cell attachment and proliferation but the DNA content and the alkaline phosphatase enzyme activity was higher for the plasma-treated ones.

Treating complete textile products is also possible. An example of this is plasma treatment to obtain hydrophobic properties on apparel [76].

Overall, the treatment of more or less finished textile products, be it high value added medical components or basic apparel pieces, is much less common. For one, the object to be treated typically has a strong '3D' character, which makes plasma treatment not evident. For another, it mostly concerns individual pieces, so that a batch-treatment process is required.

6.4
Specific Requirements for the Textile Industry

As mentioned before in this chapter, the potential advantages and application possibilities of plasma are plentiful. In spite of that, and of the numerous (research) efforts undertaken in the past, the industrial use of plasma-based processes in the textile industry is still modest, as will be discussed in more detail in Section 6.6. Clearly, there are reasons why plasma processing is not yet more widely integrated. Verschuren *et al.* deals specifically with the unique challenges posed by textile treatments [77]. Several reasons are pinpointed why plasma treatment of textiles is so difficult.

In the following we will identify a set of reasons which are intrinsically linked to the specific properties of textile materials: the chemical composition, the surface cleanliness, the 3D structure, the large surface area, and the affinity for moisture and air.

6.4.1
Chemical Composition

Textile materials have a natural (e.g., wool, cotton), artificial (e.g., viscose), or synthetic origin. The artificial and synthetic materials have a well-defined and reproducible composition, but for natural fibers the composition is very complex and it depends on a variety factors: e.g., geographical origin, seasonal variations, harvesting conditions.

Another aspect that can be of importance is the sensitivity toward the exact composition of the material. It has been reported [78] that using a similar plasma treatment on two types of polyester gave different results: using X-ray photoelectron spectroscopy (XPS) about 4% of C=O bonds were observed on a PET non-woven, whereas on a PET film no C=O bonds were found at all.

6.4.2
Surface Cleanliness

Plasma treatment is a surface treatment, influencing only the top layer. Contamination of the surfaces, which normally does not influence the standard textile process/properties, can nevertheless be detrimental for a plasma treatment.

Next to the heterogeneous mix of basic textile materials (see above), plasma treatment is further challenged by the presence of finishes, which may be difficult to remove. The result of a finish is that the chemical structure of the surface to be treated may be completely different from that of the bulk. The chemical nature of the fiber surface may strongly influence diverse properties like the bonding strength of the textile material in a composite structure or the aging of the plasma treatment.

Hence, introducing a plasma-based production step may require a new approach for the textile manufacturing process involved ahead of the plasma treatment. As an example, we refer to the time-of-flight secondary ion mass spectrometry (ToF-SIMS) images shown in Section 5.3.2.5, namely Figures 5.12 and 5.13, for an untreated and a treated PET surface, respectively. Here a plasma coating was applied to render the fabric permanently hydrophilic. The images clearly show a different effect of the treatment on the fibers for the weft and warp direction, although both are made of PET and for all other means can be considered identical. This difference was attributed to the different fabrication steps the warp and weft yarns underwent, which lead to a small but significant difference in surface chemistry for the two types of yarns (see also Section 5.3.2.5).

6.4.3
Three-dimensional Structure of Textiles

When considering the interaction of a plasma with the textile substrate, the latter has to be considered as a porous, 3D structure. It is not evident for the active plasma species to penetrate into this structure, and thus, to ensure proper treatment throughout the textile. This is due to the interplay between the life-time of the active species in the plasma and their main free path length. Both properties are strongly influenced by the gas pressure.

This means that especially for textile treatments with low pressure plasmas, the pressure will be a crucial parameter. Several studies regarding this topic have been conducted. In one of them, Poll *et al.* [79] found as optimum pressure range 0.1–10 kPa. They attribute this to different parameters, namely the characteristic geometrical distances in fabrics and the mean free path of the active particles in the gas phase for energy transfer to the textile surface.

Verschuren *et al.* [77, 80] designed some experiments to investigate the penetration of a plasma treatment at a low pressure of 53 Pa. As they report, at these pressures the textile material forms a clear barrier for a gas-containing plasma-generated active species, resulting in an 'edge effect'. They also showed how basic variations in the textile structure influenced the penetration and relate this penetration to the textile porosity.

Another experiment that shows very clearly the limited penetration depth of a low pressure plasma into a non-woven is reported by Musschoot *et al.* [81]. Here atomic layer deposition (ALD) was performed on a thick (35 mm) PET non-woven. Layers deposited using thermal ALD resulted in a conformal deposition throughout the non-woven. They also tested plasma-enhanced ALD on the non-woven material. In this case, the ALD process was assisted by an oxygen-based RF plasma at 0.3 Pa. Here, deposition was only possible in the outermost few millimeters of the non-woven, clearly showing the limited plasma penetration at these pressures.

De Geyter *et al.* [78, 82] studied the penetration of a sub-atmospheric DBD plasma discharge into three layers of non-wovens stacked on top of the grounded electrode. In the investigated pressure range (0.3–7 kPa) it was found that at higher pressures the middle layer was the least influenced by the plasma treatment while at the lowest pressures this was the case for the layer closest to the grounded electrode. This shows that the penetration of plasma into these 3D structures is rather complex and strongly pressure dependent.

On the other hand, for atmospheric plasma treatment the pressure is fixed and cannot be used to control the deposition. In order to investigate the penetration of atmospheric pressure plasma jet treatment a twist free PET filament tow as well as ultra-high modulus polyethylene (UHMPE) filament tows with 0, 1, 2, and 3 twists per centimeter were used as model systems [83]. It was reported that, as the twist level increased, the plasma treatment became less effective due to the tightened yarn structure (as measured by e.g., surface chemical composition, surface wettability, or interfacial shear strength to epoxy). Clearly, this limited penetration depth is a basic disadvantage compared with wet techniques like, for example, padding.

6.4.4
Large Surface Area

The basic building blocks of textile materials are the individual fibers or filaments and, as a result, the surface area to be treated is much larger for a textile substrate than for a flat film. We will illustrate this by a simplified model: consider 1 cm^2 of film and 1 cm^2 of woven fabric, as shown in Figure 6.5a–c.

For a flat, thin sample with a sufficiently smooth surface (e.g., a film where the roughness does not play a role), the total surface area S_{tot} of the square piece is 2 cm^2 (back and front). To calculate the surface area of the woven fabric we assume the following: (i) uniformity in warp and weft direction; (ii) 100 yarns per centimeter (yarn diameter $d_y = 100\,\mu m$); (iii) fiber diameter $d_f = 10\,\mu m$; (iv) 30 fibers per yarn. With these numbers, we find that the piece of fabric contains in both warp and weft direction each $30 \times 100 = 3000$ fibers, resulting in 6000 fibers in total. For simplicity, we neglect the bending of the fibers due to the weaving pattern, so that

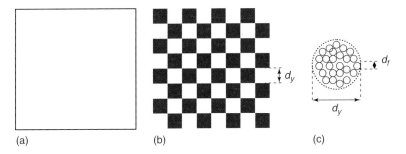

(a)   (b)   (c)

Figure 6.5 Model for calculating the total surface area, for (a) a square sample (side length 1 cm) of film, (b) a woven fabric (the yarn width d_y is indicated), and (c) the yarn cross-section (indicating how a yarn is built up by fibers with diameter d_f). Drawing not to scale [3].

the fiber length l can be set equal to the side of the square: $l = 1 \text{ cm} = 10^4 \, \mu\text{m}$. Given the formula for the surface area S_f of a cylinder ($S_f = \pi dl$), we find for the surface area of a fiber: $S_f \approx 3 \times 10^{-3} \text{ cm}^2$. For the total surface area of the piece of fabric, we still have to multiply by the number of fibers, so that we find: $S_{tot} \approx 18 \text{ cm}^2$. This is almost one order of magnitude larger than the total surface area of the corresponding flat film specimen (2 cm²). In reality the exact values for the surface area will vary strongly with, for example, the type of weaving, the density of the yarns, or the yarn diameter. Moreover, because of the limited penetration, the effectively treated surface area can be (substantially) lower than the total available surface area.

6.4.5
Moisture Regain and Air Adsorption

Natural and artificial fibers and some synthetic fibers contain an important amount of water, for example, 15% for wool, up to 8% for aramid. Because of their structure, natural fibers also incorporate a certain amount of air. Both effects will complicate the plasma treatment.

During atmospheric plasma treatment, most of the water and air contained in the textile substrate will desorb and, as a result, change the plasma composition near the surface. This phenomenon is increased by the high specific surface area. Sometimes, the resulting disturbance of the plasma can reduce or even inhibit the positive effects of the treatment, especially if the atmosphere used is very different from (humid) air.

For low pressure plasma treatment, the pumping capacity must be very high to enable good degassing, especially of the natural fibers. A pre-treatment drying step can be required.

6.5
Case Studies

In this section, we consider three case studies related to the part of the ACTECO project (see Preface) that dealt with tuning the surface tension of textile materials. The first case study (Section 6.5.1) discusses different measurement techniques to evaluate the surface energy. The second case study (Section 6.5.2) gives some examples of hydrophilic treatments and reviews results obtained during the ACTECO project. The third case (Section 6.5.3) relates to hydrophobic and oleophobic treated textiles.

6.5.1
Assessing the Surface Energy of Textiles

When designing plasma processes to tune the surface energy, it is of utmost importance to be able to reliably assess the surface energy of the treated textile matcrials. Therefore, in this section we discuss some options. First, we introduce the standard surface energy measurement methods based on goniometry and tensiometry (Section 6.5.1.1). Then follows an evaluation of the tensiometric methods in order to determine which settings can be beneficial for measuring hydrophilic textiles (Section 6.5.1.2). Afterwards, we look at some typical normalized textile methods that are in use for assessing the surface energy of textiles (Section 6.5.1.3).

6.5.1.1 Introduction to Methods for Evaluating the Surface Energy and Wetting of Textiles

In general, contact angle measurements serve as a good initial technique to characterize the surface energy and to evaluate the wettability of a solid by a liquid. The *contact angle* θ is defined geometrically as the angle formed by a liquid at the three-phase boundary where a liquid, gas, and solid intersect (see also Figure 5.32). If the angle θ is less than $90°$ the liquid is said to 'wet' the solid, with a zero contact angle represents complete wetting. If θ is larger than $90°$ it is said to be 'non-wetting'.

In literature several methods are proposed to measure the contact angle between a solid (e.g., a film) and a liquid. All underlying theories consider the material as an ideal surface: flat, rigid, perfectly smooth, and chemically homogeneous. Since textile substrates do not fulfill these requirements, the method to be used will depend on the degree of wetting, that is, the hydrophilic and hydrophobic/oleophobic behavior.

In general, two different approaches are commonly used to measure contact angles of nonporous solids (like textile materials): goniometry and tensiometry. The goniometric method involves the observation of a sessile drop of the test liquid on a solid substrate, for example, the drop method. It consists of simply putting a droplet on the sample to be analyzed and to visually determine the contact angle. More information about this topic can also be found in Section 5.7. Tensiometry (e.g., the Wilhelmy method or Washburn method) involves measuring the forces of interaction as a solid is brought into contact with a test liquid. The technique chosen depends principally on the geometry and location of the surface or coating to be studied. Goniometry is very suitable for solid substrates with (relatively) flat and nonporous areas whereas tensiometry is, for example, typically used to evaluate the contact angle of fibers.

Here, we further discuss the Wilhelmy method and the Washburn method.

6.5.1.1.1 Wilhelmy Method   For this method, a sample is hung on the balance hook of the tensiometer and brought into contact with the test liquid as shown in the figure below (Figure 6.6).

When the fiber contacts the liquid, a change in weight is detected because of buoyancy and the apparatus registers this as zero depth of immersion. As the solid is pushed into the liquid, the forces on the balance (i.e., the weight) are recorded. These forces are the net result of the actual wetting force, the weight of the probe and the effect of the buoyancy of the sample. The apparatus can be tared to offset the influence of the weight of the probe and can also remove the effect of the

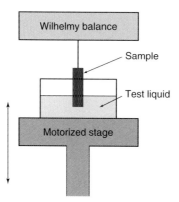

Figure 6.6 Schematic representation of the set-up for the Wilhelmy method.

buoyancy force by extrapolating the graph back to zero depth of immersion. The remaining component force is the wetting force F which is defined as: $F = \gamma_{LV} P \cos\theta$. In this expression, γ_{LV} is the surface tension of the test liquid, P is the perimeter of the cross-section of the sample at the height of the water surface and θ represents the contact angle between the solid test piece and the test fluid. The perimeter P can be determined for a certain sample type by doing a calibration measurement with a completely wetting liquid.

Thus, at any depth, data is received which can be used to calculate the contact angle. This contact angle, which is obtained from data generated as the sample advances into the liquid, is called the advancing contact angle (AA). The sample is immersed to a pre-defined depth and then the process is reversed. As the probe retreats from the liquid data collected is used to calculate a receding contact angle (RA). On almost all surfaces the AA will exceed the RA. The difference between these two values, known as the *contact angle hysteresis*, is a subject of great interest in wetting studies reported in literature, see, for example, Müller *et al.* [84] and references therein.

A drawback is that the Wilhelmy method actually measures $\cos\theta$ and not θ directly. For low $\cos\theta$ values, a small variation in $\cos\theta$ results in a very large variation in θ. This means that, inherently, the technique will be not very accurate. An example: for $\theta < 30°$ an error of 10% on the estimation of $\cos\theta$ may lead to an error of $25°$ for θ itself.

6.5.1.1.2 Washburn Method   For this method, the samples are suspended from a microbalance in a set-up similar to that used for the Wilhelmy method (Figure 6.6) and are put in contact with the wetting liquid. In contrast to the Wilhelmy method the sample is not immersed into the wetting liquid. Instead, the sample is kept in a fixed position as soon as contact is made. Thus, the liquid is absorbed by the (porous) sample, leading to an increase in its weight. The evolution of the weight according to time is recorded. Two liquids are used: hexadecane for calibration, as it leads to total wetting of the sample, and water for the actual measurement.

According to theory, when a porous solid is brought into contact with a liquid the rise of the liquid into the pores of the solid will obey the following relationship (assuming a unimodal network):

$$t = [\eta/C\rho^2\gamma \cos\theta]M^2 \tag{6.1}$$

with

$t =$ time after contact
$\eta =$ viscosity of liquid
$C =$ material constant characteristic for solid sample
$\rho =$ density of liquid
$\gamma =$ surface tension of liquid
$\theta =$ contact angle
$M =$ mass of liquid adsorbed by solid.

Following this relation, the plot of the mass M versus the square root of the time t should yield a straight line with slope $\eta/C\rho^2\gamma \cos\theta$. If the viscosity, the density and the surface tension are known from separate experiments, there are only two unknowns left in this term, θ and C. To determine the material constant C, one does a calibration measurement with a very low surface tension liquid (e.g., n-hexane). In this case, θ may be assumed to be zero (i.e., $\cos\theta = 1$) so that one can determine C (t and M are known from the measurement).

Similar to the Wilhelmy method, the Washburn method also has the drawback that one actually measures $\cos\theta$ and not θ directly. For low $\cos\theta$ values, this implies that the technique will be not very accurate, see previous discussion.

The Washburn method is derived in the assumption that the investigated sample forms a perfect porous network that can be represented by a parallel capillary model for which the effects of fluid inertia and gravity are negligible. Moreover, the wetting of the sample via the porous structure must be very reproducible to ensure that the constant C can be considered as constant for different samples. Looking at fabric samples, the challenge is to ensure that each tested sample has the same size, shape and orientation, and macro and micro-organization. As this is not obvious, the possibility of applying the Washburn model has to be evaluated (Section 6.5.1.2).

6.5.1.2 Evaluation of Methods for Measuring Hydrophilic Properties

The aim of the study reported here was to identify which of the three methods presented in the previous section was the best to evaluate hydrophilic properties of a PET fabric treated by plasma. First of all, it is mentioned that due to the open structure of the fabric, the drop method could immediately be discarded, as any droplet put on the fabric is absorbed into the fabric within tenths of a second.

6.5.1.2.1 Wilhelmy Method   First, screening tests were carried out using the Wilhelmy method (dynamic tensiometer). The measurements were carried out on fabric samples having the same size (2 cm × 3 cm), weight (about 65 mg), shape and orientation. Several measurements were performed at different laboratories.

Based on these screening tests, it was agreed to use the following measurement procedure:

- Sample size: width 2 cm, length in the range to 2–4 cm
- Detection speed: 100–200 $\mu m\,s^{-1}$
- Detection threshold: 2500 μg or manual
- Measurement speed: 100–150 $\mu m\,s^{-1}$
- Penetration depth: 10 mm.

Using these settings, a very good uniformity in the measurement results could be obtained. As an illustration, the values measured on an untreated polyester fabric: the AA was 88($\pm$3)° and 90($\pm$8)° and the RA was 49.5($\pm$3.2)° and 51.3($\pm$3.5)°, respectively, for two different laboratories performing the measurements. Hence, the results for both the AA and the RA the values show excellent agreement between the two laboratories.

6.5.1.2.2 Washburn Method   Second, the Washburn method was tested to evaluate its objectivity and its potential to measure the hydrophilic properties of fabrics. Again, care was given that the measurements were carried out on fabric samples having the same shape and orientation but this time the size was chosen to be 2 cm × 6 cm. We consider here two different samples: the 'reference' PET and the 'treated' samples, which underwent a low pressure plasma treatment with an $O_2/Ar/NH_3$ mixture.

In Figure 6.7 the evolution of the weight versus the square root of time is reported for four different measurements:

1) wetting of the reference sample with water – three repeated measurements represented by curves (1), (2), and (3);

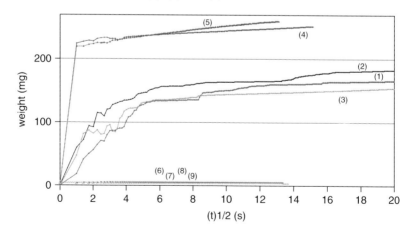

Figure 6.7 Plot of the weight gain versus the square root of time as measured for the reference and treated sample with the Washburn method. The explanation of the different curves is given in the text.

2) wetting of the treated sample with water – two repeated measurements represented by curves (4) and (5);
3) wetting of the reference sample with hexadecane – two repeated measurements represented by curves (6) and (7);
4) wetting of the treated sample with hexadecane – two repeated measurements represented by curves (8) and (9).

From this figure we can deduce some general trends:

- The weight increase of the treated samples in contact with water is larger than for the reference sample, confirming the hydrophilic character of the treatment.
- The weight increase of all samples is substantially larger for contact with water than for hexadecane.
- For contact with hexadecane no difference between the reference and the treated samples can be observed.
- For the reference samples in contact with water the three different measurement results (1), (2), and (3) deviate substantially among each other.
- Also for the treated samples in contact with water there is quite some spread between the measurement curves (4) and (5).
- The graphs do not show a linear behavior, this is especially visible for the measurement related to the reference sample in water due to the choice for the scale of the horizontal axis.

We focus now on the last observation of the list above. According to the Washburn method, this nonlinear shape of the curves shows that the porous structure of the tested samples does not represent a unimodal network (see also previous section), but rather a polymodal network. The conclusion is that the tested PET fabric may not be considered a homogeneous network meaning that for each tested sample the porosity is different. Consequently, one should perform a calibration per sample. Clearly, this is practically not feasible and thus the Wilhelmy method was chosen as the favored method. However, the recording of the weight vs. the time in the Washburn method does allow comparing the wettability of the samples on a semi-quantitative basis.

6.5.1.2.3 Summary of Evaluation   Because of the open structure of the PET fabrics that were plasma treated, the drop method could not be applied. Therefore, two methods based on tensiometry, the Wilhelmy and the Washburn method, were evaluated. A summary of the advantages and disadvantages of the methods are given in Table 6.8. The Washburn method was found not to yield sufficiently reproducible results. Hence, the Wilhelmy method was chosen as preferred test method to evaluate the contact angle of the textile surface.

6.5.1.3   **Tests and Standards for Evaluating Hydrophobic/Oleophobic Properties**
Here, we look at some typical normalized methods that are in use to assess the surface energy of textiles. These methods do not give a direct value for the surface energy. Instead, they give a normalized performance level, for example, regarding

Table 6.8 Overview of the advantages and disadvantages of the different methods used for determining the contact angle of fabrics.

Method	Advantages	Disadvantages
Drop	Easy	Droplet absorbed extremely fast into the fabric
	Accurate (on nonporous surfaces)	
Wilhelmy	Suitable for porous hydrophilic samples	$\cos\theta$ is measured, not accurate for small θ Calibration required (per sample type)
Washburn	Suitable for porous hydrophilic samples	$\cos\theta$ is measured, not accurate for small θ Calibration required (per sample)

(a)   (b)

Figure 6.8 The apparatus for performing the spray test (EN 24920) (a) Initial set-up (b) Spray of water.

the water or oil repellency. In fact, such levels typically correspond with a certain range of surface energy values.

6.5.1.3.1 Water Repellency: Spray Test The spray test (EN 24920) simulates rainfall on a pre-tensioned and slanted piece of fabric, referred to as a *swatch*. The test device (see Figure 6.8a) allows a gentle spray of water to fall on the swatch

Standard spray test ratings

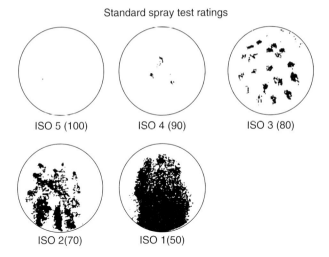

Figure 6.9 Images of the ISO references for evaluation of the spray test EN 24920.

Table 6.9 Overview of the ISO levels for the spray test (EN24920).

Level	Description
5	No sticking or wetting of upper surface
4	Slight random sticking or wetting of upper surface
3	Wetting of upper surface at spray points
2	Partial wetting of whole of upper surface
1	Complete wetting of whole of upper surface

at an angle of 45°. The standardized procedure is to first spray 250 ml of water during 30 seconds (Figure 6.8b). Subsequently, the test specimen is tapped twice to loosen surface-held and any remaining water. Finally the wetted area on the swatch is compared with a series of standards rated 1–5, which are defined by images (Figure 6.9) and a description (Table 6.9).

6.5.1.3.2 Water/Alcohol Repellency   To assess the water/alcohol repellency, the AATCC Test Method 193-2004 can be used. This method grades the fabric based on its repellency to different mixtures of water and isopropyl alcohol. The test liquid 'level 0' corresponds to pure water; the liquid corresponding to 'level 8' consists of 40% water and 60% isopropyl alcohol.

The standardized procedure is as follows. One starts with the lowest level liquid, that is, with pure water. Five drops of the liquid have to be placed across the width of the test specimen. Once a drop is placed on the sample, a grade is assigned to the drop, ranging from A to D. For determining the grade of the drop, the description in Table 6.10 is used. If not all drops have grade D, one has to repeat the procedure

Table 6.10 Classification of drop grade for the water/alcohol repellency test.

Drop grade	Description
A	Passes; clear well-rounded drop
B	Borderline pass; rounding drop with partial darkening
C	Fails; wicking apparent and/or complete wetting
D	Fails; complete wetting

with a test liquid of higher level. This has to be iterated until only grade D drops are observed or level 8 has been reached.

6.5.1.3.3 Oil Repellency   The test method EN ISO 14419 (also referred to as *AATCC Test Method 118-2002*) is used to evaluate the fabric's oleophobicity by putting drops of different liquid hydrocarbons, with a well-defined surface tension, on the surface. The test is done with eight different standard hydrocarbon compounds, ranging from 'Number 1', nujol oil (a paraffin oil) to 'Number 8', *n*-heptane, a completely nonpolar solvent. An overview of the test oils and their surface tension is given in Table 6.11.

The standardized procedure is as follows. First, five drops of a standard test liquid (typically the one corresponding to level 8) are placed with a pipette smoothly on the surface of the fabric and observed for wetting, wicking, and contact angle. The fabric should be able to repel the oil for more than 30 seconds to get the grade of the standard oil sample. Three of the five drops have to stay in drop form on the fabric, otherwise a lower grade has to be assigned. The oil repellency grade is the highest numbered test liquid which does not wet the fabric surface. According to this standard, the best value that can be achieved is 8.

Table 6.11 Overview of the oil repellency grades.

Grade	Test liquid	Surface tension (mN m^{-1})
1	Nujol oil	32
2	65% nujol oil/35% *n*-hexadecane	30.4
3	*n*-Hexadecane	27.5
4	*n*-Tetradecane	26.5
5	*n*-Dodecane	25.4
6	*n*-Decane	23.8
7	*n*-Octane	21.6
8	*n*-Heptane	20.1

Table 6.12 Overview of perfluoro groups and the typical oil repellency level that can be reached with them.

Perfluoro groups (acrylic polymers)	Oil repellency grade
$-CF_3$	0
$-CF_2-CF_3$	3–4
$-(CF_2)_2-CF_3$	6–7
$-(CF_2)_4-CF_3$	7–8
$-(CF_2)_6-CF_3$	7–8
$-(CF_2)_8-CF_3$	8

Table 6.12 gives an overview of the typical oil repellency levels that can be obtained with a variety of perfluoro groups.

6.5.2
Hydrophilic Properties Imparted by Plasma

The purpose of the experiments is to bring durable and highly hydrophilic properties to PET fabrics. Such a hydrophilicity can be a functionalization on its own but within the ACTECO project, it was intended to obtain a permanent primer layer to improve the adhesion properties of subsequent finishes.

A brief analysis of the literature was made to identify the molecules most likely to provide hydrophilic properties and to improve adhesion. The main families of gases and monomers identified were:

- for hydrophilic properties:
  - gases : O_2, N_2, NH_3, H_2/N_2, Ar, Air
  - monomers : acrylic acid, allylamine, acrylamide, methacrylamide (with Ar, water, O_2)
- for adhesion:
  - gases : O_2, N_2, NH_3, Ar/N_2, Ar, Air, CO_2, $Ar/O_2/NH_3$, H_2
  - monomers: acrylic acid, allyl alcohol, allylamine, maleic anhydride, acrylonitrile, also hexamethyl disilazane (HMDS), glycidil methacrylate

As expected, there is some overlap in the gases and monomers used for the two applications.

Hydrophilic properties were developed using plasma processes at both low and atmospheric pressure. First, we will discuss in more detail the low pressure plasma results, followed by an example of the results obtained at atmospheric pressure.

6.5.2.1 Plasma Experiments at Low Pressure
For the low pressure experiments reported here, a low pressure microwave powered system was used, which generates uniform glow discharges (see also Section 2.2.1.2).

First a selection of the most suitable precursors was made, based on the outcome of the above-mentioned literature search. As a result, the following precursor gases were investigated: O_2, N_2, NH_3, Ar, air and a mixture of $O_2/Ar/NH_3$. Also the following monomers, used in vaporized state, were tested: acrylic acid (liquid), methacrylamide (liquid), maleic acid anhydride (solid), and butylmethacrylate (liquid). Other molecules that were used in the vapor state were alcohols and chlorosulfonic acid.

Note that in literature also the use of, for example, allyl alcohol, allylamine, and acrylamide is reported for obtaining hydrophilic properties. However, for this work these were not selected due to their high toxicity.

6.5.2.1.1 First Screening of Precursors The results obtained with the different precursors and a PET fabric as substrate are given in Figure 6.10. Using the Wilhelmy method both the AA and the RA were measured (see also Sections 6.5.1.1 and 6.5.1.2). They are represented in Figure 6.10 by the dark and light colored columns, respectively. Also the water absorption was evaluated, by recording the weight of absorbed water during a complete cycle of immersion/emersion of the samples (represented by the full line and marker in Figure 6.10).

From these results, and also considering the measurement error, the following observations were made for the AA:

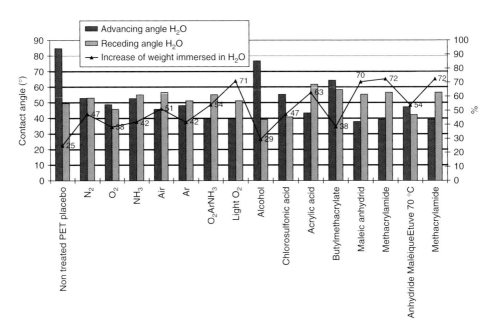

Figure 6.10 Overview of the results of the first screening test with a wide variety of precursors (horizontal axis). The AA (dark bars) and the RA (pale bars) refer to the left axis (contact angle) and the weight increase due to absorption of water (black line and symbols) refer to the right axis (%).

- The hydrophilic character of the treatments is evident as the AA of the treated PET fabric is clearly lowered.
- The effect is largest for the samples treated with $O_2/Ar/NH_3$, 'light O_2' (i.e., plasma based on O_2 but with low intensity), acrylic acid, maleic anhydride, and methacrylamide, where values for the AA of close to $40°$ could be reached, that is, a decrease of about $45°$ was realized.
- The lowest effect is observed for alcohol (decrease of only $7°$), and in a less extent for butylmethacrylate.
- With the other treatments, the contact angle decreased about $30–35°$ compared with the PET blanco.

The RAs of the treated PET are not really affected by the different treatments: they are all within $±10°$ of the RA of the untreated PET. Hence, the RA values will not be considered further.

Concerning the data obtained for the weight change of samples fully immersed in water, the influence of the treatments performed with $O_2/Ar/NH_2$, light O_2, acrylic acid, maleic acid anhydride, and methacrylamide is obvious: the weight increase is at least the double of that of the untreated sample. These results are in line with the outcome of the AA measurements.

Hence, the first important result is that the precursors $O_2/Ar/NH_3$, light O_2, acrylic acid, maleic anhydride, and methacrylamide improve the hydrophilic properties in the studied plasma conditions most.

6.5.2.1.2 Aging of the Samples   The aging behavior of the treatments was evaluated by comparing the values after one day and after one month for the AA and for the weight increase due to water absorption. These parameters were chosen because they were the most strongly influenced by the plasma treatments. For the aging, the samples were kept in a conditioned atmosphere at a temperature of $23°C$ and a relative humidity of 50%.

The results obtained are presented in Figure 6.11. The columns represent the AA measured after one day (dark) and after one month (light) of treatment. The lines represent the weight increase due to absorption of water also after one day (dark) and after one month (light).

The results for the AA show that there is an effect of aging: for all but one treated sample (alcohol) the AA after one month is higher than after one day. However, the increase is only about $10°$ and thus relatively small. Also the weight increase was only altered to a limited extent by the aging, being roughly 10% smaller. The only exception was the treatment with light O_2: a small increase of $15°$ is observed and the weight also decreased more strongly, in spite of it being one of the best conditions after one day.

Concerning the four other best initial conditions ($O_2/Ar/NH_3$, acrylic acid, maleic anhydride, and methacrylamide), the AA remain low (increase from about 40 to $50°$). The weight increase for these four treatments is also only very limitedly affected and there is practically no aging observed for the $O_2/Ar/NH_3$ and acrylic acid treatments. The treatments with NH_3, air, and N_2 also seem to remain

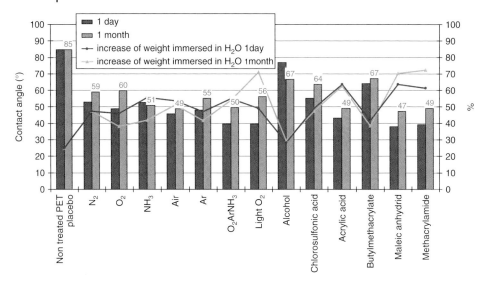

Figure 6.11 Overview of the results of the first screening test with a wide variety of precursors (horizontal axis). The AA (columns, left axis) and the weight increase due to absorption (lines, right axis) are shown after one day (dark bars) and after one month (pale bars).

unaltered after one month of aging. Because of this, for NH_3 and air treatments the AA and the weight increase are close to those of the best treatment condition for immediate results (light O_2) after one month of aging.

After six months of aging at ambient temperature and humidity, a new set of contact angle measurements and absorption tests were performed on the five best results. In summary, the result showed that for the AA the values after one day were in the range 38–53°, after one month in the range 47–51°, and after six months in the range 48–55°. For the weight increase, the following values were obtained: after one day 42–72%, after one month 55–64%, and after six months 59–63%. The conclusion is that the hydrophilic properties remain quite stable during the observed period between one and six months, even at ambient conditions.

6.5.2.2 Plasma Experiments at Atmospheric Pressure (Aldyne System)

Hydrophilic properties could also be obtained using atmospheric plasma conditions and implementing an innovative silane chemistry to deposit SiO_x coatings. The PET was treated with the Aldyne system (see also Section 2.3.1.2). With this system, a homogeneous atmospheric pressure plasma by DBD in controlled nitrogen atmosphere can be generated. As precursor SiH_4 was used.

Different plasma conditions were applied and the hydrophilic properties of the treated PET were evaluated by XPS and time-of-flight static secondary ion mass spectrometry (ToF-SSIMS). The XPS measurements revealed that the best plasma parameters were able to yield up to 20% of silicon at the surface. With ToF-SSIMS it could be shown that the coating was very uniform but a difference was observed between the weft and warp directions. This is illustrated in Figures 5.12 and 5.13. This result was attributed to a small surface contamination present on the fabric prior to plasma treatment, see also Section 6.4.2.

The samples were also evaluated for their affinity toward water. It appeared that the water absorption could be increased by about 50%, which is of the same order of magnitude as the results obtained by low pressure plasma (see results presented in Figure 6.10 in the previous section).

6.5.3
Hydrophobic/Oleophobic Properties Imparted by Plasma

Within ACTECO, one part of the textile related research was dedicated to imparting hydrophobic/oleophobic properties. The purpose of the experiments was to obtain durable hydrophobic and oleophobic properties to different types of PET fabrics. The term *durable* refers here to *washing and abrasion resistance*. The performance level was determined using the water repellency test and the oil repellency test (see Section 6.5.1.3). Both tests were performed on as-deposited samples and also after washing and abrasion testing.

The plasma depositions were performed with the microwave reactor described in Section 2.2.1.2. First, a plasma pre-treatment was performed, for example, in an argon atmosphere. Then, the precursor was applied by manual spraying. This process took place outside the vacuum chamber. Finally, the sample was put back into the vacuum chamber and a plasma treatment was performed to induce cross-linking, which is needed for a durable coating. For this latter plasma treatment the gas consisted of two components: one basic gas for generating the discharge and an additive to promote the cross-linking.

Tests were done on PET fabric intended for two different applications. The first regarded a pure PET fabric (Type 1) which was a relatively thin (weight about $100 \, \text{g m}^{-2}$ – plain weave) and typically used for professional work wear. For this application, the main requirement set was to maintain a level 6 for oil repellency after 10 washing cycles (washing according to test ISO 6330). The second application regards fabrics used for upholstery. In this case two fabrics were investigated: a pure PET fabric – Type 2 and a PET/cotton fabric, both Jacquard woven. For this application, the abrasion resistance is critical, which was evaluated by performing Martindale testing (ISO 12947/2).

6.5.3.1 Preliminary Experiments
The experimental set-up started with a screening phase used for the selection of the precursors. At first, the following precursors were investigated: a mixture of fluoro-gases based on CF_4, a liquid silicon monomer used in the vapor state and a perfluoro-monomer used in liquid state (long perfluoro-chain). The first results obtained with these precursors allowed for the following conclusions:

- All treated textile surfaces were very water repellent.
- The oil repellency test showed that good resistance to oil could only be reached with the perfluoromonomer precursor.
- A strong decrease in performance was observed after washing or abrasion of the samples.

According to these preliminary results, five new fluoro-hydrocarbon based precursors with different lengths of the perfluoro-chain were studied: three perfluoroacrylates and two perfluoroalkanes. The best results were achieved with the perfluoroacrylates.

The second phase was based on these trials and consisted of optimization of the process. Therefore, only one perfluoroacrylate (AC8F17) was selected for more extensive testing. A design of experiments was applied for optimization of all parameters considered relevant for the entire deposition process, which resulted in an extensive list of parameters:

- nature of the substrate
- pre-treatment: duration, gas choice, no pre-treatment
- quantity of the fluoro-monomer sprayed onto the substrate
- cross-linking plasma treatment: basic plasma gas choice, gas pressure, relative ratio of monomer/basic gas, electrical power, duration.

6.5.3.2 Washing Durability

First we would like to report the oil repellency levels obtained on the PET fabric – Type 1 immediately after the final plasma treatment and after 10 washings at 40 °C, see Table 6.13. For treatment 1, the basic treatment procedure, the precursor AC8F17 was sprayed onto the fabric, after pre-treatment of the fabric with argon. Then C_3F_6 was added to the plasma and the discharge was run at low pressure and low power. The duration of the plasma treatment was 5 minutes. For the other treatments (2–6), only one parameter at a time is changed compared with the basic treatment procedure:

- Treatment 2: quantity of AC8F17 was halved
- Treatment 3: no use of C_3F_6 in the plasma
- Treatment 4: no pre-treatment with argon
- Treatment 5: reduction of the duration of the plasma treatment (halved)
- Treatment 6: combination of all the changes of Treatments 2–5, that is, halved quantity of AC8F17, no use of C_3F_6, no pre-treatment, halved plasma treatment time.

As is clear from Table 6.13, some conditions result in an excellent oleophobic performance immediately after the plasma treatment, for example, the conditions 1, 2, 4, and 5 where the highest oil repellency level of 8 could be achieved. However, for some settings, relevantly lower levels are obtained. This becomes even more evident after 10 washings: for some treatments the performance disappears very strongly, for example, for Treatment 3 or 6. As other experimental conditions

Table 6.13   Oil repellency levels obtained on PET fabric for work wear applications for the as-deposited samples and after washing ('W').

Treatment	1	2	3	4	5	6
As-deposited	8	8	6	8	8	5
Washings at 40°C						
1 W	6	6	4.5	6	6	1
1 W + ironing	6.5	6.5	5.5	6.5	6.5	3.5
5 W	5	5	3.5	5	5	2.5
5 W + ironing	6.5	6.5	5.5	6.5	6.5	3.5
10 W	4.5	4.5	2	5.5	5.5	0.5
10 W + ironing	6.5	6.5	6.5	6.5	6.5	1.5

Please see text for further information on the individual treatments.

resulted in a quite good washing resistance, it is clear that the lower performance of Treatments 3 and 6 is due to the omission of C_3F_6 in the plasma. For the best result a level of 5.5 without ironing and of 6.5 with ironing was obtained. The better performance reached after ironing is well-known [83] and is due to an alignment of the long-chained fluorocarbons because of the applied heat. The achieved results matched quite well the requirements imposed by the end-user (level 6 after 10 washings without ironing).

6.5.3.3 Abrasion Durability

In this section we will report some results regarding upholstery applications (PET fabric – Type 2 and PET/cotton fabric). This time, oil repellency levels are reported (Table 6.14) on the as-deposited fabrics and after abrasion testing, performed by standard Martindale testing. Two different pressures were applied, namely 9 kPa (corresponding with the testing method for clothing application) and 12 kPa (corresponding with the testing method for upholstery).

All samples were obtained by the same optimized plasma treatment. After the plasma treatment, the samples showed excellent performance, reaching the maximum level of 8 or even higher. As mentioned before, the best value defined by the standard test method is level 8. However, to be able to distinguish among these very good results, two extra test liquids were introduced, which have an even lower surface energy as the *n*-heptane used for level 8 (see also Table 6.11). These liquids were *n*-hexane (18.4 mN m^{-1}) and *n*-pentane (15.8 mN m^{-1}), which enabled introducing level 9 and 10, respectively. This way, an extra level of '8.5' could be measured, which, of course, is *not* a standardized value. For the PET/cotton fabric, the abrasion test has been repeated twice for each testing condition in order to evaluate the reproducibility, which appears to be good.

Table 6.14 Oil repellency levels obtained on fabrics for upholstery applications for as-deposited samples and after abrasion testing.

Substrate	PET – type 2			PET/cotton		
Weight (kPa)	12	9	12	12	9	9
As deposited	8.5	8	8	8.5	8.5	8
Cycles ($\times 1000$)	Abrasion testing					
2	6.5	5.5	8	7.5	7	7.5
4	5	5.5	8	7.5	7	7.5
8	5	5.5	7.5	7.5	7	6
12	3.5	5.5	7	6	6	6
16	6.5	5.5	7	6	6	6
20	6.5	5.5	7	6	7	7

The term *'weight'* refers to the weight applied and the 'cycles' refer to the number of cycles performed during the abrasion testing.

Also after abrasion the performance remains quite good and the 'level 6' requirement imposed by the application is clearly met for the majority of the samples.

6.5.3.4 Summary of Oleophobic Properties

After analysis of all the experimental data, the best plasma conditions to achieve oleophobic properties with good to excellent resistance to washing and abrasion are characterized as follows:

- Pre-treatment: long duration, preferred gas is argon.
- Optimized quantity of perfluoromonomer, depending on the nature of the substrate and the targeted performance level.
- Cross linking plasma treatment: low gas pressure, high amount of cross-linking additive, low electrical power, short duration.

In conclusion, the experiments performed allowed to get an insight in the influence of the different treatment parameters on the final performance of the coatings. This knowledge forms the basis for a possible transfer of the developed process from laboratory scale to industrial scale.

6.6
Transferring Plasma Technology to Industrial Processes

Plasma treatment of textiles has been considered for over 20 years. At first, low pressure plasmas were the only ones available for materials processing, and thus, they were tried on textile materials. Several good results could be obtained on laboratory scale but they typically required a treatment time of several minutes. This is not compatible with most of the common in-line treatments of textiles. Hence, in spite of the potential, low pressure plasma treatment of textiles has not so far become a widely used technique, although successful examples of its use exist.

Recently, enormous technological progress has been made toward the realization of atmospheric pressure plasmas for materials processing. Such processes can be realized in-line and be integrated into existing production lines.

As already mentioned several times in this chapter, the range of applications that can potentially benefit from a plasma treatment, be it at reduced or atmospheric pressure, is large. This potential not only already sparked a lot of research, resulting in a vast amount of literature available, but also resulted in the filing of several patents regarding 'plasma' and 'textile', both terms are quoted to indicate they are considered in the broadest sense. It has been reported that about 2% of all textile related patents involve the use of 'plasma' [3]. On the other hand, almost 5% of the plasma-related patents mention 'textile' as an application domain in one way or the other. This shows that there is also an interest in textile treatment from the plasma equipment side.

Nevertheless, in spite of the many advantages of plasma treatment and of the wide range of application possibilities, this technology is still in the stage where only first-adopters are acting. It is our understanding that this cannot be attributed to one single cause. Rather, it seems to be due to a mixture of limitations that stem from (i) the plasma technology itself, (ii) textile sector related aspects, and (iii) fundamental aspects inherently connected with plasma processing of textiles.

The first group, the limitations of the plasma technology, consists of items like the reproducibility of the process, the stability during long-term industrial operation (24 hours on 24 hours, seven days a week), the use of, for example, noble gases or precursors which can become very expensive. Since these limitations are not limited to the textile sector alone but also exist for other applications; they will not be considered further here.

The other two groups will be discussed in more detail in the next two subsections.

6.6.1
Textile Sector Related Issues

The textile sector related causes, which hamper the integration of plasma into the textile production process, can be identified as (as also discussed elsewhere [3]):

- **Existing equipment:** Nowadays, the textile industry manages the required coating and finishing steps with existing equipment. Very often, that existing equipment consists of simple (and cheap) mechanical methods, for example, finishing by padding or coating using (Mayer) bar. Clearly, it is not evident to replace such types of equipment by much more expensive plasma equipment, even when this has clear benefits regarding, for example, the water and energy use (see also Section 6.1 and especially Chapter 12 on this topic).

- **Sector organization:** The majority of the textile enterprises in Europe are small and medium enterprises (SMEs). This means that their capacity for investing in more expensive equipment is relatively limited.
- **General market conditions:** From a European perspective, the textile sector as a whole can be considered to be threatened because of the concurrence with developing countries. Such an environment does not really favor decisions to start more risky investment plans.

6.6.2
Fundamental Aspects Regarding Industrialization

In Section 6.4 some specific aspects related to the plasma treatment of textiles were discussed. The first item considered was the large variety in chemical composition of the materials to be treated. Indeed, very often different substrate materials need to be treated simultaneously because textile materials are often mixtures, for example, fabrics consisting of PET and cotton, or yarns consisting of wool and polyamide. The effect of the plasma treatment on the different materials will not be identical, thus complicating the processes to be optimized. If additionally fabrics or yarns of variable compositions are treated the whole optimization process may have to be performed for each type of textile.

As mentioned in Section 6.4.2, surface cleanliness is not always guaranteed. Obviously, surface cleanliness is already an issue today in the textile industry. However, the definition of what 'clean' means can differ strongly. Introducing plasma-based processes might require a stricter characterization or quality control of the cleanliness of the material to be treated. If this is not performed, plasma treatment may appear to be 'non-reproducible', that is, yielding good results at one time and not at another. It is not always easy to prove that it is *not* the plasma treatment that causes this phenomenon, but the difference in (surface) contamination of the textile material being treated, even if this material appears identical according to the standard testing required for common manufacturing methods.

Further, practically all of the aspects mentioned in Section 6.4 (3D structure, large surface area, moisture, and air interference) have a negative impact on the maximum line speeds that can currently be obtained for plasma treatment. Because of this, the throughput (i.e., the amount of square meters that can be treated per time unit) of a plasma treatment is still often lower than those encountered in textile manufacturing and forms a bottleneck for large scale industrial application. To give an example, it is quite clear that integrating a plasma treatment during or after the extrusion process of fibers could be very beneficial. However, for state-of-the-art extrusion equipment, line speeds of a couple of thousands of meters per minute are common. To our knowledge, no existing plasma treatment can cope with such high line speeds, unless perhaps when economically unrealistic equipment is considered. There are some extrusion lines, though, where the line speed is comparatively low, for example, for monofilament extrusion (order of $100\,\mathrm{m\,min^{-1}}$ or well below). Very often, such monofilaments are also specialty products, so that they are a good candidate for in-line plasma treatment. One example can be found in a patent application [21], where an in-line low pressure plasma treatment of hollow monofilament fibers (used for blood filtration) is described.

On the other hand, textile finishing on fabrics occurs at much lower speeds, typically 10–20 m min^{-1}. Such speeds are much lower than those encountered in the plastic web converting industry where speeds of some hundreds of meters per minute are standard. However, one should take into account that the effective surface area to be treated for textiles is typically a factor 10 larger than a smooth (plastic) film (see Section 6.4.4). Hence, the effective surface area to be treated per time unit is comparable. But, on top of this, a textile substrate is a 3D object toward the active plasma species, even when it is a relatively flat woven fabric. As discussed in Section 6.4.3, this means that the plasma will have a limited penetration into the textile structure. This effect can only be circumvented by having a plasma with higher intensity or by increasing the contact time with the plasma discharge. Hence, also for this situation, in spite of the relative low processing speed, plasma treatment at high line speeds is not practical.

The conclusion is that a textile suitable for a plasma treatment should be relatively light weight and have an open structure, implying relatively little surface area to treat and allowing penetration of the plasma species. If such a substrate is treated at modest line speeds and has no surface contamination it will be even better suited for plasma treatment. Looking at the different textile substrates being processed, good candidates fulfilling these requirements may be found in the non-woven world. This can explain the recent interest in plasma treatment from this industry.

6.7
Summary

Compared with current traditional finishing processes, plasma treatment has the crucial advantage of reducing the usage of chemicals, water, and energy. It also offers the possibility of obtaining a wide variety of typical textile finishes (e.g., hydrophilic, oleophobic, antibacterial) without changing the key textile properties (e.g., hand, softness and flexibility).

Integration of the plasma processes at different stages of the textile production chain is possible, for example, at sliver, yarn, or fabric level, or even on garments or other already tailored textiles, and has been investigated with often interesting results.

Case studies regarding the plasma treatment of PET and PET/cotton fabrics are presented. The respective materials are used in applications like professional work wear and upholstery, and the intention was to tailor their surface energy. The analysis methods for this property as well as the obtained results are discussed, showing that very high oil-repellent coatings can be obtained that are both durable against washing and abrasion.

Further, the possibilities for industrial applications, which are still rather limited, are discussed. Several hurdles exist at various levels: with the plasma technology itself, with textile sector related issues and because of bottlenecks due to the very nature of the plasma treatment of textiles itself. Examples of the latter are the 3D aspect of the textile substrates and the large effective surface area to be treated. These two intrinsic textile properties pose major challenges for large-scale industrial and economically viable plasma treatment.

Nevertheless, the undeniable assets of plasma technology for textile treatment are clear and it seems only a matter of time before the technique will become more and more integrated in the textile production chain.

References

1. Malik, M.A., Ubaid-ur-Rehman, M.A., Ghaffar, A., and Ahmed, K. (2001) Water purification by electrical discharges. *Plasma Sources Sci. Technol.*, **10** (1), 82–91.
2. Temmerman, E. and Leys, C. (2005) Surface modification of cotton yarn with a DC glow discharge in ambient air. *Surf. Coat. Technol.*, **200** (1-4), 686–689.
3. Buyle, G. (2009) Nanoscale finishing of textiles via plasma treatment. *Mater. Technol.*, 24 (1), 46–51.
4. Morent, R., De Geyter, N., Verschuren, J., De Clerck, K., Kiekens, P., and Leys, C. (2008) Non-thermal plasma treatment of textiles. *Surf. Coat. Technol.*, **202** (14), 3427–3449.
5. Cai, Z., Qiu, Y., Zhang, C., Hwang, Y., and McCord, M. (2003) Effect of atmospheric plasma treatment on desizing of PVA on cotton. *Text. Res. J.*, **73** (8), 670–674.
6. Roth, J.R., Chen, Z., Sherman, D.M., Karakaya, F., Tsai, P.P.Y., Kelly-Wintenberg, K., and Montie, T.C. (2001) Increasing the surface energy and sterilization of nonwoven fabrics by exposure to a one atmosphere uniform glow discharge plasma (OAUGDP). *Int. Nonwoven J.*, **10** (3), 34–47.
7. Canal, C., Erra, P., Molina, R., and Bertrán, E. (2007) Regulation of surface hydrophilicity of plasma treated wool fabrics. *Text. Res. J.*, **77** (8), 559–564.
8. Shishoo, R. (ed.) (2007) *Plasma Technologies for Textiles*, Woodhead Publishing, Ltd, Cambridge.
9. Buyle, G. (2007) Plasma coating on textile materials. Presentation at 4th

European Coating Congress, November 21–22, 2007, Ghent.
10. *http://www.nucryst.com/platform_technology.htm* (accessed on 26 July 2009).
11. Malek, R.M.A. and Holme, I. (2003) The effect of plasma treatment on some properties of cotton. *Iran. Polym. J.*, **12** (4), 271–280.
12. Canal, C., Molina, R., Bertran, E., and Erra, P. (2004) Wettability, ageing and recovery process of plasma-treated polyamide 6. *J. Adhes. Sci. Technol.*, **18** (9), 1077–1089.
13. Negulescu, I.I., Despa, S., Chen, J., Collier, B.J., Despa, M., Denes, A., Sarmadi, M., and Denes, F.S. (2000) Characterizing polyester fabrics treated in electrical discharges of radio-frequency plasma. *Text. Res. J.*, **70** (1), 1–7.
14. Wang, C. and Qiu, Y. (2007) Two sided modification of wool fabrics by atmospheric pressure plasma jet: influence of processing parameters on plasma penetration. *Surf. Coat. Technol.*, **201** (14), 6273–6277.
15. Wu, H. and Li, V. (1999) Fiber/cement interface tailoring with plasma treatment. *Cem. Concr. Compos.*, **21** (3), 205–212.
16. Ma, Z. and Qi, H. (2007) Polypropylene fiber modified by surface cross-linking in dielectric barrier discharge. *Surf. Coat. Technol.*, **201** (Special Issue 9-11), 4935–4938.
17. Thorson, W.J. (1971) Improvement of cotton spinnability, strength, and abrasion resistance by corona treatment. *Text. Res. J.*, **41** (5), 455–458.

18. Rad, A.Y., Ayhan, H., and Piskin, E. (1998) Adhesion of different bacterial strains to low-temperature plasma-treated sutures. *J. Biomed. Mater. Res. A*, **41** (3), 349–358.

19. Loh, I.-H., Lin, H.-L., and Chu, C.C. (2004) Plasma surface modification of synthetic absorbable sutures. *J. Appl. Biomater.*, **3** (2), 131–146.

20. Warren, J., Mather, R., Neville, A., and Robson, D. (2005) Gas plasma treatments of polypropylene tape. *J. Mater. Sci.*, **40** (20), 5373–5379.

21. Krause, B. *et al.* (2007) Continuous Method for Production of a Regioselective Porous Hollow Fibre Membrane. US2007296105A1.

22. Nikiforov, A.Y. and Leys, C. (2006) Surface treatment of cotton yarn by underwater capillary electrical discharge. *Plasma Chem. Plasma Process.*, **26** (4), 415–423.

23. Wakida, T., Tokino, S., Niu, S., Lee, M., Uchiyama, H., and Kaneko, M. (1993) Dyeing properties of wool treated with low-temperature plasma under atmospheric pressure. *Text. Res. J.*, **63** (8), 438–442.

24. Wakida, T., Cho, S., Choi, S., Tokino, S., and Lee, M. (1998) Effect of low temperature plasma treatment on color of wool and nylon 6 fabrics dyed with natural dyes. *Text. Res. J.*, **68** (11), 848–853.

25. Molina, R., Jovančić, P., Jocić, D., Bertran, E., and Erra, P. (2003) Surface characterization of keratin fibres treated by water vapour plasma. *Surf. Interface Anal.*, **35** (2), 128–135.

26. Dai, X.J., Hamberger, S.M., and Bean, R.A. (1995) Reactive plasma species in the modification of wool fibre. *Aust. J. Phys.*, **48**, 939–951.

27. Kan, C.W., Chan, K., and Yuen, M.C.W. (2003) Surface characterisation of low-temperature plasma treated wool fibre. *Autex Res. J.*, **3** (4), 194–205.

28. Ryu, J., Wakida, T., and Takagishi, T. (1991) Effect of corona discharge on the surface of wool and its application to printing. *Text. Res. J.*, **61** (10), 595–601.

29. Mori, M. and Inagaki, N. (2006) Relationship between anti-felting properties and physicochemical properties of wool

30. Höcker, H. (2002) Plasma treatment of textile fibers. *Pure Appl. Chem.*, **74** (3), 423–427.

31. Danish, N., Garg, M.K., Rane, R.S., Jhala, P.B., and Nema, S.K. (2007) Surface modification of Angora rabbit fibers using dielectric barrier discharge. *Appl. Surf. Sci.*, **253** (16), 6915–6921.

32. Chen, Y., Lin, H., Ren, Y., Wang, H., and Zhu, L. (2004) Study on Bombyx mori silk treated by oxygen plasma. *J. Zhejiang Univ. Sci.*, **5** (8), 918–922.

33. Xu, X., Wang, Y., Zhang, X., Jing, G., Yu, D., and Wang, S. (2006) Effects on surface properties of natural bamboo fibers treated with atmospheric pressure argon plasma. *Surf. Interface Anal.*, **38** (8), 1211–1217.

34. Li, Z.-F., Netravali, A.N., and Sachse, W. (1992) Ammonia plasma treatment of ultra-high strength polyethylene fibres for improved adhesion to epoxy resin. *J. Mater. Sci.*, **27**, 4625–4632.

35. Tissington, B., Pollard, G., and Ward, I.M. (1992) A study of the effects of oxygen plasma treatment on the adhesion behaviour of polyethylene fibres. *Compos. Sci. Technol.*, **44** (3), 185–195.

36. Woods, D.W. and Ward, I.M. (2004) Study of the oxygen treatment of high-modulus polyethylene fibres. *Surf. Interface Anal.*, **20** (5), 385–392.

37. Zhang, C., Gopalaratnam, V.S., and Yasuda, H.K. (2000) Plasma treatment of polymeric fibers for improved performance in cement matrices. *J. Appl. Polym. Sci.*, **76** (14), 1985–2127.

38. Demuth, O., Amouroux, J., and Goldman, M. (1983) Traitement de surface de fibres textiles (polyethylene terephtalate) par décharge couronne. *Ann. Chim. Fr.*, **8**, 349–362 (only in French).

39. Inagaki, N., Tasaka, S., and Kawai, H. (1992) Surface modification of Kevlar® fiber by a combination of plasma treatment and coupling agent treatment for silicone rubber composite. *J. Adhes. Sci. Technol.*, **6** (2), 279–291.

40. Zhu, L., Teng, W., Xu, H., Liu, Y., Jiang, Q., Wang, C., and Qiu, Y. (2008) Effect

of absorbed moisture on the atmospheric plasma etching of polyamide fibers. *Surf. Coat. Technol.*, **202** (10), 1966–1974.

41. Strobel, M., Lyons, C.S., and Mital, K.L. (eds) (1994) *Plasma Surface Modification of Polymers: Relevance to Adhesion*, VSP Books, Utrecht.

42. Cech, V. (2007) Plasma-polymerized organosilicones as engineered interlayers in glass fiber/polyester composites. *Compos. Interfaces*, **14** (4), 321–334.

43. Wei, Q., Mather, R., Fotheringham, A., and Yang, R. (2002) ESEM study of wetting of untreated and plasma treated polypropylene fibers. *J. Ind. Text.*, **32** (1), 59–66.

44. Wei, Q., Mather, R., Wang, X., and Fotheringham, A. (2005) Functional nanostructures generated by plasma-enhanced modification of polypropylene fibre surfaces. *J. Mater. Sci.*, **40** (20), 5387–5392.

45. Fuchs, H., Bochmann, R., Poll, H.U., and Schreiter, S. (1999) Plasmabehandlung cellulosischer Fasermaterialien. *Textilveredlung*, **5,6**, 23–27 (only in German).

46. Belin, R. (1976) Effect of corona treatment on the cohesion between fibres and on their wettability. *J. Text. Inst.*, **67** (7-8), 249–252.

47. Sun, D. and Stylios, G.K. (2004) Effect of low temperature plasma treatment on the scouring and dyeing of natural fabrics. *Text. Res. J.*, **74** (9), 751–756.

48. Sun, D. and Stylios, G.K. (2006) Fabric surface properties affected by low temperature plasma treatment. *J. Mater. Process. Technol.*, **173** (2), 172–177.

49. Shahidi, S., Ghoranneviss, M., Moazzenchi, B., Dorranian, D., and Rachidi, A. (2005) Water repellent properties of cotton and PET fabrics using low temperature plasma of Argon. Proceedings of XXVIIth ICPIG Conference, 18–22 July, 2005, Eindhoven, The Netherlands.

50. Ghoranneviss, M., Shahidi, S., Moazzenchi, B., Anvari, A., Rashidi, A., and Hosseini, H. (2007) Comparison between decolorization of denim fabrics with Oxygen and Argon glow discharge.

Surf. Coat. Technol., **201** (Special Issue 9-11), 4926–4930.

51. Höcker, H., Thomas, H., Küsters, A., and Herrling, J. (1994) Färben von plasmabehandelter Wolle. *Melliand Text.*, **6**, 506–512 (only in German).

52. Kan, C.W. and Yuen, C.W.M. (2005) Effect of low temperature plasma treatment on wool fabric properties. *Fibers Polym.*, **6** (2), 1229–9197.

53. Wong, K., Tao, X., Yuen, C., and Yeung, K. (2000) Topographical study of low temperature plasma treated flax fibers. *Text. Res. J.*, **70** (10), 886–893.

54. Wong, K., Tao, X., Yuen, C., and Yeung, K. (1999) Low temperature plasma treatment of linen. *Text. Res. J.*, **69** (11), 846–855.

55. Ren, C.S., Wang, D.Z., and Wang, Y.N. (2006) Graft co-polymerization of acrylic acid onto the linen surface induced by DBD in air. *Surf. Coat. Technol.*, **201** (6), 2867–2870.

56. Bhat, N. and Nadiger, G. (1978) Effect of nitrogen plasma on the morphology and allied textile properties of tasar silk fibers and fabrics. *Text. Res. J.*, **48** (12), 685–691.

57. Park, D.J., Lee, M.H., Woo, Y.I., Han, D.-W., Choi, J.B., Kim, J.K., Hyun, S.O., Chung, K.-H., and Park, J.-C. (2008) Sterilization of micro-organisms in silk fabrics by microwave-induced argon plasma treatment at atmospheric pressure. *Surf. Coat. Technol.*, **202** (22-23), 5773–5778.

58. Leroux, F., Perwuelz, A., Campagne, C., and Behary, N. (2006) Atmospheric air-plasma treatments of polyester textile structures. *J. Adhes. Sci. Technol.*, **20** (9), 939–957.

59. Zhongfu, R., Xiaoliang, T., Hongen, W., and Gao, Q. (2007) Continuous modification treatment of polyester fabric by Ar-O2(10:1) discharge at atmospheric pressure. *J. Ind. Text.*, **37** (1), 43–53.

60. Temmerman, E., Akishev, Y., Trushkin, N., Leys, C., and Verschuren, J. (2005) Surface modification with a remote atmospheric pressure plasma: de glow discharge and surface streamer regime. *J. Phys. D: Appl. Phys.*, **38** (4), 505–509.

61. Selli, E., Mazzone, G., Oliva, C., Martini, F., Riccardi, C., Barni, R.,

Marcandallic, B., and Massacra, M.R. (2001) Characterisation of poly(ethylene terephthalate) and cotton fibres after cold SF6 plasma treatment. *J. Mater. Chem.*, **11**, 1985–1991.

62. Cheng, C., Zhang, L.Y., and Zhan, R.J. (2006) Surface modification of polymer fibre by the new atmospheric pressure cold plasma jet. *Surf. Coat. Technol.*, **200** (24), 6659–6665.

63. O'Hare, L.-A., O'Neill, L., and Goodwin, A.J. (2006) Anti-microbial coatings by agent entrapment in coatings deposited via atmospheric pressure plasma liquid deposition. *Surf. Interface Anal.*, **38** (11), 1519–1524.

64. Scholz, J., Nocke, G., Hollstein, F., and Weissbach, A. (2005) Investigations on fabrics coated with precious metals using the magnetron sputter technique with regard to their anti-microbial properties. *Surf. Coat. Technol.*, **192** (2-3), 252–256.

65. Brown, J.R., Chappell, P.J.C., and Mathys, Z. (1991) Plasma surface modification of advanced organic fibres. *J. Mater. Sci.*, **26**, 4172–4178.

66. Cuong, N.K., Saeki, N., Kataoka, S., and Yoshikawa, S. (2002) Hydrophilic improvement of PET fabrics using plasma-induced graft polymerization. *Turk. J. Chem.*, **23** (4), 202–208.

67. Radetic, M., Jovancic, P., Puac, N., and Petrovic, Z.L. (2007) Environmental impact of plasma application to textiles. *J. Phys. Conf. Ser.*, **71**, 012017.

68. Karahan, H.A., Özdoğan, E., Demir, A., Koçum, I.C., Öktem, T., and Ayhan, H. (2009) Effects of atmospheric pressure plasma treatments on some physical properties of wool fibers. *Text. Res. J.*, **79**, (14) 1260–1265. doi: 10.1177/0040517508095595.

69. Šimor, M., Ráhel', J., Černák, M., Imahori, Y., Štefečka, M., and Kando, M. (2003) Atmospheric-pressure plasma treatment of polyester nonwoven fabrics for electroless plating. *Surf. Coat. Technol.*, **172** (1), 1–6.

70. Hwang, Y., McCord, M., An, J., Kang, B., and Park, S. (2005) Effects of helium atmospheric pressure plasma treatment on low-stress mechanical properties of polypropylene nonwoven fabrics. *Text. Res. J.*, **75** (11), 771–778.

71. Virk, R., Ramaswamy, G., Bourham, M., and Bures, B. (2004) Plasma and antimicrobial treatment of nonwoven fabrics for surgical gowns. *Text. Res. J.*, **74** (12), 1073–1079.

72. Yang, M., Chen, K., Tsai, J., Tseng, C., and Lin, S. (2002) The antibacterial activities of hydrophilic-modified nonwoven PET. *Mater. Sci. Eng. C*, **20** (1-2), 167–173.

73. Ingram, W.O., (2007) Method and Apparatus For Making Carpet. WO2007130118.

74. Maguirea, P.D., McLaughlin, J.A., Okpalugo, T.I.T., Lemoine, P., Papakonstantinou, P., McAdams, E.T., Needham, M., Ogwu, A.A., Ball, M., and Abbas, G.A. (2007) Mechanical stability, corrosion performance and bioresponse of amorphous diamond-like carbon for medical stents and guidewires. *Diamond Relat. Mater.*, **14** (8), 1277–1288.

75. Tuzlakoglu, K. (2008) Fiber-based structures from natural origin polymers for tissue engineering approaches, PhD Dissertation. University Minho, Portugal.

76. http://www.naox.eu/ (accessed on 26 July 2009).

77. Verschuren, J., Kiekens, P., and Leys, C. (2007) Textile-specific properties that influence plasma treatment, effect creation and effect characterization. *Text. Res. J.*, **77** (10), 727–733.

78. De Geyter, N. (2008) Plasma modification of polymer surfaces in the subatmospheric pressure range. PhD Dissertation, University of Ghent, Belgium.

79. Poll, H.U., Schladitz, U., and Schreiter, S. (2001) Penetration of plasma effects into textile structures. *Surf. Coat. Technol.*, **142-144**, 489–493.

80. Verschuren, J. and Kiekens, P. (2005) Gas flow around and through textiles structures during plasma treatment. *Autex Res. J.*, **5** (3), 154–161.

81. Musschoot, J., Deduytsche, D., De Keyser, K., Dendooven, J., Haemers, J., Van Meirhaeghe, R.L., D'Haen, J., Buyle, G., and Detavernier, C. (2010) Conformality of thermal and plasma

enhanced atomic layer deposition on fibrous substrates. to be submitted.

82. De Geyter, N., Morent, R., and Leys, C. (2006) Penetration of a dielectric barrier discharge plasma into textile structures at medium pressure. *Plasma Sources Sci. Technol.*, **15** (1), 78–84.

83. Wang, C.X., Xu, H.L., Liu, Y., and Qiu, Y.P. (2008) Influence of twist and filament location in a yarn on effectiveness of atmospheric pressure plasma jet treatment of filament yarns. *Surf. Coat. Technol.*, **202** (12), 2775–2782.

84. Müller, B., Riedel, M., Michel, R., De Paul, S.M., Hofer, R., Heger, D., and Grützmacher, D. (2001) Impact of nanometer-scale roughness on contact-angle hysteresis and globulin adsorption. *J. Vac. Sci. Technol. B*, **19** (5), 1715–1720.

85. Dufour, F., Jordan, C., and Viallier, P. (2008) Wash fastness of fluorocarbon finishes on polyester fabrics. *J. Soc. Dyers Colour.*, **114** (9), 258–263.

7
Preventing Biofilm Formation on Biomedical Surfaces

Virendra Kumar, Hubert Rauscher, Frédéric Brétagnol, Farzaneh Arefi-Khonsari, Jerome Pulpytel, Pascal Colpo, and François Rossi

7.1
Bacterial Adhesion to Biomaterials: Biofilm Formation

'Fouling' is the deposition and accumulation of undesirable materials on the surfaces of devices or tools, which poses an adverse effect on their lifetime, activity, performance, and efficiency. 'Biofouling' is a process of adhesion and accumulation of biomolecules such as proteins, carbohydrates, and adhesion, accumulation and growth of micro-organisms such as prokaryotic cells and higher organisms onto material surfaces [1, 2]. Recent studies have shown that a wide range of persistent medical device-related infections may be related to the ability of infectious micro-organisms to form 'biofilms' [1].

Biofouling, which is the consequence of undesired accumulation of biomolecules and micro-organisms and subsequent biofilm formation on surfaces, is one of the most frequent and severe complications related to medical devices such as catheters or wound dressings, and leads to increasing cost of medical treatment, pain, discomfort and inconvenience, and even fatality to the patient. The problem of biofilm formation is also important in the field of food and beverage processing, pharmaceutical and cosmetics manufacturing, and in many other industrial sectors. Biofilm formation starts with a nonspecific, reversible attachment of bacteria to substrate surfaces, followed by irreversible bacterial adhesion to surfaces and finally to the production of the extracellular polymer matrix (EPM) that encases the adherent bacteria in a 3D matrix called *biofilm*. The EPM helps bacteria to firmly adhere to a substrate and protects bacteria from immune host responses and antimicrobial agents. The process of biofilm formation is governed by a number of factors, which are related to the biomaterial surface, micro-organisms, and the surrounding medium. Different approaches have been suggested to prevent biofilm formation on biomaterial surfaces, starting from sterilization of biomedical implants and devices, minimization of pre-surgery contamination, use of antimicrobial-releasing biomaterials, to the surface-engineering, and 'antibiofilm' approach, which will be discussed in the subsequent sections. This chapter will focus on understanding the interactions of micro-organisms with the material surface leading to biofilm

Plasma Technology for Hyperfunctional Surfaces. Food, Biomedical and Textile Applications.
Edited by Hubert Rauscher, Massimo Perucca, and Guy Buyle
Copyright © 2010 WILEY-VCH Verlag GmbH & Co. KGaA, Weinheim
ISBN: 978-3-527-32654-9

formation and various factors influencing the biofilm formation process, followed by an overview of various plausible strategies, including plasma technology, that are useful to prevent biofilm formation on biomaterial surfaces.

7.1.1
'Biofilm' and Its Implications in the Biomedical Field

A 'biofilm' is a spatially and metabolically three dimensionally structured microbial community encased in the EPM, developed at a material surface. Depending on the bacterial species involved, the biofilm may be composed of 10–25% bacterial cells and 75–90% EPM, also called '*Slime*' [2]. The EPM not only helps bacterial cells to adhere to a substrate and to trap nutrients from the contacting medium due to charged and neutral polysaccharide groups, but also provides a shelter or safety cover, which protects bacteria from immune host responses, predators, and antimicrobial agents [2]. Therefore, once biofilm formation takes place, it becomes very difficult to cure the infection by using a conventional antibiotic approach.

The adhesion of micro-organisms to surfaces leading to biofilm formation has a huge impact on numerous applications, such as biomedical devices (e.g., urinary catheters, intra-ocular lenses, and dialysis equipment), food packaging, marine infrastructure, water treatment, and distribution systems and heat exchangers in nuclear power plants. The implications of biofilm formation in the biomedical field are much more severe, as medical device-related biofouling and nosocomial infections are very difficult to treat once an antibiotic resistant biofilm has developed. This can make device removal necessary, in some cases it is even fatal for the patient. In most cases, it impairs the desired device function [1, 3].

According to a June 2007 report of the European Centre for Disease Prevention and Control [4], the most important disease-related threat in Europe is posed by micro-organisms, which are resistant to antibiotics. Every year around three million people in the European Union catch a healthcare-associated infection, of which approximately 50 000 people die. Similarly, nosocomial infections associated with biofilm formation are the fourth most common cause of death in the USA leading to more than $5 billion in added medical cost per annum [5]. Therefore, it is clear that indwelling device-related infections are a major cause of morbidity and mortality in hospitalized patients, adding considerably to medical costs, and thus, they are an extra economic and social burden for patients and their families.

7.1.2
Mechanism for Bacterial Adhesion to Surfaces

In its most basic form, the process of bacterial adhesion can broadly be divided into two stages depending on the types of interaction involved and the strength of the bacterial adhesion to the surface: the primary or docking phase and the secondary or locking phase [6, 7].

The primary stage in the bacterial adhesion process is the so-called 'docking phase' where a micro-organism comes into close proximity of the material surface by different driving forces, such as Brownian motion, sedimentation, and convective mass transport from bulk liquid toward material surface [8]. Once the organism is close to a surface, the final step of adhesion depends on the net resultant of various attractive and repulsive short range interactions such as van der Waals, electrostatic, hydrophobic, and steric interactions. This stage is reversible and it is dictated by a number of physico-chemical variables that define the interaction between the bacterial cell surface and the surface of interest [9]. During this initial contact, bacteria still show Brownian motion and can be easily detached by fluid shear forces, for example, rinsing [9, 10].

The second crucial stage of bacterial adhesion is the irreversible adhesion or locking phase, which includes molecularly mediated binding, such as specific ligand–receptor interaction between the adhesion of the bacterial surface and the material surface [11]. Adhesin is a substance (a surface macromolecule, commonly lectins, proteins, polysaccharides) present on the surface structures of bacteria, such as pili, fimbriae, and fibrillae. For example, *Staphylococcus epidermidis* produces a polysaccharide intercellular adhesin (PIA), which is responsible for cell-to-cell adhesion and subsequent biofilm formation [12]. At the end of this stage, adhesion becomes irreversible in the absence of physical or chemical intervention, and the micro-organism is attached firmly to the surface. At this point, much stronger physical or chemical forces are required to remove the bacteria from the surface, for example, scraping, scrubbing, or chemical cleaners. Figure 7.1 shows the importance of various interaction types involved during the initial attachment of bacteria to the surfaces as a function of the distance between the surface and the bacteria [13].

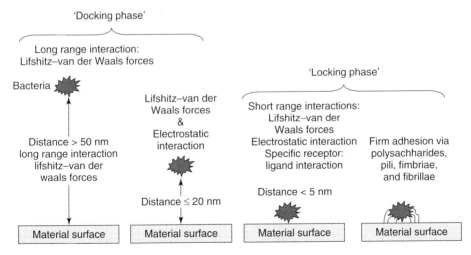

Figure 7.1 Importance of different interactions involved in the initial stages (from left to right) of bacterial adhesion as a function of the separation distance between the bacterial cell and the material surface.

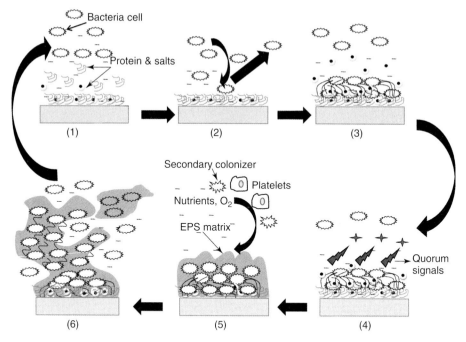

Figure 7.2 Mechanism and different processes involved in biofilm formation. The six steps are, in clockwise sequence, 1. Formation of the 'conditioning film'. 2. Reversible adhesion of bacteria. 3. Irreversible adhesion of bacteria. 4. Release of quorum signals for up-regulation of virulence factors and secretion of extracellular polymers. 5. Production of extracellular polymeric matrix, consumption of nutrients, O_2, secondary colonizers, mammalian cells, and platelets; multiplication of the cells, and formation of a 3D structure called '*biofilm*'. 6. Further development of the mature biofilm with an expressed macrostructure and finally disintegration and liberation of free-floating (planktonic) cells.

7.1.3
Biofilm Formation – a Multistep Process

Understanding the mechanisms of biofilm formation has been of fundamental importance in designing new biomaterials capable to prevent biofilm growth on their surfaces. Biofilm formation is a multistep process involving a number of interlinked physical, biological, and chemical processes, as shown in Figure 7.2 [14].

During the first stage of biofilm formation, molecules – both organic and inorganic – present in the bulk fluid are carried toward the surface by diffusion, turbulent flow and quickly adsorb onto the material surfaces via physico-chemical interactions (step 1 in Figure 7.2). The accumulation of these molecules at the solid–liquid interface of surfaces is commonly known as '*conditioning film*', which modifies the surface properties of the bare material surface [15] and mediates the bacterial adhesion to substrates. It also provides a higher concentration of nutrients at the surface compared with the liquid phase [16]. The next stage is the transport of bacterial cells to the conditioned material surfaces via a combination of transport mechanisms (i.e., diffusion, convection, sedimentation, and motility), followed by reversible adhesion of bacteria to the solid surface on the conditioning film, the so-called 'docking phase' (step 2 in Figure 7.2) [9, 11]. Once bacteria come into contact with the material surface, irreversible adhesion of bacteria on the surface takes place by specific receptor–ligand interaction between the bacterial surface adhesins and the ligand sites on the conditioned material surface (step 3 in Figure 7.2) [6].

Within minutes of irreversible attachment, adherent cells up-regulate the secretion of certain cell signaling molecules that manipulate phenotypic responses, through a process termed '*quorum sensing*' [17–19] (step 4 in Figure 7.2), which leads to stronger cellular response to a molecular stimulus, due to an increase in the number of receptors on the cell surface, that is, up-regulation of virulence factors of the pathogenic bacteria [20] and secretion of extracellular polymers [21] to bind the microcolonies together into a robust biofilm. Virulence factors are molecules produced by a pathogen that specifically cause disease, or that influence their host's function in order to allow the pathogen to thrive. The microbial colonies grow outward to form a 3D biofilm, by (i) secreting insoluble gelatinous exopolysaccharides (EPSs) or EPM, by (ii) consumption of soluble nutrients, O_2, and other bacterial species and mammalian cells, platelets, and (iii) by attachment of secondary colonizer micro-organisms (step 5 in Figure 7.2). Under favorable conditions, this stage of the build-up of a mature biofilm continues for an extended time. If the environment ceases to support the bacterial load, the equilibrium is shifted and favors dissociation of individual cells from the biofilm to seek more favorable habitats. Under such unfavorable conditions, the biofilm enters into its last stage, disintegration, degradation, and liberation of bacterial cells in the form of free-floating (planktonic) cells (step 6 in Figure 7.2), which continue the step 2 of the biofilm formation cycle [22].

7.1.4
Factors Influencing Biofilm Formation

The interaction of bacteria with material surfaces and the progress of the biofilm development are influenced by many factors related to the biomaterial surface, the type of micro-organism, the contacting medium, and the environmental conditions, as depicted in Figure 7.3. It is important to understand the role of these different factors in the bacterial adhesion process from the initial step, so that necessary action may be taken in order to prevent formation of a biofilm.

7.1.4.1 Role of the Conditioning Film
When a foreign material is placed in a physiologic fluid, organic molecules (proteins) as well as inorganic salts present in the bulk fluid are deposited quickly to the material surface and form a so called 'conditioning film', which significantly modifies the surface properties of the native surface. The adhesion of proteins influences the bacterial adhesion in two possible ways; first by altering

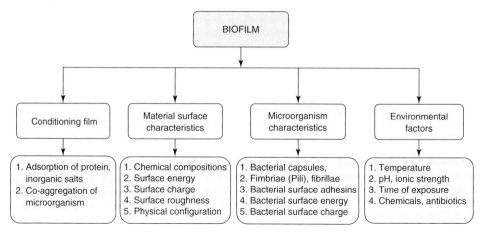

Figure 7.3 Factors influencing biofilm formation on material surfaces.

the physico-chemical properties of the material and bacterial surface, such as the surface free energy (SFE) or electrostatic charges; and secondly, by providing ligand or specific receptor sites for the surface adhesions of micro-organisms [5, 15, 23, 24]. Hence, proteins can affect the bacterial adhesin in the 'docking phase' as well as in the 'locking phase'.

Experimentally, it was found that the bacterial adhesion process is governed by the nature of the protein present in human body fluid and by the type of bacterial strains used for the study. However, in general, albumin proteins are reported to inhibit bacterial adhesion on material surfaces irrespectively of the type of material [25–27], whereas fibrinogen promotes the bacterial adhesion process [26, 28]. Fibronectin, on the other hand, shows inhibiting as well as promoting behavior toward bacterial adhesion, depending upon the bacterial strains [28–30].

It has also been found that the bacteria once adhered to the surface, also mediate and help the adhesion of other bacterial communities by a cooperative effect called 'co-aggregation process', which is mediated by cell-to-cell recognition. This way, organisms in the biofilm can recognize and adhere to genetically distinct bacteria by means of adhesins. For instance, during the process of dental plaque formation, fusobacteria (Gram-negative bacteria) are attached to the primary colonizers, that is, streptococci (Gram-positive bacteria). Subsequently, several other types of bacteria also attach, generally known as *secondary colonizers* [31]. Corpe also reported that bacteria of the genus *caulobacter spp.* appeared to attach to glass surfaces at a higher rate in the presence of *pseudomonas spp.* than in their absence [32].

7.1.4.2 Material Surface Characteristics

The role of material surface properties in influencing different interactions during various phases of biofilm development is of crucial importance for the prevention of biofilm formation, since the device material characteristics may be modified, but not the bacterial characteristics or the physiological environmental condition of a body fluid, particularly in the case of medical implants. Furthermore, the physico-chemical properties such as surface energy, surface charge, polarity, and so on, of the material surface, which affect bacterial adhesion, depend on its chemical composition. For example, polydioxanone sutures exhibited the lowest affinity toward the adherence of *Escherichia coli* and *Staphylococcus aureus*, whereas dexon sutures had the highest affinity toward these two bacteria [33].

Metal surfaces are generally hydrophilic in nature due to polar hydroxyl groups on the outer oxide layer of their surfaces. Depending on the pH of the solution, metal surfaces may have different charges [34]. On the other hand, polymers used in medical devices are in general more hydrophobic than metals. Gristina *et al.* have reported that *S. epidermidis*, which has hydrophobic surface characteristics, preferentially adheres to polymer surfaces, whereas *S. aureus*, which has a hydrophilic surface preferably adheres to metal surfaces [35]. This is one of the reasons why *S. epidermidis* often causes polymer implant infection while *S. aureus* is often the major pathogen in metal implant infections.

A surface with low SFE has a lower tendency to interact with the surrounding material objects. Hence, in aqueous solution minimal interaction possibilities between a low SFE inert surface and a bacterium are expected, which should lead to minimal bacterial adhesion as well as to easy removal of attached bacteria [36]. From a thermodynamic viewpoint, it has also been predicted that bacteria with a SFE higher than water have also little interaction tendency with low SFE surfaces, for instance in dental plaque [37, 38]. Over the past few decades, it was found that a minimal long-term adhesion of a diverse range of biological systems, including proteins, tissues, microbes, algae, and invertebrates, is associated with surfaces having initial surface tensions between 20 and 30 mJm^{-2} [39]. On the other hand, Absolom *et al.* [38] reported that bacteria with low SFE will preferentially attach to low SFE surfaces by hydrophobic interaction [40]. On the contrary, there are reports on the inhibition of protein and bacterial adhesion by high SFE surfaces, for example, surfaces covered with poly(ethylene oxide) (PEO) [also known as poly(ethylene glycol) (PEG). PEG and PEO are chemically synonymous and we use the two terms interchangeably. Historically PEG has been used for oligomers and polymers with a molecular mass below 20,000 g mol^{-1}, whereas PEO refers more to polymers with a molecular mass above 20,000 g mol^{-1}] layers, poly(methyl methacrylate) (PMMA), or plasma treated polymer surfaces with hydrophilic groups [41–48].

The surface charge of material surfaces is influenced by the chemical property of the contacting medium such as pH and ionic strength [49]. The majority of bacteria carry a net negative surface charge at physiological temperatures and pH [50], and therefore biomaterials with negatively charged surfaces are expected to discourage bacterial adhesion by Coulomb repulsion, while positively charged surfaces are expected to promote it.

In general, experimental results show that also a rough material surface enhances bacterial adhesion [51], which is attributed in part to the greater surface area offered by rough surfaces [6], and to the roughness dependent wettability of the surfaces [52, 53]. Interestingly, Taylor *et al.* postulated the concept of 'optimum roughness' at which the bacterial adhesion reaches its peak value, whereas it decreases at even higher roughness [54]. The concept of 'threshold roughness' has been suggested by Bollen *et al.*, as those authors found that a surface roughness below $R_a = 0.2\mu m$, the so called 'threshhold R_a', did not show any influence on the microbiological adhesion onto hard intra-oral surfaces [55].

The 3D morphology of a material surface, such as mono-filamental, braided, porous, or grid like surfaces, also influences the bacterial adhesion process. Merritt *et al.* found that implant site infection rates are much higher for porous and irregular surfaces [56–58] than for plain ones. Further, bacteria are found to adhere more to grooved and braided materials compared to flat ones, probably partially due to a increased surface area [59, 60]. Edwards *et al.* have theoretically described that bacteria preferentially adhere to irregularities or grooves with sizes that are comparable to the size of bacteria, since this maximizes the interaction area between bacteria and the material surface [61].

7.1.4.3 Micro-organism Characteristics

For a given material surface, different bacterial species and strains exhibit different adhesion behavior, mainly due to the variation in the physico-chemical characteristics of bacteria strains. At the initial stage of the adhesion process, surface physico-chemical properties of the bacteria, such as hydrophobicity and surface charge, play an important role. However, at a later stage, bacterial surface adhesins and specific structures of the micro-organisms, appendages, bacterial capsules, fimbriae (also called *pili*) and fibrillae, cell wall components, and extra cellular lipopolysaccharides play a significant role [62]. Almost all types of bacterial cells contain a discrete covering layer outside the bacterial cell, composed of polysaccharides and proteins, which is called '*capsule*', and governs the surface properties of bacteria in a biological fluid [63]. Bacterial strains with hydrophilic capsules exhibit a lower adhesion tendency as compared to those with hydrophobic ones [64]. Fimbriae, or pili, are a group of nanostructured rigid, straight, filamentous polymer appendages on a bacterial surface, composed of identical protein subunits called *pilin* [65]. Fimbriae mediate bacterial adhesion through surface adhesins and fimbriae-dependent surface hydrophobicity [66].

Protein-binding receptors present on the bacterial surface, termed *bacterial surface adhesions*, typically known as 'microbial surface components recognizing adhesive matrix molecules' (MSCRAMMs), are responsible for irreversible adhesion to the conditioned surfaces through specific binding with adsorbed proteins on the material surface [24, 67]. For example, *S. aureus* has fibronectin binding proteins (FnbpA and FnbpB), fibrinogen-binding (Fbe) proteins (ClfA and ClfB) and elastin binding proteins (IsdA, IsdB, IsdC, IsdH, Ebh, Emp, EbpS). Moreover, IsdA and Emp (extracellular matrix protein-binding protein) display a broad-binding specificity for extracellular matrix (ECM) and plasma proteins [24]. Likewise, *S. epidermidis* also has a Fbe adhesin [68], a heparin-binding protein IsaB [69], and collagen binding GehD lipase [70]. *S. aureus* and *S. epidermidis* may also synthesize the PIA during the course of biofilm formation.

The surface energy of bacteria, which depends on the growth medium, the bacterial age, and the bacterial surface structure, is an important physical factor for bioadhesion to material surfaces [71]. It has been found that hydrophobic bacteria, for example, *S. epidermidis*, preferably adhere more on hydrophobic material surfaces whereas hydrophilic bacteria, for example, *S. aureus*, adhere better onto hydrophilic surfaces. This was further verified experimentally, by treating *S. epidermidis* cells with pepsin or extraction with aqueous phenol to reduce their hydrophobicity. It was found that the bacterial cells after treatment exhibited reduced adhesion to hydrophobic fluoropolymer surfaces [72, 73].

The net charge on the bacterial surface depends on the overall surface isoelectric point of the micro-organism and the pH and ionic strength of the contacting medium. The surface charge of bacteria varies according to bacterial species and depends on the growth medium, bacterial age, and bacterial surface structure [71]. Electrostatic interaction chromatography (ESIC) and zeta potential measurements showed that the majority of bacteria (which are negatively charged at physiological pH [64, 74]), are found to adhere much faster to positively charged surfaces as compared to negatively charged polymers, due to the electrostatic interaction [74]. The positively charged bacteria *Stenotrophomonas maltophilia* in contrast exhibited a much stronger adhesion to negatively charged surfaces such as glass, as compared to the negatively charged bacterium *Pseudomonas putida mt-2* [75]. Using similar reasoning, some research groups, for example, Dickson and Koohmaraie [15] and Van Loosdrecht *et al.* [10], have reported positive correlations between bacterial surface charge and their tendency to attachment.

7.1.4.4 Environmental Factors

Local environmental factors such as pH, ionic strength, temperature, exposure time, presence of chemicals and antibiotic agents, and so on, play a significant role for the adhesion behavior of bacteria on biomaterial surfaces exposed to biological fluids. Those factors affect bacterial adhesion by influencing the physical interactions in the initial phase of adhesion by changing the surface physico-chemical properties of bacteria or/and materials, influencing gene regulation or quorum signals during biofilm formation, and by affecting temperature-dependent reaction rates of enzymes, which control the development of physiological and biochemical systems of bacteria [76]. A temperature increase above $37\,^{\circ}$C causes significant reduction in the adhesion of *Porphyromonas gingivalis* and *Streptococcus mutans* due to the down-regulation of fimbrial expression [77], while thermal stress induces polysaccharide intercellular adhesin expression in *S. epidermidis* [78]. The ionic strength of the medium influences the bacterial adhesion to surfaces by interfering with their mutual electrostatic interactions. A variation in pH may change the ionization state of functional groups (i.e., carboxyl and amino groups) present on the bacteria surface and, consequently, the electrostatic forces between the bacteria and the substrate [79]. The bacteria–surface adhesion has been reported to be the highest at a pH of the medium close to the isoelectric point of the bacteria, that is, at the point of zero charge, due to absence of electrostatic repulsion forces [49, 80]. Higher ionic strength of the solution results in a higher bacteria–metal surface adhesion force due to screening of the repulsive electrostatic interaction, thus facilitating the hydrophobic interaction [49].

The presence of the antibiotics cephalothin, clindamycin, and vancomycin in sub-inhibitory concentration has been reported to decrease *S. epidermidis* adhesion by 30–80%. [81]. Park *et al.* reported that an increase of the phosphate ion concentrations (between 0 and 0.5 mM) decreased *S. aureus* adhesion to iron-coated sand, possibly due to reversing their surface charge from positive to negative by adsorbed phosphate ions. However, it was also reported that increasing the phosphate concentration in the range between 0.5 and 2.0 mM promotes bacterial attachment, possibly due to compression of the electrical double layers between bacteria and phosphate-adsorbed/negatively charged surfaces by free phosphate ions [82]. Pre-treatment of *S. epidermidis* with pepsin decreases its adhesion to a polymer surface [72]. *S. pyogenes* adhesion can be prevented by pre-treatment with trypsin and pepsin or HCl solutions [83]. Satou *et al.* reported that the number of bacteria adhering to different substrate surfaces increased with time until they reached a saturation level that was specific for each type of surface [84].

7.2
Biofilm Prevention Strategies

The conventional approach for prevention of biofouling and microbial infections is based on the use of compounds that kill or inhibit the growth of the microbes. A more recent strategy for the prevention of infections is based on the knowledge that the interface between the biomaterial and the surrounding body fluid is the actual battleground where accidental contamination can first develop into colonization and, subsequently, into the establishment of a clinically relevant infection. So the most convenient way to ward off the development of a biofilm is to interfere with the early phases of microbial adhesion by modifying the surface properties of the biomaterial. Plasma processing of materials offers an efficient method to modify material surfaces for the development of antifouling surfaces. The latest and most ambitious approach is based on a strategy of interfering with the biological processes, which are involved in a later phase of biofilm formation.

7.2.1
Pre-surgery Precautionary Approach

This approach is based on personal hygiene to avoid an external pre-surgery biocontamination of implants, which is the primary cause of implant-related infections. It includes (i) sterilization of biomedical implants; (ii) thorough routine cleaning or detachment of the biofoulers; (iii) improvement of the operating standards; (iv) minimizing the possibility of contamination during surgery; (v) reducing the establishment of an infection by peri-operative antibiotic prophylaxis, and (vi) confining pathogenic strains by patient isolation [85]. According to existing medical guidelines, all devices that are intended to penetrate the human body or that come into direct contact with the patient's first immune defence system and tissues have to be sterilized in order to prevent infections caused by pathogenic micro-organisms. There are several possibilities for sterilization, for example, autoclaving, gamma-ray irradiation, ethylene oxide (EO) exposure, and plasma discharge treatment of the implants [86].

7.2.2
Antimicrobial-releasing Biomaterials

This is a common approach to control the bacterial infection, and it includes the impregnation and deposition of antimicrobial/antibiotic agents to the device material. The antimicrobial agents leach out in the course of time and kill microbes or inhibit and prevent the growth of surface attached bacteria [87, 88]. A large number of antibiotic agents such as cefzolin [89], minocycline–rifampin [90], teicoplanin [91], silver [92], chlorhexidine–silver sulfadiazine [93], usnic acid, lysostaphin [94], vancomycin [95], gendine [88], gentamicin [96], and so on, have been employed for this purpose. Some of these drug-delivery coatings have already been demonstrated to be effective in preventing infections *in vivo*. However, among these, only the minocycline–rifampin and chlorhexidine–silver sulfadiazine combinations have been shown to reduce infection rates during short-term use in clinical trials [87]. For long-term applications effective solutions are still lacking. One of the major reasons for this limitation may be the incompatibility between the antimicrobial agents and the biomedical materials which can prevent a sufficient loading and proper long-term release of the antibiotic agents. The release of an antibiotic agent below the lethal dose for micro-organisms leads to serious consequences, such as, development of antibiotic-resistant bacterial strains, accelerated biofilm formation and induced virulence factor expression [97, 98]. For instance, exposure to sub-inhibitory concentrations of imipenem leads to an increased *Pseudomonas aeruginosa* biofilms volume and alginate polymer matrix production due to induced gene coding for alginate biosynthesis [97]. Similarly, Hoffman *et al.* reported that sub-inhibitory concentrations of aminoglycoside antibiotics such as tobramycin induced biofilm formation of *P. aeruginosa* and *E. coli* [98]. However, in combination with a systemic therapy, antibiotic-releasing materials were found to exhibit convenient synergic efficacy.

7.2.3
Surface-engineering Approach

This approach works on three different strategies: (i) hydrophilic high surface energy surfaces; (ii) hydrophobic, inert, low surface energy surfaces; and (iii) tethered antimicrobial, self sterilizing agents bound directly to the device surfaces.

Ikada *et al.* predicted theoretically that the work of adhesion of biomolecules (the interfacial surface energy) in aqueous media approaches zero when the water contact angle θ of polymer surfaces approaches 0 or $90°$ [99]. This means that for

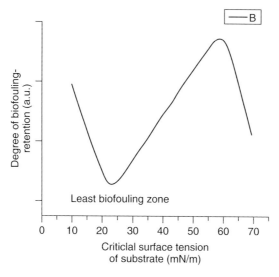

Figure 7.4 A generalized relation between the biofouling retention strength (B) and the critical surface tension of the substrate. The absolute value of the minimum fouling zone may vary and depend on the specific biological system, the time of contact, and the mechanical forces of removal [102].

a material surface there are two possibilities to approach zero work of adhesion (i.e., to be nonadhesive): one is to create a hydrophilic water-like surface, and the other one is to create an inert hydrophobic surface [23]. Moreover, Baier [100] and Dexter [101] correlated experimental results on bioadhesion with the SFE of the substrates and established a generalized correlation curve between bioadhesion and the substrate surface energy, which is nonlinear with a minimum at 20–30 mN m^{-1} (Figure 7.4) [100, 102], also known as the '*Baier curve*' [23]. The correlation curve indicated decreased biofouling retention on surfaces with low surface free energies and surfaces with very high surface energies [102].

7.2.3.1 High Surface Energy Approach

A substantial amount of work has been carried out using this strategy to develop surfaces which are inert against protein and bacteria adhesion, using hydrophilic polymers with specific chemical and structural properties [42, 48]. Following this approach, PEG coated surfaces were found to be very promising in suppressing the adhesion of proteins and a variety of bacterial strains, for instance, *S. mutans* on PEG-coated polystyrene [43], *P. aeruginosa* on plasma deposited PEG-like film [44], and *Staphylococcus epidermidis*, *S. aureus*, *Streptococcus salivarius*, *E. coli*, and *P. aeruginosa* on PEG-brush coated glass [43–45]. The PEG polymer is a linear, neutral polyether with the general structure RO–(CH$_2$–CH$_2$–O)$_n$–CH$_2$–CH$_2$–OH, where R is often H, CH$_3$, or another chemical group. Characteristic properties of PEG are its formal electrostatic neutrality, its unique electronic/hydration structure and its low interfacial free energy. These go together with entropically driven polymer steric exclusion and osmotic effects and are common aspects used to explain its unique antifouling surface characteristics [103].

Several parameters seem to be involved in the mechanism that leads to the resistance of poly(ethylene oxide) (PEO) coatings toward biomolecule [104] and cell adhesion [105, 106], including chain density, chain length, and chain conformation [107–109]. However, it appears that self-repulsion of the chains in water plays an important role for the nonfouling properties of PEO films. This property initiates a dynamic sweeping process, which limits the adsorption of proteins on the surface [110]. When a biomolecule approaches hydrated PEG brushes, the compression of extended PEG brushes hinders the free mobility of the PEG chains and creates an unfavorable entropy loss, which compensates the entropy gain from the released water molecules. This makes protein adsorption entropically unfavorable [111].

PEG coatings were reported to suffer from aging due to autooxidation, particularly in the presence of oxygen and transition metal ions [112], and thus, have limited success in preventing long-term biofilm formation [45]. PEG was reported to decompose after being exposed to air for one week at 45 °C and after one month at 20 °C. Furthermore, under *in vivo* conditions, PEG may be oxidized enzymatically to aldehydes and acids, allowing proteins and cells to attach [112]. Interestingly, Cringus-Fundeanu *et al.* reported recently that another polymer, namely a covalently attached hydrophilic polyacrylamide (PAAm) layer on a silicon wafer reduces the attractive forces and adhesion of *S. aureus* and *S. salivarius* significantly [48].

7.2.3.2 Low Surface Energy Approach

Low surface energy polymeric coatings, such as fluorocarbon coatings, can act as nonadherent surfaces to bacteria and other colonizing micro-organisms [113, 114]. The increased interest in this approach was motivated by the existence of a natural antifouling surface of gorgonian corals with a surface energy of 23–27 mN m^{-1} [115]. Busscher *et al.* reported up to 94% lower adhesion of *Streptococcus sanguis* on fluoroethylenepropylene (FEP) as compared to glass [116]. Everaert *et al.* showed that fluoroalkylsiloxane coatings on silicone rubber surfaces reduced yeast and bacterial adhesion by 50–77% as compared to original silicone rubber [117].

A concept and engineering definition of a so-called 'theta surface' has been proposed by Baier to support bioengineering solutions for biocompatibility and biofouling control [102]. '*Theta surface*' is the term used to define the characteristic expression of the outermost atomic features of a surface, which shows very little adhesion of biomolecules in different test systems such as blood, tissue, saliva, tear, and oceanographic biofluids, and which is identified by the bioengineering criterion of having a critical surface tension (CST) or SFE in the range of 22–24 mN m^{-1} [102]. Several research groups observed a reduced bioadhesion on substrates with a surface energy in the range between 20 and 30 mN m^{-1} for various surfaces and a broad range of biological agents, including proteins, tissues, microbes, algae, and invertebrates [39, 118–120].

In *in vivo* studies carried out on surfaces attached to human teeth, with a SFE from 20 to 88 mN m^{-1}, it was found that on low SFE surfaces (i) a smaller number of bacteria attached and (ii) those bacteria were attached less tightly as compared to high SFE surfaces [121]. Another *in vivo* study showed that modification of a silicone rubber voice prosthesis with perfluoroalkylsiloxane of eight fluorocarbon units resulted in reduced biofilm formation [122]. It is interesting to note that, in general, besides low adhesion of bacteria onto low SFE surfaces, it is easy to detach the adhered bacterial cells under physical shear stresses such as flow or passage of air–liquid interface [36, 116, 117, 120], which suggests a promising medical application of low SFE material surfaces in devices such as urine catheters, oral devices, and voice prosthesis, as these encounter high shear forces.

7.2.3.3 Surfaces with Bound Tethered Antimicrobial Agents

This class of prevention strategy includes chemical attachment of antimicrobial agents, such as quaternary ammonium/phosphonium groups and zwitterionic group based compounds to material surfaces. Quaternary ammonium salts (QAS) and phosphonium salts (PSs) have been used as disinfectants against a wide range of micro-organisms including Gram-positive and Gram-negative bacteria, yeasts, and moulds [123–125]. The antimicrobial activity of polymeric QAS and PS results from both ionic and hydrophobic interactions between the QAS and components of the microbial cell wall and the cytoplasmic membrane leading to malfunction in cellular processes or cell death [126].

The chemically bonded tethered QAS and PS compounds on device surfaces give rise to so-called 'self-sterilizing' surfaces and exhibit their antimicrobial action without releasing as a free molecule in the peri-prosthetic region [123, 127]. For instance, poly(vinylbenzyl trimethylammonium chloride) (PVBT) grafted to cotton fabric by Co^{60}-gamma radiation exhibited a strong antibacterial activity (up to six log cycle reduction) against *E. coli.* and *S. aureus* (Figure 7.5), and retains its antibacterial activity after several cycles of washings [123]. Poly(4-vinyl-*N*-hexyl pyridinium bromide) grafted polyurethane (PU) fibrous membranes, and phosphonium group containing polymers grafted onto polypropylene via photo-grafting showed effective antibacterial activity against *S. aureus* and *E. coli* [125, 128]. 3-(Trimethoxysilyl)-propyldimethyloctadecylammonium chloride coating covalently attached to silicone rubber substrates exhibited strong antibacterial activity against *S. aureus*, *S. epidermidis*, *E. coli* and *P. aeruginosa* [127].

Van der Mei *et al.* have recently reported the antimicrobial activity of poly(diallyl dimethylammonium chloride)-coated glass surfaces against waterborne pathogens, namely *Raoultella terrigena*, *E. coli*, and *Brevundimonas diminuta* [129]. Majumdar *et al.* have also reported the potential application of 'tethered' QAS moieties chemically bound to polysiloxane substrates as an environmental-friendly coating to control marine biofouling [130]. Tethered zwitterionic groups including phosphorycholine (PC), sulfobetaine, and carboxybetaine, have been postulated as new generation of non-biofouling materials [131–133]. Poly(sulfobetaine methacrylate) (pSBMA) and catechol-containing zwitterionic polymers (*N*-(3-sulfopropyl)-*N*-(methacryloxyethyl)-*N*,*N*-dimethylammonium betaine) (pSBMA-catechol) have been reported to resist nonspecific protein adsorption and adhesion and biofilm formation of *S. epidermidis* and *P. aeruginosa* [134, 135].

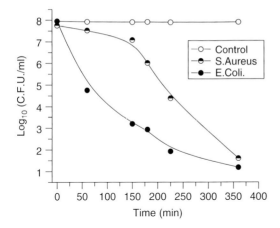

Figure 7.5 Antibacterial activity of ^{60}Co-gamma radiation
induced grafted PVBT-g-cotton against *S. aureus* and *E. coli*
as a function of time [123].

There are some limitations in the use of QAS or PS. They lack a broadband antibiotic effect, that is, they are efficient only against specific micro-organisms, and there is the possibility of attracting more bacterial cells and proteins onto the surface via electrostatic attraction. Because of the latter, the antimicrobial activity of these surfaces may easily be screened once they are covered with proteins and dead bacterial cells.

7.2.4
'Antibiofilm' Approach

Genetic studies focused on identifying the critical molecular components and processes involved in biofilm formation indicate that biofilm formation is a regulated process, with specific adhesins mediating cellular attachment to abiotic surfaces [12, 136] and other genetic elements controlling the overall microscopic architecture of the biofilm [18]. Accordingly, one general approach for synthesizing 'antibiofilm' agents is to identify compounds that impair the production or the proper assembly of the adhesins and intercellular signaling molecules giving 'quorum signals' involved in the biofilm formation process. This innovative approach also includes the development of substances that specifically inhibit bacterial virulence, called '*antipathogenic*' drugs. Unlike 'antibacterial' drugs, they do not kill bacteria or stop their growth, and so are assumed not to lead to the development of resistant strains [5]. The mode of action of 'antipathogenic' drugs is based on the inhibition of regulatory systems that govern the expression of a series of bacterial virulence factors leading to bacterial infection. Based on different modes of action under the general 'antibiofilm' approach, various strategies have been proposed by different research groups. These include a synthetic peptide vaccine and antibody therapy [137], inhibiting or negating cell–cell signaling [20, 138], negating biofilm formation by disrupting the iron metabolism [139], up-regulation of biofilm detachment promoters (rhamnolipids) [140], and enhancing macrophage phagocytosis by developing artificial 'opsonins' [141].

7.3
Role of Plasma Processing in Biofouling Prevention

Surface modification is a relatively straightforward method for regulating the interfacial properties of medical devices without disrupting the bulk properties of the device material, and hence avoids costly changes of materials. Plasma modification has various advantages compared to other techniques of surface engineering. It is an environmentally benign, solvent free dry process, has a low process cost, good process control, and no adverse effects on the bulk properties of the material [142]. Non-equilibrium plasmas have been widely used for the plasma modification of polymeric materials for various purposes, including biomedical applications [143–147].

Different plasma surface modification methodologies can be employed in order to prevent biofouling on biomedical device surfaces as depicted in Figure 7.6.

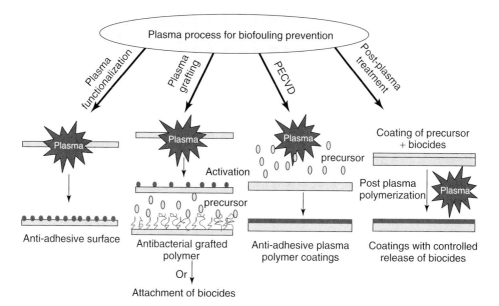

Figure 7.6 Sketch representing the application of plasma processing technology in the development of biofouling-resistant surfaces: from left to right: direct plasma surface treatment, plasma activation pre-treatment and further deposition of precursor in liquid phase, direct plasma deposition of precursor, pre-deposition of precursor, and post plasma polymerization (curing). The final result is represented by the treated (coated) substrate after the lower (last) processing step (at the bottom of sketch).

Those include (i) surface functionalization by treatment of the material surface with chemically reactive gas plasma [148, 149]: (ii) attachment of antibacterial agents to the material surface via plasma grafting [127, 128]: (iii) microbe-repelling or anti-adhesive plasma polymer coating [150, 151]: and (iv) plasma polymer coatings with controlled release of biocides [152, 153]. Besides plasma induced surface modification of medical devices, plasma processes can also be used as a sterilization tool for medical implants and devices.

7.3.1
Plasma Surface Functionalization

The physico-chemical characteristics of the material surface can be effectively modified by incorporation of different chemical groups, such as hydroxyl, peroxyl, carbonyl, carboxylic, amino or fluorocarbon groups onto the polymer surfaces by plasma functionalization using chemically reactive plasma discharges, for example, Ar/O_2, N_2, NH_3, CF_4, and so on. [143, 154–157].

CF$_4$ plasma has been applied to produce anti-adhesive fluorinated hydrophobic cellulose hemodialysis membrane surfaces in order to prevent biofilm formation [158]. PMMA intra-ocular lenses were treated by CF_4 plasma to prevent adhesion of proteins and subsequent development of inflammatory cells [159]. Recently, it has been reported that fluorination can also induce antibacterial surfaces properties [149, 157]. Ar–CF$_4$ discharge plasma treatment of wool and polyamide-6 fabrics leads to hydrophobic surfaces, which exhibit antibacterial properties against *E. coli*, *S. aureus*, and *Bacillus subtilis* [149].

Plasma treatment can also be effectively used to develop polymer surfaces with high SFE by generating hydrophilic groups on top of them, which is repellent against proteins and bacteria. He and He/O_2 plasma treatment of poly(ethylene terephthalate) (PET) surfaces and oxygen glow discharge treatment of poly(vinyl chloride) (PVC) endotracheal tubes have been demonstrated to reduce adhesion of *S. epidermidis* [41, 46, 47]. In order to overcome the protein fouling problem in protein filtration membranes, nitrogen-based plasma systems, such as N_2, NH_3, Ar/NH_3, and O_2/NH_3 have been used to incorporate hydrophilic surface chemistry onto different polymer substrates such as microporous polyethersulfone membranes [160], PP microfiltration membranes [161] and PP hollow fiber microporous membranes [162].

7.3.2
Plasma-Induced Grafting

Plasma grafting can be used to obtain antibacterial properties by two strategies: the first strategy involves the generation of reactive radicals on the polymer surfaces by the plasma discharge, which react with the unsaturated monomer of an antibacterial agent and initiate a graft polymerization reaction. In the second strategy, hydroxyl, peroxy, and/or hydroperoxy groups can be generated by treating the surface in the presence of oxygen. Those species can be used later as initiator reservoir to initiate graft polymerization of monomers under suitable experimental conditions [127, 163]. Thome *et al.* have successfully used the two plasma grafting strategies to produce antibacterial surfaces by grafting of QAS polymers, for example, dial-lyl dimethylammonium chloride (DADMAC), onto PE surfaces [163]. Gottenbos *et al.* have also used plasma discharge to generate OH groups on silicone rubber surfaces, which were further reacted with 3-(trimethoxysilyl)-propyl dimethyloc-tadecylammonium chloride to develop QAS-grafted silicone rubber surfaces with antimicrobial properties [127]. In another study, Ar plasma pre-treatment was used to generate surface oxide and peroxide groups in order to facilitate the UV-grafting of poly(4-vinyl pyridine) onto PU membranes, later quaternized with hexylbro-mide to generate quaternary ammonium group containing poly(4-vinyl-*N*-hexyl pyridinium bromide), which showed a very good antibacterial activity against *S. aureus* and *E. coli* [128]. PEG oligomers have also been covalently attached to plasma-functionalized silicon rubber surfaces to develop antifouling surfaces [150]. Antibacterial polymer surfaces can also be generated by covalent attachment of antibiotic drug molecules such as penicillin onto surfaces using both the plasma and the organic chemical route [164].

7.3.3
Plasma Polymerization

The plasma polymerization process, realized by plasma enhanced chemical vapor deposition (PECVD), is based on the polymerization of simple precursor gases (e.g., methane, ethylene, propylene, CF_4) or more complex organic monomers with unsaturated functionality (e.g., acrylic acid, allyl alcohol, and perfluoroacrylates) to produce a thin polymer coating on substrates. When exposed to plasma, hydrocarbon precursor molecules get fractured with the production of free radicals. The latter initiate a polymerization process, which leads to the growth of a polymer film on the substrate underneath. Such plasma-polymerized thin films offer many advantageous features: they can be produced pinhole-free and highly cross-linked and show strong adhesion to the surface.

Anti-adhesive and biofouling resistant polymer coatings such as PEG based hydrophilic coatings and fluorocarbon based hydrophobic coatings [42, 114, 165, 168], as well as coatings impregnated with biocides [152] can be deposited by using plasma induced polymerization of precursor molecules either by simultaneous PECVD or by post plasma treatment. PEG like coatings with very good protein repelling characteristics have been deposited by plasma polymerization of different precursor monomers with varying number of EO units namely, monoethylene vinyl ether (EO1V), diethylene glycol vinyl ether (EO2V) [165], diethylene glycol dimethyl ether (DEGDME) (diglyme) [42, 166], triethylene glycol monoallyl ether (EO3A) [169], tetraethylene glycol dimethylether (tetraglyme), and cyclic ethers such as 15-crown-5 monomers [170]. Tsai *et al.* developed a series of plasma deposited fluorocarbon thin films by radio frequency glow-discharge (RFGD) treatment using a varying C_3F_6/CH_4 ratio in the monomer feed. Platelet adhesion to the fluorocarbon surfaces was found to be lower than to PET or the methane glow-discharge-treated PET [168].

Wei *et al.* produced hexamethyldisiloxane (HMDSO) coatings with varying wettability by plasma polymerization of HMDSO followed by oxygen (O_2)-plasma treatment and found that the hydrophobic coatings discourage the adhesion of fibronectin and L929 mouse fibroblasts [171].

Hendricks *et al.* have deposited triethylene glycol dimethyl ether (triglyme) and poly(butyl methacrylate) (PBMA) based coatings onto PU surfaces embedded with antibiotics (ciprofloxacin), using radiofrequency glow discharge plasma polymerization, for antifouling and antibacterial coatings with controlled release of the antibiotic from the surface. The rate of initial bacterial cell adhesion to triglyme-coated PU and pBMA-coated PU releasing ciprofloxacin was 0.77 and 6%, respectively, as compared to the PU control without any antibiotics. However, the rate of adherent cell accumulation due to cell growth and replication was zero for the pBMA-coated PU-releasing ciprofloxacin, but no decrease was found for the triglyme-coated PU as compared to PU controls [152]. Silver-releasing, Ag/PEO-like coatings deposited by combined PECVD deposition of diethylglycol dimethyl ether (DEGDME) with the simultaneous sputtering from the silver RF electrode of an asymmetric parallel-plate reactor, showed very good antibacterial activity against *S. epidermidis* due to the combined effect of PEO (antifouling) and silver ions (antimicrobial) [153]. Wei Zhang *et al.* prepared antibacterial PVC surfaces by coatings with the antimicrobial agents triclosan (2,4, 4P-trichloro-2P-hydroxydiphenylether) and bronopol (2-bromo-2-nitropropane-1,3-diol) using a plasma process. The PVC surfaces were pre-treated with O_2 plasma to produce more hydrophilic groups and then triclosan and bronopol were coated via solvent dry method followed by post plasma treatment to ensure a better attachment of those drug molecules onto the substrate [172].

7.3.4
Plasma Sterilization

Several commonly used sterilization methods, namely autoclaving, EO treatment, gamma-ray irradiation, and electron beam irradiation have certain limitations. For instance, autoclave sterilization processes are generally not suitable for all polymeric materials due to their limited thermal stability. EO sterilization suffers from the residual toxic chemical species, which constitute an unacceptable risk in biomedical implants or devices. Gamma irradiation and electron beam irradiation are very effective methods of sterilization but have poor consumer and public acceptance, require high radiation safety equipment, and are generally very costly. Sterilization processes based on nonthermal plasma discharge treatments, have emerged as comparatively environmental friendly and cost efficient approaches [173, 174].

The nonthermal plasma contains heavy particles (molecules, atoms, free radicals, ions) and lighter species (electrons and photons) generated by excitation and ionization of gas molecules by electric discharges, which can efficiently kill or inactivate micro-organisms, resulting in a pronounced sterilization effect. Several mechanisms are thought to lead to plasma sterilization. The first mechanism involves reactive radicals (such as OH· and NO·) which are generated in a plasma and can cause irreparable surface lesions to micro-organism, leading to cell death [175]. This is termed '*etching*', in analogy to plasma etching of synthetic polymers. More recently, the mechanism of chemical sputtering obtained by the synergetic effect of ion bombardment and reactive species such as O, H, or even O_2 has been illustrated on different bacterials strains [176], proteins, and pyrogens [177]. It is also thought that a plasma can damage the bacterial membrane, which may affect the interaction between the DNA and the membrane proteins, and could cause leakage of DNA from the cell, particularly in bacteria where the DNA is anchored to the membrane [178]. The third suggested mechanism is based on DNA strand breaks and surface lesions caused by the combined effect of free radicals and UV radiation [179]. Recent studies indicate that plasma sterilization is most efficient when the plasma parameters (e.g., discharge composition, plasma power, pressure, and type of discharge), are tuned in such a way that pathogenic biomolecules, bacteria, and bacterial spores are attacked by the combined action of the active species, that is, radicals, ions, and photons, so that they can act in a synergistic way [180, 181].

Hence, for an industrial application of the plasma sterilization process two aspects should be thoroughly investigated. The first is the plasma characteristics and its operating conditions, such as the nature of the gas used to generate the plasma, the plasma power, the design of the electrodes, the effect of the plasma on the treated material, and so on. The second aspect concerns the characteristics of the micro-organisms, such as the resistance of bacterial strains to plasma and their repair mechanisms. This way, plasma technology can become a valuable alternative to established sterilization approaches.

7.4
Case Study: Plasma-deposited Poly(ethylene oxide)-like Films for the Prevention of Biofilm Formation

7.4.1
PEO Films and Plasma Deposition

As discussed previously, PEG, or PEO, is a compound that can be used both for improving biocompatibility and for reducing bacterial adhesion [182–186]. In this section, we discuss the properties of PEO-like coatings obtained by plasma polymerization of DEGDME $(CH_3OCH_2CH_2)_2O$ vapor and how their properties can be tuned by changing the deposition parameters [187, 188]. This is followed by a section describing biological properties of the plasma-polymerized PEO-like coatings, that is, cell and protein adhesion, and the influence of several sterilization processes on the surface chemistry and the bioadhesive properties of these coatings. Finally, the incorporation of Ag nanoparticles (NPs) in the plasma-polymerized PEO film is discussed as a possibility to achieve antibacterial properties.

Plasma polymerization via PECVD processing can be applied to deposit functional polymer coatings with an extreme range of surface chemistries and on a large variety of substrate materials. Moreover, the plasma deposition proceeds in one step and can be performed also on uneven 3D surfaces. The pinhole-free films can be deposited in a wide range of thicknesses from ultra-thin (1–10 nm) up to several hundreds of nanometers with good adhesion and without affecting the bulk physical properties of the treated materials. Plasma polymerization is a low pressure and (not necessarily) low temperature process induced by a glow discharge via a pure organic vapor or a mixture of an organic vapor and a reactive or nonreactive gas. During deposition, two processes occur simultaneously: the ionization of gaseous species, which induces the plasma creation, and the fragmentation/recombination of particles that lead to polymerization. These two processes have a strong influence on the properties of the deposited film and can be controlled by tuning the deposition parameters. Moreover, the films deposited by plasma polymerization have different physical properties than those of films prepared by conventional free radical polymerization of the monomer (e.g., by spin coating). These physical properties depend on the film microstructure, that is, the chemical composition and the atomic organization, as well as on the operating conditions.

The PEO-like films were prepared in a home-built, capacitively coupled plasma reactor, which is described in Section 2.3.2, along with the general deposition conditions. Fourier transform infra-red (FTIR) spectroscopy, imaging ellipsometry, and X-ray photo electron spectroscopy (XPS) were employed for the physico-chemical characterization of the films, while the adsorption of proteins was measured by quartz crystal microbalance (QCM). Cell adhesion was studied using L929 mouse fibroblasts. All experimental details can be found in [42].

7.4.2
Plasma Polymerization by Continuous Wave Plasma

7.4.2.1 Retention of the PEO Character and Film Stability
A fixed pressure of 30 Pa for pure discharge or with a mixture of DEGDME (15% DEGDME in Ar) has been found to be suitable for PEO-like growth, which gives a linear increase of the growth rate from less than 1 nm min^{-1} to around 6 nm min^{-1} (12 nm min^{-1} using pure DEGDME) for injected powers between 1 and 25 W.

Both XPS and FTIR show that for mixtures of DEGDME/Ar as well as for pure DEGDME the retention of the PEO character depends sensitively on the plasma power [189, 190]. In FTIR, the PEO character is indicated by the intensity of the C–O stretching peak (around 1106 cm^{-1}), which decreases rapidly with increasing plasma power (Figure 7.7a). Other peaks typical for PEO-like films are at 2800–3000 cm^{-1} (C–H stretching), while absorption bands between 3200 and 3500 cm^{-1} (OH or NH stretching) are absent. The decrease of the PEO character is also highlighted by the ratio of the C–O/C–H bands, which is plotted as a function of the plasma power in Figure 7.7b. Increasing the power generally induces a strong fragmentation of the glyme monomer in the plasma discharge and leads to coatings with less PEO character [190].

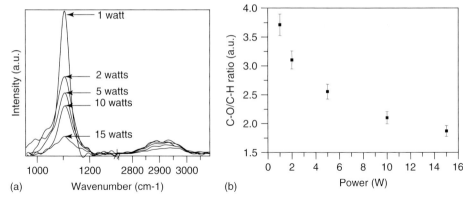

(a)

(b)

Figure 7.7 (a) Infra-red spectra of PEO-like coatings deposited on silicon wafers at different plasma powers: by increasing the processing power, the C–O peak intensity decreases. (b) Evolution of the ratio C–O/C–H determined from FTIR versus plasma power. Mixture of DEGDME and Ar (30 Pa, deposition time = 30 min): the PEO-like character decreases by increasing the processing power.

The same trend is observed from the XPS data and here in particular from the C_{1s} signal. That region can be fitted by a combination of four distinct components, respectively assigned to C–C and C–H moieties (285 eV), C–O groups usually related in the literature to the PEO character (286.5 eV) [190], O–C–O/C=O groups (288.0 eV), and a fourth component at 289.2 eV, corresponding to COOR/H groups. C_{1s} spectra of deposited films as a function of the plasma power are shown in Figure 7.8. These data and their evolution with increasing plasma power show a strong change of the relative intensities of the C–C and C–H (285 eV) and the C–O (286.5 eV) moieties with the increase of input plasma power, which indicates the loss of PEO character, whereas the other contributions are small and remain relatively constant. A maximum of around 70% PEO character (corresponding to the percentage of the peak at 286.5 eV) is obtained for the lowest plasma power whereas this character is reduced to 40% for highest plasma power. This trend is confirmed by the evolution of the O/C ratio versus plasma power [42].

For biotechnological and biomedical applications, the deposited films must be stable in different media (water, phosphate buffered saline (PBS), etc.). After immersion of samples with PEO-like films in ultra-pure water at 25 °C for five days both O/C ratio and PEO character, as analyzed by XPS, remain quite constant (variation less than 5%). However, FTIR shows that samples prepared with a power of 5 W or more exhibit new chemical moieties as indicated by a new absorption band around 1730 cm^{-1}, which can be attributed to terminal C=O moieties. A higher plasma power probably increases also the number of free radicals or unsaturated bonds in the polymers [191], which are very reactive, so that chemical rearrangement is possible when the coatings are immersed in water. These modifications in the bonding structure have also an influence on the optical properties of the film as evidenced by a slight modification of the refractive index

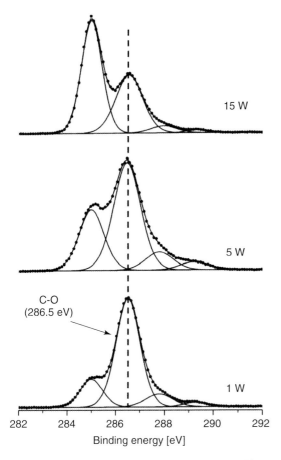

Figure 7.8 C_{1s} high resolution XPS spectra for PEO-like coatings as a function of the plasma power. By increasing the processing power, the signal at 286.5 eV decreases in intensity, confirming the abatement of the PEO-like character of the coatings.

after the water bath, which can be linked to a decrease of the film density. This in turn can be attributed either to the creation of terminal bonds (supported by FTIR measurements, which indicate the creation of C=O bonds) and/or to the decrease of the coatings' crosslinking [192]. Interestingly, when immersed in acetone using the same protocol as that for water, no significant modification of the PEO-like coatings can be detected.

7.4.2.2 Protein Adsorption

Protein adhesion, tested by quartz crystal microbalance (QCM) with bovine serum albumin (BSA) on PEO-like coatings as a function of the deposition power (1, 5, and 15 W) showed that compared to the reference (SiO_2), all the coatings lead to

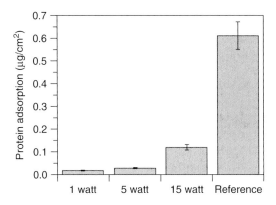

Figure 7.9 Protein adsorption (BSA) on PEO-like films deposited at different plasma powers.

a remarkable decrease of the protein adsorption with a reduction of around 75% for the films produced at 15 W and more than 95% for the films produced at lower power (Figure 7.9). Even the coatings with a small fraction of PEO groups ($\sim$ 40% at 15 W) lead to a strong reduction of the protein adsorption. It should also be underlined that the protein adsorption differs only little between samples produced at 1–5 W in spite of a significant difference in the PEO character ($\sim$70% for 1 W and $\sim$ 55% for 5 W).

7.4.2.3 Cell Attachment and Proliferation

Cellular adhesion tests have been performed using fibroblast L929 cells on nontreated polystyrene Petri dishes. The percentage of attached cells was determined by using the Alamarblue method after one and two days of proliferation (for all the tests, the amount of cells on the nontreated polystyrene dishes after one day of incubation was used as normalization factor and set at 100%). Results of cell culture experiments on coatings made with a mixture of DEGDME and Ar are summarized in Figure 7.10.

Coatings deposited at 1 W plasma power greatly reduce the number of attached cells ($\geq$ 95%), even after two days of incubation, that is, they show a strong cell repulsive behavior. In contrast, coatings prepared at powers of 5 and 15 W exhibit very good cell adhesion properties and the number of adherent cells is similar to the control. After two days of incubation, cell proliferation continues to increase, which indicates that these coatings are not cytotoxic. In order to find out whether the antifouling character of the 1 W coating was related to a cytotoxic effect, the supernatant was removed after one day of incubation and seeded on nontreated Petri dishes. Then, after one day, the cells were counted. The increase of the number of cells after this procedure confirms the non-cytotoxicity of these polymer coatings. These results confirm that there is a strong correlation between the PEO character of the films and their cell repulsive/adhesive properties. For pure discharges of DEGDME, the results are similar, that is, only coatings produced at 1 W exhibit good cell repellent properties. Furthermore, a comparison of protein

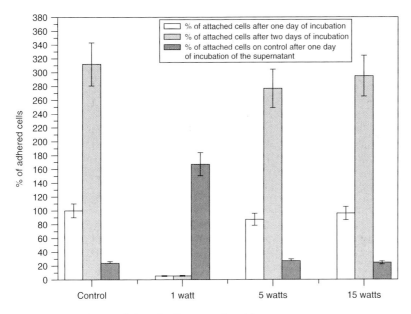

Figure 7.10 Evolution of the cell adhesion and proliferation on PEO-like films at various plasma powers. (Mixture of DEGDME and Ar, 30 Pa, deposition time = 30 min.)

and cell adsorption shows that the study of the adsorption of only one type of protein is not sufficient for predicting the cell-adhesive behavior and further sets of experiments are required in order to clearly distinguish the role of the protein adsorption in the cell adhesion process. Indeed, films produced at 1–5 W lead to a strong reduction of the protein adsorption (> 95%) but give rise to a very different behavior of cells. These results show that a minimum of 70% of PEO character should be reached in order to have a good cell repulsive effect in PEO-like coatings produced by plasma.

Finally, the anti-adhesive properties of the coatings were confirmed by direct observation of the cells at the surface of the coatings by optical microscopy (Figure 7.11). These photos reveal different cell morphologies after 24 hours of incubation for coatings produced at low and high power. In the case of the coating deposited at 1 W, isolated cells or cell clusters without attachment could be observed on the surface. The cells have adopted a globular shape, which minimizes the cell/substrate interaction. In contrast, coatings produced at 15 W exhibit a strong cell/substrate adhesion.

It is also important to know the minimum thickness of the coating which is necessary to obtain cell repulsive properties and the effect of the thickness on the cell repulsive properties. For this purpose, coatings have been produced at 1 W, but with different deposition times, leading to layer thicknesses between 5 and 40 nm. Their cell repellent properties have been studied by analyzing cell growth. The

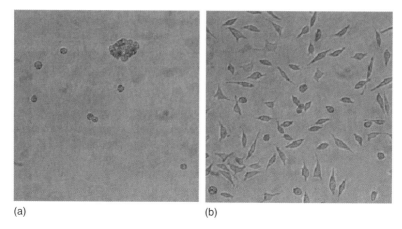

(a)   (b)

Figure 7.11 (a) Optical image of the cell repulsive effect of the PEO-like coating produce at 1 W. (Mixture of DEGDME and Ar, 30 Pa, deposition time = 30 min.) (b) Deposition power 15 W. Image sizes: 550 µm × 550 µm.

results obtained this way reveal that coatings with a thickness of more than 10 nm are cell repellent whereas thinner coatings allow cell attachment on the surface.

7.4.2.4 Aging

The aging of the produced coatings has been studied as a function of the storage time. For this purpose, the samples were stored under ambient conditions in darkness by wrapping them with aluminum foil. For this purpose, a series of coatings has been deposited under identical conditions, and for four months their cell repellent properties have been tested in monthly intervals. Under these storage conditions the samples did not show aging effects, which also holds true for samples prepared in pulsed plasma mode (see below).

7.4.3
Plasma Polymerization in Pulsed Mode

Only films with a high level of PEO-like character (>70%) exhibit good protein and cell repellent properties, as discussed in the previous section. For that purpose it is necessary to employ very low power plasma (typically 1 W). The control of such a low power is difficult in plasma deposition and hence the process does not have the robustness necessary for an industrial application. In order to reach a high level process control, pulsed plasma deposition has also been studied. In a pulsed plasma discharge the RF power is modulated by varying the time when the plasma is on (t_{on}) and switched off (t_{off}), typically in the range of milliseconds. The modulation of the plasma is defined by the duty cycle (DC), which is the relationship between t_{on}, t_{off}, and the pulse duration ($t_{on} + t_{off}$), that is, DC = ($t_{on}/(t_{on} + t_{off})$). The equivalent power (P_{eq}) injected during the plasma treatment is given by

Table 7.1 Summary of the Properties of PEO-like Coatings Deposited at Different Pulsed Plasma Conditions Using (i) a Mixture of 15% DEGDME/85% Ar or (ii) Pure DEGDME (pressure = 30 Pa, Deposition Time = 80 min for (i) and (ii)).

t_{on} (ms)	t_{off} (ms)	DC	P_{peak} (W)	% PEO-like character determined by XPS	Cell adhesion
(i) Mixture of 15 % DEGDME in Ar					
1	200	0.5%	5	52	Similar to control
3	27	10%	5	61	Similar to control
5	45	10%	5	64	Similar to control
10	**100**	**10%**	**5**	**72**	**Repellent**
10	100	10%	10	58	Similar to control
10	100	10%	20	54	Similar to control
10	200	5%	5	64	Similar to control
(ii) Pure DEGDME					
10	190	5%	5	66	Similar to control
3	27	10%	5	73	Repellent
5	45	10%	5	68	Repellent
10	90	10%	5	72	Repellent
20	180	10%	5	71	Repellent
10	40	20%	5	73	Repellent
10	20	33%	5	68	Similar to control

$P_{eq} = DC \times P_{peak}$ where P_{peak} is the applied power during t_{on}. Thus, by using low values of DC, high values of P_{peak} can be used while the equivalent power remains relatively low.

During the 'plasma on' phase the processes at the surface are driven by the reactive species (ions, radicals, electrons, etc.) and the substrate is subjected to a combination of surface modification, deposition, and ablation. On the other hand, in the 'plasma off' phase, the density of the reactive species decreases strongly. However, since the radicals, having lifetimes from a few milliseconds to several seconds, are remaining, they are able to continue a kind of classical polymerization reaction in the gas phase and at the surface.

Table 7.1 shows the results obtained for several deposition conditions in pulsed mode for a mixture of DEGDME and Ar as well as for pure DEGDME. The XPS measurements have been carried out after submersion of the samples in ultrapure water for 12 hours in order to check the stability of the coatings in aqueous media. The PEO-like character of the film is evaluated from the C−O component (286.5 eV) of the C_{1s} spectra.

The PEO-like content decreases when increasing the P_{peak} as a result of more pronounced monomer fragmentation and competitive surface deposition/ablation effects. Furthermore, even at constant and low P_{peak}, the PEO-like character of the film can be different and depends on the DC values as well as on the t_{on} and t_{off} periods. When t_{on} is less than 10 ms or t_{off} is around 200 ms and the deposition is done from a mixture of DEGDME and Argon, coatings which do not have enough retention of PEO-like character are produced. These findings are due to the fact that during t_{on} the plasma discharge is ignited by a high initial voltage, followed by a moderate operating voltage. If there are enough remaining charge carriers the plasma can be easily re-ignited.

However, if t_{off} is too long (200 ms) most of the charge carriers are lost, such that each plasma pulse requires a high ignition voltage similar to continuous wave plasma which counteracts the desired effects of soft plasma deposition. Also, t_{off} < 100 ms favors the ablation process more than the polymerization. The best conditions to reproducibly obtain coatings with good nonfouling properties were found in a pulsed plasma discharge of a mixture of DEGDME and Ar with t_{on} and t_{off} 10 and 100 ms, respectively, at a nominal power of 5 W. Moreover, the PEO-like coatings obtained under these conditions exhibit a rather strong adhesion to the substrate, even after sonication in acetone. Furthermore, ellipsometry and XPS measurements of the samples after immersion in water for up to 30 days show that the coating does not swell or undergo chemical changes. For deposition from pure DEGDME at a nominal power of 5 W a DC of 10% (with little dependence on t_{on}) was found to be the most suitable for producing coatings with good antifouling properties.

7.4.4
Sterilization of PEO-like Films

If PEO-like films are to be used as a coating layer on medical devices or on tools for surgery they have to withstand sterilization processes. This sterilization step is crucial, because bacterial contamination of biomaterial surfaces during surgery is the primary cause of implant related infections [86]. However, previous studies have shown that sterilization can affect the surface properties of polymers [193–195]. It is therefore important to know whether sterilization processes change the surface chemistry and/or affect the bioadhesive properties of thin plasma polymerized PEO coatings.

Therefore, chemical modifications of the surface and changes in the bioadhesive properties of PEO coatings deposited by pulsed plasma discharge from pure DEGDME ($t_{on} = 10$ ms, $t_{off} = 100$ ms, nominal power = 5 W) with a thickness of ~20 nm as described in the previous section were analyzed after application of the most common sterilization and decontamination processes [196]:

- autoclaving (i.e., steam sterilization for 20 minutes at 121 °C)
- gamma ray irradiation (normal and double dose, 35 and 70 kGy)
- ethylene oxide (ETO) treatment (normal cycle with simple and double dose and low temperature cycle)
- treatment by two types of Ar/H$_2$ low pressure plasma: (i) inductively coupled plasma (ICP) [197], operated in 20 : 1 Ar/H$_2$ mixture at a pressure of 10 Pa and 50 W deposited power and (ii) a microwave (MW) plasma reactor [198], operating

with a 90 : 10 mixture of Ar and hydrogen at a pressure of 15 Pa and at total gas flow of 100 sccm. The surfaces were treated for 30 seconds.

It has been discussed earlier that plasma treatment can be used to destroy bacterial spores [195] or to inactivate bacterial endotoxins [197, 198]. The chemical modifications of the surface induced by these techniques were analyzed by XPS, whereas the bioadhesive properties were studied using L929 fibroblast cells as model.

Sterilization of the films by autoclaving, ETO and γ-ray treatment lead to only slight modifications of the surface chemistry of the layer, with a reduction of both the oxygen content and the PEO-like character, and some reticulation of the layer [196]. However, these modifications do not alter the bioadhesive properties of the plasma polymerized film, which remains cell repellent. In contrast, it was observed that the applied low pressure plasma sterilization techniques are not suitable for sterilization of this class of thin coating since they lead to etching as well as chemical modification of the layer and, as a consequence, to a loss of the initial cell repellent properties of the coating [196]. These results show that plasma polymerization technology is suitable for the production of robust nonadhesive thin films that can be sterilized by two standard methods.

7.4.5
Composite Films: Ag Nanoparticles in a PEO-like Matrix

For certain biomedical applications, such as medical devices and textiles for wound dressings, it is desirable that the coatings should not only show an antifouling effect but also show an antibacterial effect. The PEO-like coatings discussed in the previous section show good antifouling (protein and cell repellent) properties, but have no antibacterial properties. It is, therefore, desirable to further engineer the PEO-like films toward those properties.

Silver is known to have extraordinary inhibitory and bactericidal properties since ancient times. While being relatively nontoxic to human cells, silver possesses antibacterial properties for a broad spectrum of bacterial strains that are found in industrial processes as well as in the human body [199–201]. However, Ag ions or salts have only limited usefulness as an antimicrobial agent for several reasons, including interfering effects of salts and the requirement for a sufficiently high release of Ag ions from the metal form. These kinds of limitations are thought to be overcome by the use of Ag NPs [202]. In fact, Ag NPs have been tested in various fields of biological sciences, for example, drug delivery, wound treatment, binding with HIV gp120 protein [203], in water treatment and as an antibacterial compound against both Gram-positive and Gram-negative bacteria [204–210].

Therefore, samples with plasma-deposited PEO were treated to achieve antibacterial properties by incorporation of silver NPs in addition to anti-adhesive properties with regard to proteins.

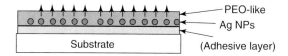

Figure 7.12 Sketch of a sample with a composite coating, consisting of an adhesive layer, Ag nanoparticles, and a PEO-like cover layer.

7.4.5.1 Synthesis of Ag Nanoparticles and Deposition on Surfaces

It has been demonstrated earlier that PEO-like films and Ag NPs can be deposited at the same time using dedicated equipment and under careful control of the deposition conditions to fabricate a PEO/Ag nanocomposite [211]. On the other hand, it is also advantageous to fabricate Ag NP/PEO-like coatings by using a strategy that allows adjusting the nature of the coating and to synthesize and deposit the Ag NPs independently from each other. The latter strategy, which was followed here, consists of several fabrication steps:

- Deposition of an adhesive layer, which allows adhesion of the Ag NPs independent of the substrate. This step can be omitted if the NPs already adhere to the substrate material itself.
- Fabrication of the NPs using chemical processes that allow adjusting the size of the NPs according to the needs of the final coating.
- Deposition of the NPs from liquid suspension. This allows controlling the density and size of the NPs by using stock solutions of different NPs at selected dilutions. Furthermore, this approach is suitable to deposit NPs also on non-flat surfaces, such as tools or textile fabrics.
- Plasma deposition of the coating of interest, which allows controlling the surface chemistry.

The structure of such a composite coating is sketched in Figure 7.12. The rate of diffusion of Ag out of the system, and therefore the antibacterial effect, will finally depend on the thickness of the coating layer. The thickness of the coating will also influence the time scale on which the coating is antibacterially active.

A good method to deposit Ag NPs on the surface without creating big aggregates and to achieve a homogeneous distribution of stably adherent Ag NPs on the surface is to synthesize the NPs by wet chemistry, and then to deposit them on the surface from a colloidal solution. Following this approach, Ag NPs of two different sizes were synthesized using the polyol process [212]. To obtain a colloidal solution of small, stable, and monodispersed Ag particles, poly(vinyl pyrrolidone) (PVP, average molecular weight 10 kDa), and $AgNO_3$ salt were dissolved in ethylene glycol at room temperature, and the solution was heated to 120 °C at a heating rate of 1 °C per minute. The particles were redispersed in ethanol, and analyzed by dynamic light scattering. The Ag NPs had a medium diameter of 10 nm with a narrow size distribution.

Larger Ag NPs with a mean diameter of 80 nm were prepared using the injection method. PVP (55 kDa) and Ag salt were dissolved in ethylene glycol at room temperature, and another solution of PVP and ethylene glycol was prepared in parallel. These two solutions were simultaneously injected into 160 °C hot ethylene glycol at a rate of 0.375 ml min^{-1}. After an aging time of 46 hours and washing in acetone the solution was redispersed in ethanol.

Aqueous dispersions of these NPs were first tested for their antibacterial activity, using the macrodilution broth method on *S. aureus* AATCC 6538 bacteria. It turned out that dispersions of small NPs (10 nm, stock concentration 6.35 gl^{-1} stock particle concentration 1.155×10^{18} particles per liter) had a clear antibacterial effect in solution. They were able to reduce the concentration of bacterial colony forming units (CFUs), starting from $\sim 10^5$ to 3.7×10^3 CFU ml^{-1} after 24 hours contact time at a 20-fold dilution from stock solution, and there was a reduction below the counting limit for the twofold dilution of Ag NP dispersion. Starting from $\sim 10^7$ CFU ml^{-1}, the concentration was also reduced below the counting limit by the twofold dilution. On the other hand, dispersions of large (80 nm) NPs at comparable mass concentrations had only a minor antibacterial effect in comparison to that observed for the 10 nm NPs. Therefore, for depositions only dispersions with Ag NPs of 10 nm mean diameter were used.

For the characterization of the general properties of the composite layers, Thermanox coverslips were used as substrates to test the adhesion of the Ag NPs and their distribution on the surface. These coverslips were dipped into Ag NP dispersions and pulled out again under controlled speed. Atomic force microscopy measurements made after drying of the solvent showed that a homogeneous distribution of adhering Ag NPs can be achieved. The particles are well dispersed and there is a low concentration of larger aggregates (Figure 7.13).

These samples were also analyzed by XPS to determine their Ag coverage. Comparison of the XPS spectra of the substrate before dipping into the Ag NP solution and after dipping (Figure 7.13b) indicates the deposition of Ag by the dipping process, which, together with the AFM images, proves the deposition of Ag NPs.

The results also show that (i) the resulting surface concentration of the Ag NPs depends on the concentration of the Ag NPs in the solution and (ii) there is a direct correlation between the results obtained with the two characterization methods (Figure 7.13c).

7.4.5.2 Composite AgNP/PEO Surfaces and Their Antibacterial Activity

Preliminary tests of the antibacterial properties of the combined Ag NP/PEO-like samples were made with selected samples, and carried out by the Acteco partner Biomatech. Those tests indicated a slight to moderate antibacterial activity of the coatings, which were also able to reduce biofilm formation. This is illustrated with the following two examples.

The first example shows the reduction of bacterial adhesion. The samples were prepared by dipping the PEO-covered samples into Ag NP dispersions of (i) 25-fold and (ii) 50-fold dilution from stock. After that, they were covered by a ~ 10 nm thick PEO-like film by plasma-assisted chemical vapor deposition (CVD), so that the samples had the following sequence of layers: PEO/AgNP(25 × or 50 ×)/

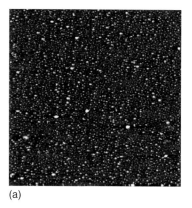

(a)

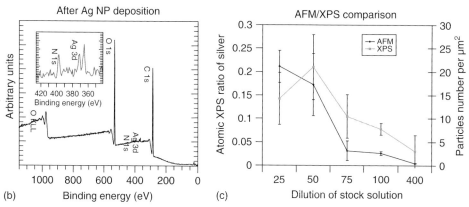

(b)   Binding energy (eV)   (c)   Dilution of stock solution

Figure 7.13 (a) AFM image (15 μm × 15 μm) of Ag nanoparticles deposited on the surface of a Thermanox slide. (b) XPS spectrum after deposition of Ag nanoparticles. (c) Correlation between results obtained by AFM and XPS.

PEO/Thermanox. Bacterial adhesion of *S. epidermis* RP62A (ATCC 35984) was tested after 2 and 24 hours of contact. The samples were inoculated with 3×10^3 CFU, incubated for the desired time and compared with nontreated Thermanox slides. The results of these experiments are shown in Figure 7.14a.

The second example shows the reduction of biofilm formation. For these samples the composite Ag NP/PEO-like coatings were prepared on poly(acrylic acid) (PAA) covered Thermanox slides. The PAA film served as adhesive layer for the Ag NPs. The Ag NPs were deposited by dipping the samples into (i) fivefold and (ii) 10-fold dilution from stock. After that, they were covered by a ~10 nm thick PEO-like film using plasma-assisted CVD, as described earlier, so that the layer sequence was PEO/AgNP(5 × or 10 ×)/PAA/Thermanox. Biofilm formation of *S. epidermis* RP62A (ATCC 35984) was tested after 72 hours contact time. The results on biofilm formation on those samples are shown in Figure 7.14b.

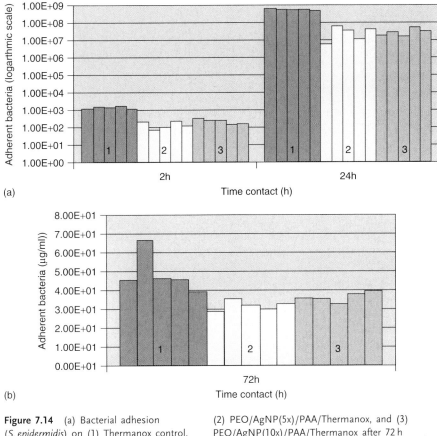

(a)

(b)

Figure 7.14 (a) Bacterial adhesion (*S. epidermidis*) on (1) Thermanox control, (2) PEO/AgNP(25x)/PEO/Thermanox, and (3) PEO/AgNP(50x)PEO/Thermanox after 2 and 24 h contact time. (b) Biofilm formation (*S. epidermidis*) on (1) Thermanox control, (2) PEO/AgNP(5x)/PAA/Thermanox, and (3) PEO/AgNP(10x)/PAA/Thermanox after 72 h contact time. Each test was done on five samples, represented by five bars of each sample group.

Regarding bacterial adhesion, a significant difference was determined between the control samples and the Ag treated Thermanox after 2 and 24 hours of contact. Both treatments showed statistically relevant anti-adhesion properties, and the Mann and Whitney analysis gave a value of $p = 0.008$. The bacterial adhesion was not significantly different between the two sets of samples. The samples tested for biofilm reduction showed a statistically significant reduction of biofilm formation at 72 hours of between 26 and 35% with respect to Thermanox control samples.

These results can only serve as starting point for further, more detailed studies of the antibacterial effect of composite Ag NP and PEO-like coatings. Particularly, the antibacterial efficacy has to be improved by several orders of magnitude and it also has to be tested for other species of bacteria before this type of coating can be used for biomedical purposes.

7.5
Summary

Biofouling and more specifically, 'biofilm formation' on biomaterial surfaces is a very complex process involving a number of physical, chemical, and biological processes, which are in turn governed by various parameters, such as the nature of the biomaterial surface, the type of micro-organism, and the contacting biological medium. Therefore, it is not easy to propose a general theory of bacterial adhesion, which leads to the formation of a biofilm, despite the large amount of research work that has been done to understand the mechanisms of bacterial adhesion and biomaterial infection. Moreover, some reports with contradictory results on the influence of different factors on the adhesion behavior of micro-organism further complicate the subject. The lack of experimental consistency is partly due to the great variation in experimental design with respect to the nature of the bacteria, the substrates, sterility, contacting media, and other experimental conditions. However, advances in the understanding of biofilm formation, coupled with the surface engineering of biomaterials, provide different potential strategies to prevent or significantly reduce biofilm infections in biomedical devices. Some of these prevention strategies include deposition of antimicrobial-releasing coatings, generation of anti-adhesive coatings, tethered antimicrobial surfaces, and impairing the biological processes involved in biofilm formation. Plasma processing technology, which has been used effectively for surface modification of material surface with great success, plays an important role in development of antifouling material surfaces.

As a case study, deposition of PEO-like coatings by plasma polymerization of DEGDME under low pressure was studied in detail as a function of the plasma parameters in a dedicated plasma reactor. Fouling and nonfouling surfaces have been produced by RF plasma polymerization of DEGDME monomer by varying the injected power. For a given precursor concentration and a constant pressure, the plasma power plays a crucial role for the properties of the coatings. For low power (1 W), high retention of the monomer precursor chemistry was achieved, which leads to the formation of coatings with a very good protein and cell repulsive behavior. However, higher plasma power (≥ 5 W) gives rise to a significant fragmentation of the monomer precursor and leads to coatings with a low concentration of PEO groups and so promote cell adhesion and growth. Moreover, PEO-like coatings exhibit a high stability toward water and acetone. Plasma polymerization in pulsed mode with a DC of 10% in pure DEGDME has been found to be very efficient for producing antifouling coatings. Sterilization of PEO-like films by autoclaving, ETO and γ-ray treatment lead to slight modifications of the surface chemistry of the layer, but does not alter its cell-repellent properties. In contrast, low pressure plasma sterilization is not suitable for sterilization of PEO-like coating since it leads to both etching and chemical modification of the layer and to a loss of the initial cell-repellent properties.

Plasma polymerization technology is suitable for the production of robust nonadhesive thin films that can be sterilized by several standard methods, which makes this technique very promising for the coating of surfaces for biomaterials applications. For instance, composite coatings consisting of 10 nm Ag NPs in PEO-like matrix showed an antibacterial effect toward *staphylococcus epidermis* and are able to reduce biofilm formation.

References

1. Hall-Stoodley, L., Costerton, J.W., and Stoodley, P. (2004) *Nat. Rev. Microbiol.*, **2**, 95.
2. Costerton, J.W., Cheng, K.J., Geesey, G.G., Ladd, T.I., Nickel, J.C., Dasgupta, M., and Marrie, J.T. (1987) *Annu. Rev. Microbiol.*, **41**, 435.
3. Mah, T.F. and O'Toole, G.A. (2001) *Trends Microbiol.*, **9**, 34.
4. European Centre for Disease Prevention and Control, Annual Epidemiological Report on Communicable Diseases in Europe, *http://ecdc.europa.eu/en/publications/ Publications/0706_SUR_Annual_ Epidemiological_Report_2007.pdf*
5. Bryers, J.D. (2008) Medical biofilms, *Biotechnol. Bioeng*, **100**, (1) 1–18.
6. An, Y.H. and Friedmann, R.J. (1998) *J. Biomed. Mater. Res.*, **43**, 338.
7. Marshall, K.C. (1985) Mechanisms of Bacterial Adhesion at Solid-Water Interfaces in *Bacterial Adhesion: Mechanisms and Physiological Significance* (eds D.C. Savage and M. Fletcher), Plenum Press, New York, p. 133.
8. Yang, J., Bos, R., Belder, G.F., Engel, J., and Busscher, H.J. (1999) *J. Colloid Interface Sci*, **220**, 410.
9. Marshall, K.C., Stout, R., and Mitchell, R. (1971) *J. Gen. Microbiol*, **68**, 337.
10. Van Loosdrecht, M.C.M., Lyklema, J., Norde, W., Schroa, G., and Zehnder, A.J.B. (1987) *Appl. Environ. Microbiol*, **53**, 1898.
11. An, Y.H., Dickinson, R.B., and Doyle, R.J. (2000) Mechanisms of Bacterial Adhesion and Pathogenesis of Implant and Tissue Infections in *Handbook of Bacterial Adhesion: Principles, Methods, and Applications* (eds Y.H. An and R.J. Friedman), Humana Press Inc., Totowa, p. 1.
12. Heilmann, C., Schweitzer, O., Gerke, C., Vanittanakom, N., Mack, D., and

13. Götz, F. (1996) *Mol. Microbiol*, **20**, 1083–1091.
14. Busscher, H.J. and Weerkamp, A.H. (1987) *FEMS Microbiol. Rev*, **46**, 165.
15. Nikolaev, Y.A. and Plakunov, V.K. (2007) *Microbiology*, **76**, 125.
16. Dickson, J.S. and Koohmarare, M. (1989) *Appl. Environ. Microbiol*, **55**, 832.
17. Kumar, C.G. and Anand, S.K. (1998) *Int. J. Food Microbiol*, **42**, 9.
18. Bjarnsholt, T. and Givskov, M. (2007) *Philos. Trans. R. Soc. Lond. B Biol. Sci.*, **362**, 1213.
19. Davies, D.G., Parsek, M.R., Pearson, J.P., Iglewski, B.H., Costerton, J.W., and Greenberg, E.P. (1998) *Science*, **280**, 295.
20. Hodgkinson, J.T., Welch, M., and Spring, D.R. (2007) *ACS Chem. Biol*, **2**, 715.
21. Otto, M. (2006) *Curr. Top. Microbiol. Immunol*, **306**, 251.
22. Davies, D.G. and Geesey, G.G. (1995) *Appl. Environ. Microbiol*, **61**, 860.
23. Webb, J.S., Thompson, L.S., James, S., Charlton, T., Tolker-Nielsen, T., Koch, B., Givskov, M., and Kjelleberg, S. (2003) *J. Bacteriol*, **185**, 4585.
24. Vladkova, T. (2007) *J. Univ. Chem. Technol. Metall*, **42**, 239.
25. Clarke, S.R. and Foster, S.J. (2006) *Adv. Microbiol. Physiol*, **51**, 187.
26. An, Y.H. and Friedman, J.R. (1996) *J. Hosp. Infect*, **33**, 93.
27. Brokke, P., Dankert, J., Carballo, J., and Feijen, J.J. (1991) *Biomater. Appl*, **5**, 204.
28. An, Y.H., Stuart, G.W., McDowell, S.J., McDaniel, S.E., Kang, Q., and Friedman, R.J.J. (1996) *Orthop. Res*, **14**, 846.
29. Kuusela, P., Vartio, T., Vuento, M., and Myhre, E.B. (1985) *Infect. Immunol*, **50**, 77.

29. Naylor, P.T., Ruch, D., Brownlow, C., Webb, L.X., and Gristina, A.G. (1989) *Trans. Orthop. Res. Soc*, **14**, 561.

30. Espersen, F., Frimodt-Moller, N., Corneliussen, L., Rosdahl, V.T., and Clemmensen, I. (1990) *Acta Pathol. Microbiol. Immunol. Scand*, **98**, 471.

31. Rickard, A.H., Gilbert, P., High, N.J., Kolenbrander, P.E., and Handley, P.S. (2003) *Trends Microbiol*, **11**, 94.

32. Corpe, W.A. (1974) Periphytic Marine Bacteria and the Formation of Microbial Films on Solid Surfaces in *Effect of the Ocean Environment on Microbial Activity* (eds R. Colwell and R. Morita), University Park Press, Baltimore, p. 397.

33. Chu, C.C. and Williams, D.F. (1984) *Am. J. Surg*, **147**, 197.

34. Fukuzaki, S., Urano, H., and Hagata, K. (1995) *J. Jpn. Soc. Food. Sci. Technol.-Nippon Shokuhin Kagaku Kogaku Kaishi*, **12**, 700.

35. Gristina, A.G., Hobgood, C.D., and Barth, E. (1987) Biomaterial Specificity, Molecular Mechanisms and Clinical Relevance of *S. Epidermidis and S. Aureus* Infections in Surgery in *Pathogenesis and Clinical Significance of Coagulase-Negative Staphylococci* (eds G.Pulverer, P.G. Quie, and G. Peters), Gustav Fischer Verlag, Stuttgart, p. 143.

36. Baier, R.E. (1980) Substrate Influences on Adhesion of Microorganisms and Their Resultant New Surface Properties in *Adsorption of Micro-Organisms to Surfaces* (eds G.Bitton, K.S. Marshall), John Wiley & Sons, Inc., New York, USA, p. 59.

37. Weerkamp, A.H., Quirynen, M., Marechal, M., Van der Mei, H.C., Van Steenberghe, D., and Busscher, H.J. (1989) *Microb. Ecol. Health Dis*, **2**, 11.

38. Absolom, D.R., Lamberti, F.V., Policova, Z., Zingg, W., Van Oss, C.J., and Neumann, A.W. (1983) *Appl. Environ. Microbiol*, **46**, 90.

39. Callow, M.E. and Fletcher, R.L. (1994) *Int. Biodeterior. Biodegrad.*, **34**, 333.

40. Norde, W. (2003) Water in *Colloids and Interfaces in Life Sciences* (ed. W.Norde), Marcel. Dekker Inc., New York, p. 47.

41. Balazs, D.J., Triandafillu, K., Chevolot, Y., Aronsson, B.-O., Harms, H., Descouts, P., and Mathieu, H.J. (2003) *Surf. Interface Anal.*, **35**, 301.

42. Brétagnol, F., Lejeune, M., Papadopoulou-Bouraoui, A., Hasiwa, M., Rauscher, H., Ceccone, G., Colpo, P., and Rossi, F. (2006) *Acta Biomater.*, **2**, 165.

43. Holmberg, K., Bergstrom, K., Brink, C., Osterberg, E., Tiberg, F., and Harris, J.M. (1993) *J. Adhes. Sci. Technol.*, **7**, 503.

44. Johnston, E.E., Bryers, J.D., and Ratner, B.D. (1997) *Polym. Prep. ACS Div. Polym. Chem.*, **38**, 1016.

45. Roosjen, A., Van der Mei, H.C., Busscher, H.J., and Norde, W. (2004) *Langmuir*, **20**, 10949.

46. Amanatides, E., Mataras, D., Katsikogianni, M., and Missirlis, Y.F. (2006) *Surf. Coat. Technol.*, **200**, 6331.

47. Katsikogianni, M., Amanatides, E., Mataras, D., and Missirlis, Y.F. (2008) *Colloids Surf., B Biointerfaces*, **65**, 257.

48. Cringus-Fundeanu, I., Luijten, J., Van der Mei, H.C., Busscher, H.J., and Schouten, A.J. (2007) *Langmuir*, **23**, 5120.

49. Husmark, U. and Ronner, U. (1990) *J. Appl. Bacteriol*, **69**, 557.

50. Rijnaarts, H.H.M., Norde, W., Lyklema, J., and Zehnder, A.J.B. (1999) *J. Colloid Interface Sci*, **14**, 179.

51. Boyd, R.D., Verran, J., Jones, M.V., and Bhakoo, M. (2002) *Langmuir*, **18**, 2343.

52. Wenzel, R.N. (1936) *Ind. Eng. Chem*, **28**, 988.

53. Rosales, J.I., Marshall, G.W., Marshall, S.J., Watanabe, L.G., Toledanol, M., Cabrerizo, M.A., and Osoriol, R. (1999) *J. Dent. Res*, **78**, 1554.

54. Taylor, R.L., Verran, J., Lees, G.C., and Ward, A.J. (1998) *J. Mater. Sci. Mater. Med*, **9**, 17.

55. Bollen, C.M.L., Papaioannou, W., Van Eldere, J., Schepers, E., Quirynen, M., and Van Steenberghe, D. (1996) *Clin. Oral Implants Res*, **7**, 201.

56. Merritt, K., Shafer, J.W., and Brown, S.A. (1979) *J. Biomed. Mater. Res*, **13**, 101.

57. Locci, R., Peters, G., and Pulverer, G. (1981) *Zbl. Bakt. Hyg. 1. Abt. Orig. B*, **173**, 285.

58. Quirynen, M., Van der Mei, H.C., Bollen, C.M., Schotte, A., Marechal, M., Doornbusch, G.I., Naert, I., Busscher, H.J., and Van Steenberghe, D. (1993) *J. Dent. Res*, **72**, 1304.

59. Bos, R., Van der Mei, H.C., Gold, J., and Busscher, H.J. (2000) *FEMS Microbiol. Lett*, **189**, 311.

60. Medilanski, E., Kaufmann, K., Wick, L., Wanner, O., and Harms, H. (2002) *Biofouling*, **18**, 193.

61. Edwards, K.J. and Rutenberg, A.D. (2001) *Chem. Geol*, **180**, 19.

62. Morra, M. and Cassinelli, C. (1997) *J. Biomater. Sci. Polym. Ed*, **9**, 55.

63. Sutherland, I.W. (1983) *CRC Crit. Rev. Microbiol*, **10**, 173.

64. Hogt, A.H., Dankert, J., and Feijen, J. (1985) *J. Gen. Microbiol*, **131**, 2485.

65. Jones, G.W. and Isaacson, R.E. (1983) *CRC Crit. Rev. Microbiol*, **10**, 229.

66. Lindahl, M., Faris, A., Wadstrom, T., and Hjerten, S. (1981) *Biochim. Biophys. Acta*, **677**, 471.

67. Klemm, P. and Schembri, M.A. (2000) *Int. J. Med. Microbiol*, **290**, 27.

68. McCrea, K.W., Hartford, O., Davis, S., Eidhin, D.N., Lina, G., Speziale, P., Foster, T.J., and Hook, M. (2000) *Microbiology*, **147**, 1535.

69. Fallgren, C., Utt, M., and Ljungh, A. (2001) *J. Med. Microbiol*, **50**, 547.

70. Bowden, M.G., Visai, L., Longshaw, C.M., Holland, K.T., Speziale, P., and Hook, M. (2002) *J. Biol. Chem*, **277**, 43017.

71. Dankert, J., Hogt, A.H., and Feijen, J. (1986) *CRC Crit. Rev. Biocompat*, **2**, 219.

72. Hogt, A.H., Dankert, J., De Vries, J.A., and Feijen, J. (1983) *J. Gen. Microbiol*, **129**, 1959.

73. Van Loosdrecht, M.C.M., Lyklema, J., Norde, W., Schraa, G., and Zehnder, A.J.B. (1987) *Appl. Environ. Microbiol*, **53**, 1893.

74. Gottenbos, B., Grijpma, D.W., Van der Mei, H.C., Feijen, J., and Busscher, H.J. (2001) *J. Antimicrob. Chemother*, **48**, 7–13.

75. Jucker, B.A., Harms, H., and Zehnder, A.J.B. (1996) *J. Bacteriol*, **178**, 5472.

76. Ahmed, N.A.A.M., Petersen, F.C., and Scheie, A.A. (2008) *Oral Microbiol. Immunol*, **23**, 492.

77. Amano, A., Premaraj, T., Kuboniwa, M., Nakagawa, I., Shizukuishi, S., Morisaki, I., and Hamada, S. (2001) *Oral Microbiol. Immunol*, **16**, 124.

78. Rachid, S., Ohlsen, K., Witte, W., Hacker, J., and Ziebuhr, W. (2000) *Antimicrob. Agents Chemother*, **44**, 3357.

79. Hamadi, F., Latrache, H., Zekraoui, M., Ellouali, M., and Bengourram, J. (2009) *Mater. Sci. Eng. C*. **29**, (4), 1302–1305.

80. Sheng, X., Ting, Y.P., and Pehkonen, S.O. (2008) *J. Colloid Interface Sci*, **321**, 256.

81. Pascual, A., Fler, A., Westerdaal, N.A.C., and Verhoef, J. (1986) *Eur. J. Microbiol*, **5**, 518.

82. Park, S.-J., Lee, C.-G., and Kim, S.-B. (2009) *Colloids Surf., B Biointerfaces*, **68**, 79.

83. Ofek, I., Simpson, W.A., Whitnack, E., and Beachy, E.H. (1985) *Infect. Immunol*, **47**, 341.

84. Satou, N., Satou, J., Shintani, H., and Okuda, K. (1988) *J. Gen. Microbiol*, **134**, 1299.

85. Strachan, C.J.L., (1995) *J. Hosp. Infect*, **30** (Suppl), 54.

86. Arciola, C.R., Alvi, F.I., An, Y.H., Campoccia, D., and Montanaro, L. (2005) *Int. J. Artif. Organs*, **28**, 1119.

87. Eiff, C.V., Jansen, B., Kohnen, W., and Becker, K. (2005) *Drugs*, **65**, 179.

88. Chaiban, G., Hanna, H., Dvorak, T., and Raad, I. (2005) *J. Antimicrob. Chemother*, **55**, 51.

89. Kamal, G.D., Pfaller, M.A., Rempe, L.E., and Jebson, P.J. (1991) *J. Am. Med. Assoc*, **265**, 2364.

90. Raad, I., Darouiche, R., Hachem, R., Mansouri, M., and Bodey, G.P. (1996) *J. Infect. Dis*, **173**, 418.

91. Jansen, B., Jansen, S., Peters, G., and Pulverer, G. (1992) *J. Hosp. Infect*, **22**, 93.

92. Ahearn, D.G., Grace, D.T., Jennings, M.J., Borazjani, R.N., Boles, K.J., Rose, L.J., Simmons, R.B., and Ahanotu, E.N. (2000) *Curr. Microbiol*, **41**, 120.

93. Maki, D.G., Stolz, S.M., Wheeler, S., and Mermel, L.A. (1997) *Ann. Intern. Med*, **127**, 257.

94. Shah, A., Mond, J., and Walsh, S. (2004) *Antimicrob. Agents Chemother*, **48**, 2704.

95. Pai, M.P., Pendland, S.L., and Danziger, L.H. (2001) *Ann. Pharmacother*, **35**, 1255.

96. Donelli, G. and Francolini, I. (2001) *J. Chemother*, **13**, 595.

97. Bagge, N., Schuster, M., Hentzer, M., Ciofu, O., Givskov, M., Greenberg, E.P., and Høiby, N. (2004) *Antimicrob. Agents Chemother*, **48**, 1175.

98. Hoffman, L.R., D'Argenio, D.A., MacCoss, M.J., Zhang, Z., Jones, R.A., and Miller, S.I. (2005) *Nature*, **436**, 1171.

99. Ikada, Y., Suzuki, M., and Tamada, Y. (1984) Polymer Surfaces Possessing Minimal Interaction with Blood Components in *Polymers as Biomaterials* (eds W.S. Shalaby, A.S. Hoffman, B.D. Ratner, and T.A. Herbett), Plenum Press, New York, p. 135.

100. Baier, R.E. (1973) Influence of the Initial Surface Condition of Materials on Bioadhesion in *Proceedings of the 3rd International Congress on Marine Corrosion and Fouling* (eds R.F.Acker, B.F. Brown, J.R. DePalma, W.P. Iverson), Northwestern University Press, Evanston, p. 633.

101. Dexter, S.C. (1979) *J. Colloid Interface Sci*, **70**, 346.

102. Baier, R.E. (2006) *J. Mater. Sci: Mater. Med*, **17**, 1057.

103. Hoffman, A.S. (1999) *J. Biomater. Sci. Polym. Ed*, **10**, 1011.

104. Sherman, M.R., Williams, L.D., Saifer, M.G.P., French, J.A., Kwak, L.W., and Oppenheim, J.J. (1997) Conjugation of High-Molecular Weight Poly(ethylene glycol) to Cytokines: Granulocyte-Mocrophage Colony-Stimulating factors as Model Substrates in *Poly(ethylene glycol) Chemistry and Biological Applications* (eds J.M. Harris and S. Zalipsky), Am. Chem. Soc., Washington, DC, p. 155.

105. Lee, J.H., Kopecek, J., and Andrade, J.D. (1989) *J. Biomed. Mater. Res*, **23**, 351.

106. Nagaoka, S. and Nakao, A. (1990) *Biomaterials*, **11**, 119.

107. Morra, M. (2000) *J. Biomater. Sci. Polym. Ed*, **11**, 547.

108. Malmsten, M., Emoto, K., and Van Alstine, J.M. (1998) *J. Colloid Interface Sci*, **202**, 507.

109. Szleifer, I. (1997) *Biophys. J*, **72**, 595.

110. Kingshott, P. and Griesser, H.J. (1999) *Curr. Opin. Colloid Interface Sci*, **4**, 403.

111. Hermans, J. (1982) *J. Chem. Phys*, **77**, 2193.

112. Ostuni, E., Chapman, R.G., Holmlin, R.E., Takayama, S., and Whitesides, G.M. (2001) *Langmuir*, **17**, 5605.

113. Quirynen, M. (1994) *J. Dent*, **22** (Suppl), S13.

114. Thorpe, A.A., Nevell, T.G., and Tsibouklis, J. (1999) *Appl. Surf. Sci*, **137**, 1.

115. Vrolijk, N.H., Targett, N.M., Baier, R.E., and Baier, A.E. (1990) *Biofouling*, **2**, 39.

116. Busscher, H.J., Sjollema, J., and Van der Mei, H.C. (1990) Relative Importance of Surface free Energy as a Hydrophobicity Measure in Bacterial Adhesion to solid Surfaces-Observations with *Streptococcus sanguis* in *Microbial Cell Surface Hydrophobicity* (eds R.J.Doyle and M. Rosenberg), American Society for Microbiology, Washington, DC p. 335.

117. Everaert, E.P.J.M., Van der Mei, H.C., and Busscher, H.J. (1998) *Colloids Surf., B*, **10**, 179.

118. Bakker, D.P., Huijs, F.M., De Vries, J., Klijnstra, J.W., Busscher, H.J., and Van der Mei, H.C. (2003) *Colloids Surf., B*, **32**, 179.

119. Zhao, Q., Wang, S., and Müller-Steinhagen, H. (2004) *Appl. Surf. Sci*, **230**, 371.

120. Meyer, A. *et al.* (2006) *Biofouling*, **22**, 411.

121. Quirynen, M., Marechal, M., Busscher, H.J., Weerkamp, A., Arends, J., and Darius, P.L. (1989) *J. Dent. Res*, **68**, 796.

122. Everaert, E.P.J.M., Mahieu, H.F., Van De Belt-Gritter, B., Peeters, A.J.G.E., Verkerke, G.J., Van der Mei, H.C., and Busscher, H.J. (1999) *Arch.*

Otolaryngol. Head Neck Surg, **125**, 1329.

123. Kumar, V., Bhardwaj, Y.K., Rawat, K.P., and Sabharwal, S. (2005) *Radiat. Phys. Chem*, **73**, 175.

124. Ohta, Y., Kondo, Y., Kawada, K., Teranaka, T., and Yoshino, N. (2008) *J. Oleo Sci*, **57**, 445.

125. Kanazawa, A., Ikeda, T., and Endo, T. (1993) *J. Polym. Chem. A: Polym Chem*, **31**, 3031.

126. Akihiko, K., Ikeda, T., and Endo, T. (1993) *J. Polym. Sci. A: Polym. Chem*, **31**, 335.

127. Gottenbos, B., Van der Mei, H.C., Klatter, F., Nieuwenhuis, P., and Busscher, H.J. (2002) *Biomaterials*, **23**, 1417.

128. Yao, C., Li, X., Neoh, K.G., Shi, Z., and Kang, E.T. (2008) *J. Membr. Sci*, **320**, 259.

129. Van der Mei, H.C., Rustema-Abbing, M., Langworthy, D.E., Collias, D.I., Mitchell, M.D., Bjorkquist, D.W., and Busscher, H.J. (2008) *Biotechnol. Bioeng*, **99**, 165.

130. Majumdar, P. *et al.* (2008) *Biofouling*, **24**, 185.

131. Chen, S., Yu, F., Yu, Q., He, Y., and Jiang, S. (2006) *Langmuir*, **22**, 186.

132. Zhang, Z., Chen, S., Chang, Y., and Jiang, S. (2006) *J. Phys. Chem. B*, **110**, 10799.

133. Shi, Q., Su, Y., Zhao, W., Li, C., Hu, Y., Jiang, Z., and Zhu, S. (2008) *J. Membr. Sci*, **319**, 271.

134. Cheng, G., Zhang, Z., Chen, S., Bryers, J.D., and Jiang, S. (2007) *Biomaterials*, **28**, 4192.

135. Li, G., Cheng, G., Xue, H., Chen, S., Zhang, F., and Jiang, S. (2008) *Biomaterials*, **29**, 4592.

136. Danese, P.N., Pratt, L.A., Dove, S.L., and Kolter, R. (2000) *Mol. Microbiol*, **37**, 424.

137. Cachia, P.J. and Hodges, R.S. (2003) *Biopolymers*, **71**, 141.

138. Otto, M. (2004) *FEMS Microbiol. Lett*, **241**, 135.

139. Kaneko, Y., Thoendel, M., Olakanmi, O., Britigan, B.E., and Singh, P.K. (2007) *J. Clin. Invest*, **117**, 877.

140. Boles, B.R., Thoendel, M., and Singh, P.K. (2005) *Mol. Microbiol*, **57**, 1210.

141. Gyimesi, E., Bankovich, A.J., Schuman, T.A., Goldberg, J.B., Lindorfer, M.A., and Taylor, R.P. (2004) *Immunol. Lett*, **95**, 185.

142. Lieberman, M.A. and Lichtenberg, A. J. (eds) (1994) *Principles of Plasma Discharges and Materials Processing*, John Wiley & Sons, Inc., New York.

143. Arefi-Khonsari, F. and Tatoulian, M. (2007) Plasma Processing of Polymers by a Low-frequency Discharge with Asymmetrical Configuration of Electrodes in *Advanced Plasma Technology* (eds R. d'Agostino, P. Favia, H. Ikegami, Y. Kawai, N. Sato, and F. F. Arefi-Khonsari), Weinheim, p. 137.

144. Gheorgiu, M.F.A.-K., Amouroux, J., Placinta, G., Popa, G., and Tatoulian, M. (1997) *Plasma Sources Sci. Technol*, **6**, 1.

145. Rejeb Ben, S., Tataoulian, M., Arefi-Khonsari, F., Fischer-Durand, N., Martel, A., Lawrence, J.F., Amouroux, J., and Le Goffic, F. (1998) *Anal. Chim. Acta*, **376**, 133.

146. Jafari, R., Arefi-Khonsari, F., Tatoulian, M., Le Clerre, D., Talini, L., and Richard, F. (2009) *Thin Solid Films*. **517** (19), 5763–5768 doi: 10.1016/j.tsf.2009.03.217.

147. Gomathi, N., Sureshkumar, A., and Neogi, S. (2008) *Curr. Sci*, **94**, 1478.

148. Poncin-Epaillard, F. and Legeay, G. (2003) *J. Biomater. Sci. Polym. Ed*, **14**, 1005.

149. Canal, C., Gaboriau, F., Villeger, S., Cvelbar, U., and Ricard, A. (2009) *Int. J. Pharm*, **367**, 155.

150. Jiang, H., Manolache, S., Wong, A.C., and Denes, F.S. (2006) *J. Appl. Polym. Sci*, **102**, 2324.

151. Hanein, Y., Pan, Y.V., Ratner, B.D., Denton, D.D., and Bohringer, K.F. (2001) *Sens. Actuators B*, **81**, 49.

152. Hendricks, S.K., Kwok, C., Mingchao Shen, T.A., Horbett, T.A., Ratner, B.D., and Bryers, J.D. (2000) *J. Biomed. Mater. Res*, **50**, 160.

153. Sardella, E., Favia, P., Gristina, R., Nardulli, M., and d'Agostino, R. (2006) *Plasma Process. Polym*, **3**, 456.

154. Kwon, O.J., Myung, S.W., Lee, C.S., and Choi, H.S. (2006) *J. Colloid Interface Sci*, **295**, 409.

155. Friedrich, J.F., Mix, R., and Kühn, G. (2005) *Surf. Coat. Technol*, **200**, 565.

156. Mühlhan, C., Weidner, S.T., Friedrich, J., and Nowack, H. (1999) *Surf. Coat. Technol*, **116-119**, 783.

157. Ishihara, M., Kosaka, T., Nakamura, T., Tsugawa, K., Hasegawa, M., Kokai, F., and Koga, Y. (2006) *Diamond Relat. Mater*, **15**, 1011.

158. Poncin-Epaillard, F., Legeay, G., and Brosse, J.-C. (1992) *J. Appl. Polym. Sci*, **44**, 1513.

159. Eloy, R., Parrat, D., Duc, T.M., Legeay, G., and Béchetoille, A. (1993) *J. Cataract. Refract. Surg*, **19**, 364.

160. Kull, K.R., Steen, M.L., and Fisher, E.R. (2005) *J. Membr. Sci*, **246**, 203.

161. Yan, M.-G., Liu, L.-Q., Tang, Z.-Q., Huang, L., Li, W., Zhou, J., Gu, J.-S., Wei, X.-W., and Yu, H.-Y. (2008) *Chem. Eng. J*, **145**, 218.

162. Yu, H.-Y., He, X.-C., Liu, L.-Q., Gu, J.-S., and Wei, X.-W. (2007) *Water Res*, **41**, 4703.

163. Thome, J., Hollander, A., Jaeger, W., Trick, I., and Oehr, C. (2003) *Surf. Coat. Technol*, **174-175**, 584.

164. Aumsuwan, N., Heinhorst, S., and Urban, M.W. (2007) *Biomacromolecules*, **8**, 713.

165. Wu, Y.J., Timmons, R.B., Jen, J.S., and Molock, F.E. (2000) *Colloids Surf., B Biointerfaces*, **18**, 235.

166. Muir, B.W., Tarasova, A., Gengenbach, T.R., Menzies, D.J., Meagher, L., Rovere, F., Fairbrother, A., McLean, K.M., and Hartley, P.G. (2008) *Langmuir*, **24**, 3828.

167. Evans, L.V. and Clarkson, N. (1993) *J. Appl. Bacteriol. Symp. Suppl*, **74**, 119S.

168. Tsai, W.B., Shi, Q., Grunkemeir, J.M.C.M., and Horbett, T.A. (2004) *J. Biomater. Sci. Polym. Ed*, **15**, 817.

169. Beyer, D., Knoll, W., Ringsdorf, H., Wang, J.H., Timmons, R.B., and Sluka, P. (1997) *J. Biomed. Mater. Res*, **36**, 181.

170. Lopez, G.P., Ratner, B.D., Tidwell, C.D., Haycox, L.L., Rapoza, R.J., and Horbett, T.A. (1992) *J. Biomed. Mater. Res*, **26**, 415.

171. Wei, J., Yoshinari, M., Takemoto, S., Hattori, M., Kawada, E., Liu, B., and Oda, Y. (2007) *J. Biomed. Mater. Res. B: Appl. Biomater*, **81B**, 66.

172. Zhang, W., Chu, P.K., Ji, J., Zhang, Y., Liu, X., Fu, R.K.Y., Ha, P.C.T., and Yan, Q. (2006) *Biomaterials*, **27**, 44.

173. Rossi, F., Kylián, O., and Hasiwa, M. (2006) *Plasma Process. Polym*, **3**, 431.

174. Moreau, M., Orange, N., and Feuilloley, M.G.J. (2008) *Biotechnol. Adv*, **26**, 610.

175. Laroussi, M. and Leipold, F. (2004) *Int. J. Mass Spectrom*, **233**, 81.

176. Raballand, V., Benedikt, J., Wunderlich, J., and von Keudell, A. (2008) *J. Phys. D: Appl. Phys*, **41**, 115207.

177. Rossi, F., Kylián, O., Rauscher, H., Gilliland, D., and Sirghi, L. (2008) *Pure Appl. Chem*, **80**, 1939.

178. Moreau, M., Feuilloley, M.G.J., Veron, W., Meylheuc, T., Chevalier, S., Brisset, J.-L., and Orange, N. (2007) *Appl. Environ. Microbiol*, **73**, 5904.

179. Moreau, S., Moisan, M., Tabrizian, M., Barbeau, J., Pelletier, J., Ricard, A., and Yahia, L.H. (2000) *J. Appl. Phys*, **88**, 1166.

180. Stapelmann, K., Kylián, O., Denis, B., and Rossi, F. (2008) *J. Phys. D: Appl. Phys*, **41**, 192005.

181. Kylián, O., Benedikt, J., Sirghi, L., Reuter, R., Rauscher, H., von Keudell, A., and Rossi, F. (2009) *Plasma Process. Polym*, **6**, 255.

182. Zhang, M.Q., Desai, T., and Ferrari, M. (1998) *Biomaterials*, **19**, 953.

183. Fushimi, F., Nakayama, M., Nishimura, K., and Hiyoshi, T. (1998) *Artif. Organs*, **22**, 821.

184. Yoshioka, H. (1991) *Biomaterials*, **12**, 861.

185. Kingshott, P., Wei, J., Bagge-Ravn, D., Gadegaard, N., and Gram, L. (2003) *Langmuir*, **19**, 6912.

186. Dong, B., Jiang, H., Manolache, S., Lee Wong, A.C., and Denes, F.S. (2007) *Langmuir*, **23**, 7306.

187. Brétagnol, F., Ceriotti, L., Lejeune, M., Papadopoulou-Bouraoui, A., Hasiwa, M., Gilliland, D., Ceccone, G., Colpo, P., and Rossi, F. (2006) *Plasma Process. Polym*, **3**, 30.

188. Brétagnol, F., Ceriotti, L., Valsesia, A., Sasaki, T., Ceccone, G., Gilliland,

D., Colpo, P., and Rossi, F. (2007) *Nanotechnology*, **18**, 135303.

189. Wu, Y.J., Timmons, R.B., Jen, J.S., and Molock, F.E. (2000) *Colloids Surf., B*, **18**, 235.

190. Sardella, E., Gristina, R., Senesi, G.S., d'Agostino, R., and Favia, P. (2004) *Plasma Process. Polym*, **1**, 63.

191. Poncin-Epaillard, F., Chevet, B., and Brosse, J.C. (1994) *J. Adhes. Sci. Technol*, **8**, 455.

192. Van Os, M.T., Menges, B., Foerch, R., Vansco, G.J., and Knoll, W. (1999) *Chem. Mater*, **11**, 3252.

193. Nair, P.D. (1995) *J. Biomater. Appl*, **10**, 121.

194. Deng, M. and Shalaby, S.W. (1995) *J. Appl. Polym. Sci*, **58**, 2111.

195. Hirata, N., Matsumoto, K.-I., Inishita, T., Takenaka, Y., Suma, Y., and Shintani, H. (1995) *Radiat. Phys. Chem*, **46**, 377.

196. Brétagnol, F., Rauscher, H., Hasiwa, M., Kylián, O., Ceccone, G., Hazell, L., Paul, A.J., Lefranc, O., and Rossi, F. (2008) *Acta Biomater*, **4**, 1745.

197. Halfmann, H., Bibinov, N., Wunderlich, J., and Awakowicz, P. (2007) *J. Phys. D: Appl. Phys*, **40**, 4145.

198. Kylián, O., Hasiwa, M., Gilliland, D., and Rossi, F. (2007) *Plasma Process. Polym*, **4**. doi: 10.1002/papp.200700093.

199. Clement, J.L. and Jarrett, P.S. (1994) *Met. Based Drugs*, **1**, 467.

200. Zhao, G. and Stevens, S.E.Jr. (1998) *Biometals*, **11**, 27.

201. Furno, F., Morley, K.S., Wong, B., Sharp, B.L., Arnold, P.L., Howdle, S.M., Bayston, R., Brown, P.D., Winship, P.D., and Reid, H.J. (2004) *J. Antimicrob. Chemother*, **54**, 1019.

202. Tiwari, D.K., Behari, J., and Sen, P. (2008) *Curr. Sci*, **95**, 647.

203. Elechiguerra, J.L., Burt, J.L., Morones, J.R., Camacho-Bragado, A., Gao, X., Lara, H.H., and Yacaman, M.J. (2005) *J. Nanobiotechnol*, **3**. doi: 10.1186/1477.

204. Jain, P. and Pradeep, T. (2005) *Biotechnol. Bioeng*, **90**, 59.

205. Son, W.K., Youk, J.H., Lee, T.S., and Park, W.H. (2004) *Macromol. Rapid Commun*, **25**, 1632.

206. Li, P., Li, J., Wu, C., Wu, Q., and Li, J. (2005) *Nanotechnology*, **16**, 1912.

207. Lok, C.-N., Ho, C.-M., Chen, R., He, Q.-Y., Yu, W.-Y., Sun, H., Tam, P.K.-H., Chiu, J.-F., and Che, C.M. (2006) *J. Proteome Res*, **5**, 916.

208. Baker, C., Pradhan, A., Pakstis, L., Pochan, D.J., and Shah, S.I. (2005) *J. Nanosci. Nanotechnol*, **5**, 244.

209. Shrivastava, S., Bera, T., Roy, A., Singh, G., Ramachandran, P., and Das, D. (2007) *Nanotechnology*, **18**, 225103.

210. Pal, S., Tak, Y., and Song, J.M. (2007) *Appl. Environ. Microbiol*, **73**, 1712.

211. Balazs, D.J., Triandafillu, K., Sardella, E., Iacoviello, G., Favia, P., d'Agostino, R., Harms, H., and Mathieu, H.J. (2005) PE-CVD Modification of Medical-grade PVC to Inhibit Bacterial Adhesion: PEO-like and Nanocomposites Ag/PEO-like Coatings in *Plasma Processes and Polymers* (eds R. d'Agostino, P. Favia, C. Oehr, M.R. Wertheimer), Wiley-VCH Verlag GmbH, Weinheim, Germany, 351–372.

212. Kim, D., Jeong, S., and Moon, J. (2006) *Nanotechnology*, **17**, 4019.

8

Oxygen Barriers for Polymer Food Packaging

Joachim Schneider and Matthias Walker

8.1
Introduction

State-of-the-art polymer materials applied for food packaging consist of a multilayer structure of different polymers, as only a combination of several polymers can provide comprehensive barrier properties toward a multitude of gases [1]. The manufacturing of packaging materials made of a multilayer structure of different polymers is very expensive compared with homo-polymeric packaging materials, particularly considering the recent steady increase in commodity prices. Moreover, homo-polymeric materials are easily recyclable in contrast to multilayer polymers. In order to achieve barrier properties toward a multitude of gases comparable with those provided by multilayer polymer food packaging materials, an additional barrier coating has to be applied to homo-polymeric food packaging [2–5]. Recyclability of the polymer material will not be affected if the barrier coating is in the nano-scale. Low-pressure microwave plasma processes are a promising tool for the deposition of such barrier coatings, as these processes have low impact on thermally sensitive materials. A description of the microwave system used for the experiments described in this chapter can be found in Section 2.2.1.

8.2
Fundamentals of Gas Diffusion through Polymers

Diffusion of gases, vapors, and liquids in polymers is an important and often the controlling factor in numerous applications, for example, in membrane separation processes, reverse osmosis, ultra-filtration, and packaging for foods and beverages. For all these applications, barrier coatings are needed and play an important role. Therefore, a better understanding of the mechanisms of diffusion is highly desirable in order to achieve significant improvements in these fields of application.

The diffusion process of gases, vapors, and liquids, respectively, through polymers depends on the structure of the polymer membrane. In case of a polymer

Plasma Technology for Hyperfunctional Surfaces. Food, Biomedical and Textile Applications.
Edited by Hubert Rauscher, Massimo Perucca, and Guy Buyle
Copyright © 2010 WILEY-VCH Verlag GmbH & Co. KGaA, Weinheim
ISBN: 978-3-527-32654-9

with pores, different mass transport mechanisms can occur, depending on the pore size and the shape of the permeating molecules. The mass transport can be described by viscous processes, by ordinary or Fickian diffusion, or by another form of diffusion, namely the Knudsen flow [6].

In nonporous polymer materials, the mass transport is generally of the activated-diffusion type and can be described by the following three steps [6]:

1) the absorption of the permeating molecules on the surface of the polymer and the infiltration of the molecules into the polymer matrix;
2) the diffusion through the polymer describable by a concentration gradient; and
3) the desorption of the permeating molecules from the opposite polymer surface.

The two Fick's laws are the fundamental laws for theoretical description of the diffusion of gases, vapors, and liquids through polymer materials (see for example, Vieth [6] or Comyn [7]). Fick's first law (Equation 8.1) states that the flux $\vec{j}$ of the permeating molecules is proportional to the concentration gradient $\vec{\nabla} c$:

$$\vec{j} = -D\,\vec{\nabla} c \tag{8.1}$$

where

$\vec{j}$ is the particle flux diffusing across an area per time unit,

D is the diffusion coefficient, and

$c(x,t)$ is the local concentration of the permeant.

D depends on the nature of the polymer-permeant system and can be constant or a function of the concentration. Integration of Equation 8.1 for the specified geometry and boundary conditions yields the total amount of permeant diffusing through a polymer.

Fick's second law (Equation 8.2) describes the non-steady state and can be written as

$$\frac{\partial c}{\partial t} = \vec{\nabla}\left(D\,\vec{\nabla} c\right) \tag{8.2}$$

In general, the diffusion coefficient depends on the concentration of the permeant in the polymer, on the spatial coordinates, on the thickness of the polymer, and on the time passed since diffusion started. Additionally, the diffusion coefficient is a function of the temperature and, due to an activated process, is obeying in a limited temperature range the relationship similar to the Arrhenius equation (Equation 8.3):

$$D = D_0\, e^{-\frac{E_D}{RT}} \tag{8.3}$$

where

E_D is the activation energy for diffusion,

R the gas constant, and

T the absolute temperature.

The activation energy is a measure for the energy applied against the cohesive forces of the polymer for formation of the gaps through which diffusion in homogeneous polymers will occur.

Here, only a concentration dependence and in some special cases a temperature dependence of the diffusion coefficient will be discussed. Furthermore, we assume a one-dimensional diffusion process, where x is perpendicular to the surface of the polymer foil. In this case, Fick's laws are given by Equations 8.4, 8.5 and 8.6:

$$j = -D(c, T)\frac{dc}{dx} \tag{8.4}$$

and

$$\frac{dc}{dt} = \frac{d}{dx}\left(D(c, T)\frac{dc}{dx}\right) \tag{8.5}$$

or

$$\frac{dc}{dt} = D(c, T)\frac{d^2c}{dx^2} + \frac{dD(c)}{dc}\left(\frac{dc}{dx}\right)^2 \tag{8.6}$$

8.2.1
Diffusion, Solubility, and Permeability of Polymers

The interactions between the permeating molecules and the polymer chains and segments are not negligible in many polymer-permeant systems. In these cases, the diffusion coefficient D is not a constant, but a function of the concentration c. Several theories and computational techniques can be found in literature which describe the diffusion mechanisms. The 'molecular' models are based on specifically postulated motions of the permeating molecules and the polymer chains and take into account the intermolecular forces [8, 9]. On the molecular scale, the diffusion of gases can be described by molecular dynamic methods [10]. A more phenomenological description of the diffusion mechanisms is based on the so-called 'free-volume' models. These models do not directly consider the molecular structure of the polymer-permeant system, but relate the diffusion coefficient to the free-volume of the system (see for example, Huang *et al.* [11] or Fels *et al.* [12]). Here, the diffusion coefficient is mainly determined by the amount of free-volume present in the polymer-permeant system. In particular, the free-volume model proposed by Fujita satisfactorily describes the diffusion of a number of organic liquids and vapors in a variety of polymers [13]. In these polymer-permeant systems, the sorbed penetrant swells and plasticizes the polymer, resulting in an increase of the free volume and in a strong increase of the diffusion coefficient.

However, if the interactions between the permeating molecules and the polymer are small, the diffusion coefficient D in the two Fick's laws is independent of the concentration c. This holds for many permanent gases (that is, a gas at a pressure and temperature far from its liquid state) at room temperature and at moderate pressure and presents a good approximation for, for example, the diffusion of oxygen (O_2) through poly(ethylene terephthalate) (PET), polypropylene (PP), and polyethylene (PE). In this chapter, we show results on the diffusion of O_2 through PP and PET, and therefore we consider the diffusion coefficient to be constant.

In a typical and often applied permeation experiment, the polymer film separates two chambers which contain the permeant at different concentrations. The molecules diffuse from the chamber at high concentration c_0 to the chamber at low concentration c_1. Then the experimental condition is Equation 8.7:

$$c_0 \gg c_1 \approx 0 \tag{8.7}$$

According to Figure 8.1, the boundary conditions can be written as Equation 8.8:

$$c(x = 0, t) = c_0$$
$$c(x = L, t) = 0 \tag{8.8}$$

where
    L is the thickness of the polymer.

The concentration profiles at any time t and at any distance x in the polymer can be obtained from the solution of Fick's second law (see Equation 8.5). A standard method for obtaining the analytical solution is to assume that the variables are separable [14]. Then, we have Equation 8.9:

$$c(x, t) = X(x)T(t) \tag{8.9}$$

where X and T are functions of x and t, respectively. Substitution of $c(x,t)$ in Equation 8.5 in combination with the boundary conditions of Equation 8.8 lead to the following (general integral) solution, Equation 8.10 (for a detailed description see for example, Crank [14]):

$$c(x, t) = c_0 \left(1 - \frac{x}{L}\right) - \frac{2}{\pi} \sum_{n=1}^{\infty} \frac{c_0}{n} \sin\left(\frac{n\pi x}{L}\right) \cdot e^{-\frac{Dn^2\pi^2 t}{L^2}} \tag{8.10}$$

A typical series of concentration–distance profiles for different times is shown in Figure 8.2.

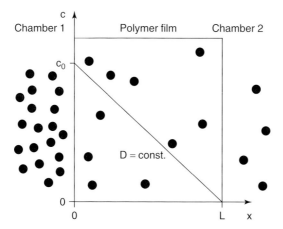

Figure 8.1 Boundary conditions in a typical permeation experiment and a typical steady-state concentration distribution for a diffusion coefficient D independent of the concentration of the permeant.

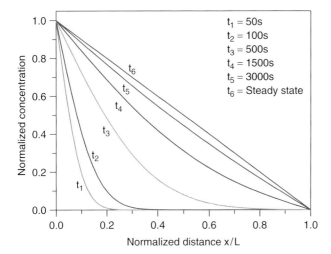

Figure 8.2 Concentration–distance profiles within a polymer foil for D = const.

It is illustrated that the profiles are convex to the origin when the diffusion coefficient is independent of the concentration.

In the steady state, that is, $t \rightarrow \infty$ or $dc/dt = 0$, the concentration distribution is given by Equation 8.11:

$$c_s(x) = c_0 \left(1 - \frac{x}{L} \right) \tag{8.11}$$

This means that the concentration decreases linearly within the polymer foil.

The flux of the permeating molecules at $x = L$ can be calculated by substituting Equation 8.10 into Fick's first law (see Equation 8.4) to give Equation (8.12):

$$j(x = L, t) = D \left(\frac{c_0}{L} + \frac{2c_0}{L} \sum_{n=1}^{\infty} \cos (n\pi) \cdot e^{-\frac{Dn^2\pi^2 t}{L^2}} \right) \tag{8.12}$$

Using Equation 8.13:

$$\cos (n\pi) = (-1)^n \tag{8.13}$$

we obtain Equation 8.14:

$$j(x = L, t) = D \left(\frac{c_0}{L} + \frac{2c_0}{L} \sum_{n=1}^{\infty} (-1)^n \cdot e^{-\frac{Dn^2\pi^2 t}{L^2}} \right) \tag{8.14}$$

When the steady state is attained, t becomes large enough to make the exponential term negligibly small. Then Equation 8.14 can be written as Equation 8.15:

$$j_s = D \cdot \frac{c_0}{L} \tag{8.15}$$

Assuming that the gas–polymer system obeys Henry's law [8], the concentration c_0 of the permeant can be related to the partial pressure p_0 by Equation 8.16:

$$c_0 = S \cdot p_0 \tag{8.16}$$

where

S is the solubility coefficient.

Then Equation 8.15 can be written as Equation 8.17:

$$j_s = D \cdot S \cdot \frac{p_0}{L} = P \cdot \frac{p_0}{L} \tag{8.17}$$

where the permeability coefficient P is given by Equation 8.18:

$$P = D \cdot S \tag{8.18}$$

Thus, the particle flux through a polymer depends both on the diffusion coefficient D and on the solubility coefficient S. Since diffusion coefficients may not be independent of the concentration, and Henry's law may not apply, the permeability coefficient P is in general not a fundamental property. However, the permeability coefficient P is the most frequently measured and reported quantity characterizing the barrier properties of a polymer.

8.2.2
Diagnostic Methods

The two methods commonly used to evaluate molecular diffusion in polymers are [6, 7]:

1) the time resolved measurement of the particle flux through the polymer (transmission method) and,
2) the kinetic study of the sorption and desorption in the polymer.

In the latter method, a piece of polymer foil is located in the permeant, that is, liquid, vapor, or gas, and the weight gain M_t is plotted against time. This is usually done with a microbalance. From the graph of M_t/M_∞ versus $t^{1/2}/L$, where M_t is the total mass gain of the polymer foil with the thickness L at the time t and M_∞ is the equilibrium condition, the diffusion coefficient D can be calculated.

In the transmission method, the polymer film separates two chambers which contain gas, vapor, or liquid at different concentrations. The molecules permeate from the chamber at high concentration c_0 to the chamber at low concentration c_1. The experimental condition $c_0 \gg c_1 \approx 0$ is applied very often and can be achieved by evacuation of the chamber at low concentration or by introduction of a carrier gas. With the 'vacuum' method, the pressure increase is plotted against time. Then the amount $Q(t)$ of the molecules permeated through the polymer foil can be obtained by integration of Equation 8.14 over time to give Equation 8.19:

$$Q(t) = \int_0^t j(x = L)dt' = \int_0^t D \left(\frac{c_0}{L} + \frac{2c_0}{L} \sum_{n=1}^{\infty} (-1)^n \cdot e^{-\frac{Dn^2 \pi^2 t}{L^2}} \right) dt' \tag{8.19}$$

The integration yields Equation 8.20 (for a detailed description see for example, Crank [14])

$$Q(t) = \frac{Dc_0}{L} \left(t - \frac{L^2}{6D} \right) - \frac{2c_0 L}{\pi^2} \sum_{n=1}^{\infty} \frac{(-1)^n}{n^2} \cdot e^{-\frac{Dn^2 \pi^2 t}{L^2}} \tag{8.20}$$

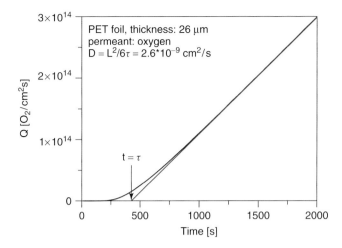

Figure 8.3 Amount of the permeated penetrant Q as a function of time. The diffusion coefficient D results from extrapolation of the steady-state line to the time axis.

The plot of Equation 8.20 for O_2 diffusion through a PET foil is shown in Figure 8.3. The extrapolation of the steady-state part of the permeation curve Equation 8.21:

$$Q_s(t) = \frac{Dc_0}{L}\left(t - \frac{L^2}{6D}\right) \tag{8.21}$$

to the time axis yields the time value called *time lag* Equation 8.22:

$$\tau = \frac{L^2}{6D} \tag{8.22}$$

so that the diffusion coefficient D can be calculated by Equation 8.23:

$$D = \frac{L^2}{6\tau} \tag{8.23}$$

Assuming that Henry's law (see Equation 8.16) can be applied, the steady-state part of Equation 8.21 can be written as Equation 8.24:

$$Q_s(t) = \frac{P \cdot p_0}{L}t \tag{8.24}$$

so that the permeability coefficient P can be determined. With the values for D and P, the solubility coefficient S can be calculated from Equation 8.18. The time lag technique has been applied successfully to a wide variety of polymer-gas systems with constant diffusion coefficient. The diffusion coefficient D cannot be determined from the time lag technique if it depends on, for example, c, x, or t. Then we only obtain a mean value for D.

The carrier gas method is another method that we used in our experiments to determine the permeation through polymer foils. A sketch of the corresponding experimental set-up is shown in Figure 8.4. As already described, the permeation

Permeant (O$_2$)

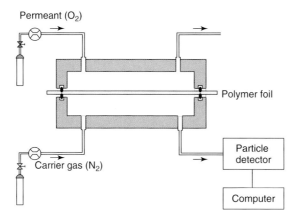

Polymer foil

Particle detector

Carrier gas (N$_2$)

Computer

Figure 8.4 Carrier gas method: experimental set-up for measurement of the permeation through polymer foils.

cell consists of two chambers separated by the polymer foil. The permeant is introduced into the upper chamber of the cell, and a carrier gas, for example, nitrogen (N$_2$), flows through the other chamber with a constant rate and sweeps the diffusing molecules to the detector. In our experiments, we used a ceramic (zirconium dioxide, ZrO$_2$) detector to determine the O$_2$ flow rate. In contrast to the 'vacuum' method, both compartments of the permeation cell are at atmospheric pressure. With this particular experimental device, the temporal behavior of the concentration of the permeant in the carrier gas is measured.

As an example, Figure 8.5 shows the O$_2$ diffusion through a PET foil. The coefficients for diffusion, permeability, and solubility can be determined from Equation 8.14. Applying the steady-state relation from Equation 8.15 at the time

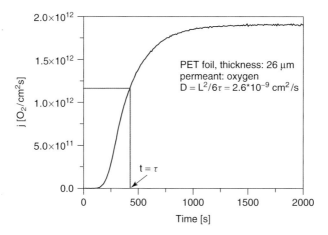

PET foil, thickness: 26 μm
permeant: oxygen
$D = L^2/6\tau = 2.6*10^{-9} \text{ cm}^2/\text{s}$

$t = \tau$

Figure 8.5 Particle flux j as a function of time. The time $t = \tau$ corresponding to $j = 0.616 \cdot j_s$ determines the diffusion coefficient D.

lag $\tau = \frac{L^2}{6D}$ we obtain Equation 8.25:

$$\frac{j(x = L, t = \tau)}{j_s(x = L, t = \tau)} = 1 + 2\sum_{n=1}^{\infty}(-1)^n \cdot e^{-\frac{n^2\pi^2}{6}} \approx 0.616 \tag{8.25}$$

The time $t = \tau$ with $j(t = \tau) = 0.616 \cdot j_s$ can be determined from the permeation slope, so that D is given by $D = \frac{L^2}{6\tau}$. We can obtain the permeability coefficient P from the steady-state particle flux $j_s = P \cdot \frac{p_0}{L}$, and the solubility coefficient S can be calculated from Henry's law (see Equation 8.16).

8.2.3
Barrier Concepts

In order to improve the barrier properties of a polymer, a number of different techniques can be applied. Here, not all of the multifarious methods described in literature can be discussed, but a short overview will be given.

The physical quantities that affect the permeability result from Equation 8.17 for the steady-state particle flux. According to this equation, j_s is inversely proportional to the thickness L of the polymer foil, so that an increase of the thickness leads to a reduction of the steady-state particle flux. When the foil thickness is increased beyond a certain value, it becomes uneconomical to further increase it in order to obtain lower permeation. Optimization, that is, minimizing the diffusion coefficient D, is another way of reducing the steady-state particle flux. This can for example, be achieved by incorporating fillers into the polymer or by using a polymer material with large areas of crystallinity. The presence of crystalline regions within the polymer effectively increases the length of the diffusion path, which leads to a reduction of the steady-state particle flux.

In other barrier concepts, a suitable barrier layer, consisting of ethylene-vinyl alcohol (EVOH), for example, is included into the polymer. However, when the relative humidity is increased, EVOH shows a strong decrease in its barrier properties concerning the permeability toward O_2. The sensitivity toward moisture is very high, because EVOH is extremely hygroscopic. Copolymerization with ethylene will significantly reduce the sensitivity of EVOH toward moisture. Inversely, the ethylene content generally lowers the barrier properties of the EVOH copolymer. In order to preserve the barrier properties of EVOH concerning its permeability toward gases when used for packaging of moist food, a water vapor barrier layer will prevent the EVOH film from being in direct contact with humidity. In addition, adhesive layers are needed which produce a good bond strength between the barrier layer and the polymer. The resulting barrier polymer consists of a multilayer structure. This kind of barrier concept is commonly used in the packaging industry [1].

Evaporated or sputtered metallic films such as aluminum films offer further possibilities of reducing the permeability of polymer materials. The thickness of the metallic film does not exceed 0.1 μm in general, mainly for economic reasons. Unlike thin metal foils, the quality of the barrier of evaporated metallic coatings depends not only on the thickness of the film, but also on the morphology of the polymer surface being coated by the metal and on the conditions during deposition. Consequently, the high gas permeation barriers of thin metal foils cannot be achieved by evaporated metallic films in general. As optically transparent barrier coatings are often required, and as metallic films cannot be used for packaging of food to be prepared in a microwave oven, oxide barrier coatings such as silicon oxide (SiO_x) or aluminum oxide (Al_2O_3) offer an excellent alternative to metallic films. In industrial practice, electron beam evaporation or magentron sputtering are principally applied for vacuum deposition of SiO_x coatings on polymer materials used for packaging applications. The high deposition rates that are achievable present the main advantage of these technologies. The main drawback is that neither electron beam evaporation nor sputtering easily facilitates the modification of the stoichiometry of the barrier film during the coating process. Reactive evaporation and sputtering techniques compensate for this disadvantage by using plasma technology to produce reactive particles: the oxygen content of SiO_x in reactive evaporation deposition of SiO with O_2 can be tuned with the aid of a plasma source producing reactive oxygen radicals. Such films satisfy all demands with respect to optical and microwave transparency and barrier properties.

Direct plasma enhanced chemical vapor deposition (PECVD) of SiO_2-like films [15–17] is another method that does not rely on evaporation or sputtering techniques. Depending on the targeted application, organosilicon compounds like hexamethyldisiloxane (HMDSO), tetramethylsilane (TMS), or tetraethoxysilane (TEOS) are commonly used in mixtures with O_2 as deposition gases. The corresponding films produce excellent results in terms of transparency and of barrier properties. Furthermore, the plasma technique offers the possibility of varying the carbon and hydrogen content of the film by modification of the process parameters [18, 19]. For example, films grown at low O_2 concentrations contain many CH-groups, and a relatively soft, organic film is polymerized. An increase of the O_2 concentration causes oxidation of the CH-groups, and the chemical composition changes to an inorganic SiO_2-like film [20]. Such films with different CH-content can be combined to a multilayer structure which compensates, for example, a thermal mismatch between the barrier coating and the polymer material [21]. As it can be seen, barrier films play an important role in packaging industry. However, there are a lot of other applications where barrier films are needed, for example, barriers toward organic vapors and liquids [22–24] or barriers for fuel cell systems [25, 26].

8.3
Case Study: Plasma Deposition of SiO_x Barrier Films on Polymer Materials Relevant for Packaging Applications

8.3.1
Materials and Measurements

8.3.1.1 Selection of Two-dimensional and Three-dimensional Polymer Substrates
PET and PP were selected as polymer materials due to their diverse fields of application in packaging industry. Most of the plastic bottles produced worldwide

are made of PET, and PET foil is also used for packaging of food [27]. PP can be found in packaging industry both in the form of foil material and as 3D containers, simple or complex in structure [28]. In contrast to PET, PP materials are much more sensitive to thermal treatment even at temperatures well below 100 °C due to their comparatively high coefficient of thermal expansion, which further increases with higher temperatures [29]. Due to this fact, the development of processes for plasma deposition of efficient barrier coatings concerning the permeability toward O$_2$ is assumed to be more difficult on PP substrates than on samples made of PET.

According to the requirements set within the ACTECO project (see Preface), injection-molded trays made of pure PP should provide barrier properties concerning the permeability toward O$_2$ equivalent to the steady-state O$_2$ particle flux of 5 cm^3 O$_2$/(m^2·24 h·bar) or even less after plasma deposition of barrier coatings.

SiO$_x$-based barrier films are very promising with regard to this target, particularly as the barrier coatings had to be both optically and microwave transparent and should facilitate metal detection from outside after filling and sealing of the trays. Starting with plasma deposition of SiO$_x$ barrier films on 2D PET and PP foil samples, respectively, the knowledge and experience attained in these tests should then be transferred to the development of an industrially relevant process for plasma deposition of efficient SiO$_x$ barrier films on 3D injection-molded trays made of pure PP.

8.3.1.2 Measurement of the Steady-state O$_2$ Particle Flux

The permeability toward O$_2$ of the SiO$_x$ films on PET foil samples and on PP substrates, respectively, was analyzed by the carrier gas method (see Section 8.2.2). The O$_2$ particle flux through the uncoated and coated PET and PP foil samples, respectively, was recorded by a self-made system using a ZrO$_2$ detector and by a MultiPerm oxygen and water vapor permeability analyzer from ExtraSolution, respectively. The measurements of the O$_2$ particle flux were carried out at the constant temperature of $T = 30.0$ °C of the measurement cell and at 0% relative humidity of the permeant O$_2$ (cf. Figure 8.4). The results of the steady-state O$_2$ particle flux presented in Section 8.3.3 for uncoated and SiO$_x$ coated PP trays were recorded by the ACTECO project partner, Biophy Research in Fuveau, France, in a Mocon O$_2$ transition rate test system OX-TRAN Model 2/61 adapted for O$_2$ particle flux measurements on entire 3D substrates. The analytical testing at Biophy was performed at the constant temperature of $T = 30$ and 23 °C, respectively, with the relative humidity of the permeant air being approximately 25%.

8.3.1.3 Measurement of the Coating Thickness

In order to get additional information on the thickness of the SiO$_x$ films, microscope slides were plasma coated along with the foil samples. They were attached to the inner and outer surfaces of the PP trays during parallel plasma deposition tests performed at the same process parameters applied for the corresponding plasma deposition of SiO$_x$ films on PP trays designated for measurement of the steady-state O$_2$ particle flux. Small strips of adhesive tape made of Kapton fixed onto the microscope slides were removed after plasma deposition. The resulting step from the plasma-coated surface of the microscope slide to its uncoated surface due to removal of the strip of adhesive tape was analyzed by use of a Perthometer C5D profilometer from Mahr Perthen.

8.3.2
SiO$_x$ Barrier Films on PET Foil

PET foil samples Hostaphan RD (26 µm in thickness) were plasma coated under low-pressure conditions in an electron cyclotron resonance (ECR) set-up driven by microwaves at 2.45 GHz. Figure 8.6 shows the schematic of the ECR set-up (see also Section 2.2.1.4). A horn antenna is applied at medium height to the side of the vacuum vessel approximately 80 cm in diameter and 50 cm in height [30, 31]. Microwave power is radiated via this horn antenna to the inside of the vacuum vessel at low-pressure conditions of 1 Pa up to approximately 50 Pa. For the microwave frequency $f_{mw} = \omega_{mw}/(2 \cdot \pi) = 2.45$ GHz, the magnetic field strength has to be $B_0 = 0.0875$ T in order to fulfill the ECR condition (cf. Equation 2.13 in Section 2.2.1.4). The magnetic field strength of the magnets was chosen accordingly to meet the ECR condition approximately 1 cm above the bottom plate, directly beyond the location of the movable magnet configuration. As a consequence, a confined plasma is formed in the sphere where the ECR condition is valid. When the ECR plasma is moved along with the horizontal movement of the magnet configuration, 2D substrates with an area of up to 40 cm × 50 cm lying directly on top of the bottom plate can be plasma treated homogeneously, that is, the deviation of the film thickness in a corresponding ECR plasma deposition process will not exceed the value of 5–10%.

8.3.2.1 SiO$_x$ Barrier Films Deposited from O$_2$: HMDSO Gas Mixtures
HMDSO was applied as precursor in combination with O$_2$ as reactive gas for ECR plasma deposition of SiO$_x$ barrier films on PET foil samples performed at the

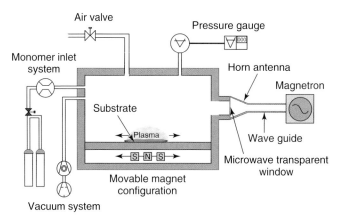

Figure 8.6 Schematic of the experimental set-up of the ECR plasma source for plasma deposition on 2D substrates [31].

working pressure of $p = 6-8$ Pa and at the constant mean microwave power of approximately 560 W in pulse mode (pulsing frequency $f = 1$ kHz, duty cycle 50%). The PET foil samples approximately 10 cm × 10 cm were located in the middle of the vacuum vessel directly on top of the bottom plate, as it is shown in Figure 8.6. During plasma deposition, the magnet configuration was in a steady horizontal movement with the cycle time of approximately 60 seconds. As a consequence, the PET foil samples were directly exposed to the plasma.

8.3.2.1.1 O$_2$ Permeation Measurements: Determination of the Diffusion Coefficient

Figure 8.7 shows the measured O$_2$ particle flux through an uncoated PET foil sample as a function of time. The graph in Figure 8.7 theoretically describing the O$_2$ particle flux through the uncoated PET foil, is calculated in the following way: the diffusion coefficient D is considered to be independent of the concentration profile c of the permeant O$_2$ in the PET foil. In a first step, Fick's second law (Equation 8.5) is solved numerically with the boundary conditions $c(x = 0) = c_0$ and $c(x = L) = 0$ and the initial condition $c(x, t = 0) = 0$ (cf. Figure 8.1). As a result, the concentration profiles $c(x, t)$ are obtained. The time-dependent O$_2$ particle flux at the position $x = L$ of the PET foil can then be calculated by substitution of the concentration profiles $c(x, t)$ into Fick's first law (Equation 8.4). The following values were applied to calculate the graph in Figure 8.7 theoretically describing the O$_2$ particle flux through the uncoated PET foil: $D = 2.63 \cdot 10^{-9}$ cm^2 s^{-1}, $c_0 = 5.35 \cdot 10^{17}$ O$_2$ molecules cm^{-3} and $S = 2 \cdot 10^{-2}$. In Figure 8.7, the calculated graph simulates the time-dependent O$_2$ particle flux through the uncoated PET foil quite well.

Plasma deposition of SiO$_x$ films is an efficient means of noticeably reducing the steady-state O$_2$ particle flux of $j_s = 1.9 \cdot 10^{12}$ O$_2$ molecules/(cm^2·s) measured for the uncoated PET foil sample. In a preliminary test, the significant decrease

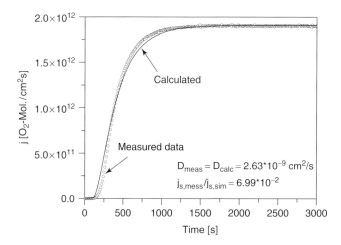

Figure 8.7 Experimental and calculated results of the O$_2$ particle flux through an uncoated PET foil sample 26 μm in thickness.

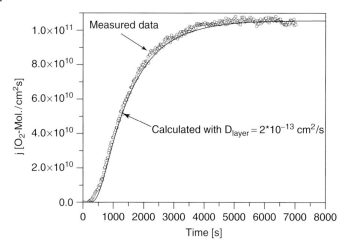

Figure 8.8 Experimental and calculated results of the O_2 particle flux through a plasma-coated PET foil sample 26 μm in thickness. The 200 nm SiO$_x$ barrier film was deposited from the O_2: HMDSO gas mixture ratio of 7 : 1.

of the steady-state O_2 particle flux down to $j_s = 1.05 \cdot 10^{11}$ O_2 molecules/(cm²·s) was obtained by a SiO$_x$ film approximately 200 nm in thickness plasma deposited from the O_2: HMDSO gas mixture ratio of 7 : 1 on the PET foil sample. Figure 8.8 shows the corresponding temporal behavior of the O_2 particle flux. The diffusion coefficient D related to the O_2 permeation through the uncoated PET foil is still valid, but as the O_2 permeability of the coated PET foil is dominated by the SiO$_x$ film, the diffusion coefficient D_{layer} of the film had to be optimized for the best fit of the calculated graph to the measured O_2 particle flux as a function of time. The calculated graph in Figure 8.8 based on the optimized diffusion coefficient $D_{layer} = 2 \cdot 10^{-13}$ cm² s^{-1} of the SiO$_x$ film is in very good agreement with the measured data.

8.3.2.1.2 O_2 Permeation Measurements: Variation of the O_2: HMDSO Gas Mixture Ratio As already mentioned, ECR plasma deposition of the SiO$_x$ barrier films was performed by applying HMDSO as precursor in combination with O_2 as reactive gas. This 'O_2: HMDSO gas mixture ratio', often referred to as '*gas mixture ratio*', is a crucial parameter. Here, we considered its variation in the range of 1 : 1 – 40 : 1. The thickness of the SiO$_x$ barrier films was kept to approximately 100 nm for all gas mixture ratios, in order to obtain directly comparable results when analyzing the quality of the barrier properties toward O_2 by determination of the O_2 permeability of the uncoated and plasma-coated PET foil samples, respectively.

The measurement values of the steady-state O_2 particle flux (i.e., the oxygen transition rate (OTR), as the steady-state O_2 particle flux will be called subsequently) are presented in Figure 8.9 as a function of the O_2: HMDSO gas mixture ratio applied for plasma deposition of the individual SiO$_x$ films on PET foil samples.

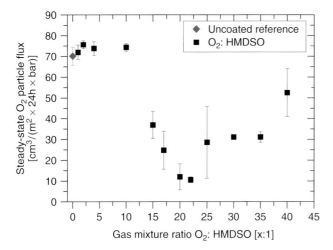

Figure 8.9 Steady-state O$_2$ particle flux (corresponding to the oxygen transition rate, OTR) as a function of the O$_2$: HMDSO gas mixture ratio, obtained for SiO$_x$ barrier films deposited on PET foil by the ECR plasma source.

The error bars in Figure 8.9 represent the arithmetic average of the OTR results determined for SiO$_x$ films deposited repeatedly at the same gas mixture ratio. The average over up to eight samples was taken. The OTR results obtained for ratios in the range of 1 : 1 up to 10 : 1 show a slight increase compared with the OTR value of $70(\pm 4)$ cm^3 O$_2$/(m^2·24 h·bar) of the uncoated PET foil samples. When increasing the ratio from 10 : 1 to 15 : 1, the OTR can be significantly reduced from $74(\pm 2)$ cm^3 O$_2$/(m^2·24 h·bar) down to $37(\pm 7)$ cm^3 O$_2$/(m^2·24 h·bar). Minimum OTR values of $12(\pm 6)$ cm^3 O$_2$/(m^2·24 h·bar) and of $11(\pm 1)$ cm^3 O$_2$/(m^2·24 h·bar) were obtained for the gas mixture ratios of 20 : 1 and 22 : 1, respectively. For ratios higher than 25 : 1, the OTR values are slowly increasing again.

8.3.2.1.3 FTIR Analysis: Chemical Composition of the Surface of the SiO$_x$ Barrier Films Deposited from Different O$_2$: HMDSO Gas Mixtures

According to Figure 8.9, the SiO$_x$ barrier films providing the best barrier properties toward O$_2$ were deposited from the gas mixture ratios of 20 : 1 and 22 : 1. In order to obtain information as to how the chemical composition of the films could be causal for their quality, small sheets of glass metalized with molybdenum were placed next to the PET foil samples during plasma deposition. The thickness of the SiO$_x$ films on the molybdenum layer was approximately 100 nm for all investigated gas mixtures. The plasma-coated sheets of glass were analyzed *ex situ* with a Bruker Vector 22 Fourier-transform infra-red (FTIR) spectrometer by use of a reflectance unit.

The FTIR spectra recorded from the SiO$_x$ films deposited at different O$_2$: HMDSO gas mixture ratio are presented in Figure 8.10. Peaks at 2960, 2905, 1260, and 845 cm^{-1} can be detected in the FTIR spectra for gas mixtures containing only

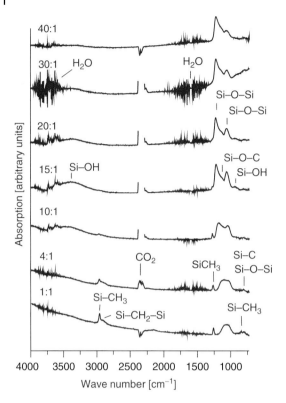

Figure 8.10 Infra-red spectra of SiO$_x$ barrier films deposited from different O$_2$: HMDSO gas mixture ratios by use of the ECR plasma source [31].

a small amount of O$_2$. These peaks can be assigned to anti-symmetric stretching vibrations of Si–CH$_3$ groups, symmetric stretching vibrations of CH$_2$ in Si–CH$_2$–Si groups (or even Si–CH$_2$–CH$_x$ groups), symmetric bending vibrations of CH$_3$ in Si–CH$_3$ groups and CH rocking vibrations in Si–CH$_3$ groups, respectively [32–34]. The intensity of these peaks is maximum in the FTIR spectrum for the gas mixture ratio of 1 : 1. Increasing the O$_2$ content is not only causal for the significant drop in peak intensity, but also for the shift of the peak positions toward higher wave numbers. As a consequence, only the symmetric bending vibrations of CH$_3$ in Si–CH$_3$ groups can be still identified at the wave number of 1279 cm^{-1} in the FTIR spectrum for the gas mixture of 15 : 1, and for ratios of 20 : 1 and even higher, all bands assignable to CH$_2$ and CH$_3$ vibrations, respectively, have disappeared from the FTIR spectra. The shift of the peak position of the symmetric bending vibrations of CH$_3$ in Si–CH$_3$ groups in the range between 1260 and 1279 cm^{-1} is shown in Figure 8.11. A major shift of the peak position from 1265 to 1278 cm^{-1} occurs when the proportion of O$_2$ in the gas mixture is increased from 4 : 1 to 10 : 1. The decreasing peak intensity together with the shift of the peak position toward higher wave numbers of the Si–CH$_3$ vibration bands in particular present a strong

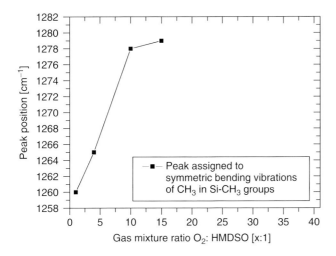

Figure 8.11 Position of the peak assigned to symmetric bending vibrations of CH$_3$ in Si–CH$_3$ groups as a function of the O$_2$: HMDSO gas mixture ratio.

indication for the decreasing number of CH$_3$ groups bonded to a single silicon atom, due to progressive oxidation of the CH$_3$ groups at higher gas mixture ratios [20, 32].

The small shoulder at 1150 cm^{-1} in the FTIR spectra for gas mixture ratios in the range between 15 : 1 and 40 : 1 related to Si–O–C stretching vibrations in particular, but also the small peak at 804 cm^{-1} assignable to both Si–C anti-symmetric stretching vibrations and bending vibrations of Si–O–Si groups, imply that the chemical composition of the SiO$_x$ barrier films deposited from ratios even higher than 15 : 1 is not free of carbon atoms [35–38].

The two intense peaks at 1228 and at 1062 cm^{-1} in the FTIR spectrum for the gas mixture ratio of 20 : 1 in Figure 8.10 related to out-of-phase and in-phase anti-symmetric stretching vibrations of Si–O–Si groups, respectively, in the transverse optical mode show a shift of the peak position similar to the bands representing vibrations of Si–CH$_3$ groups [39, 40]. The shift of the peak assigned to in-phase anti-symmetric stretching vibrations of Si–O–Si groups starts at 1038 cm^{-1} in the FTIR spectrum for the ratio of 4 : 1 (see Figure 8.12). The maximum shift of the peak position happens between the ratios of 10 : 1 (1045 cm^{-1}) and 15 : 1 (1066 cm^{-1}), indicating a significant change in stoichiometry of the SiO$_x$ film from a polymer-like to a quartz-like chemical composition [41–43]. For ratios from 15 : 1 up to 40 : 1, the peak position remains almost constant at approximately 1068 cm^{-1}.

Figure 8.13 presents the shift of the peak position of the out-of-phase anti-symmetric Si–O–Si stretching vibration: the peak position is shifting almost linearly from 1076 cm^{-1} in the FTIR spectrum for the gas mixture ratio of 1 : 1 to 1222 cm^{-1} in the FTIR spectrum for the gas mixture ratio of 15 : 1. When the O$_2$ content is further increased up to 40 : 1, the respective peak position stays almost constant at 1228 cm^{-1}. This phenomenon is ascribable to the excitation of both the longitudinal and the transverse mode phonons of the anti-symmetric Si–O–Si

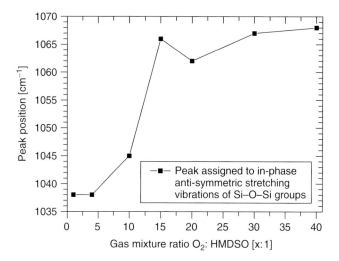

Figure 8.12 Position of the peak assigned to in-phase anti-symmetric stretching vibrations of Si–O–Si groups as a function of the O$_2$: HMDSO gas mixture ratio.

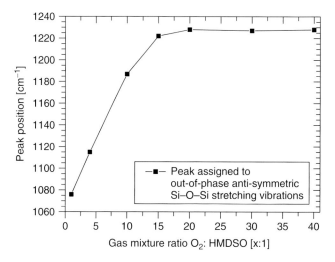

Figure 8.13 Position of the peak assigned to out-of-phase anti-symmetric stretching vibrations of Si–O–Si groups as a function of the O$_2$: HMDSO gas mixture ratio.

stretching, an effect already found by Berreman on films with a thickness of significantly less than 1 μm [39, 44–47].

The shift of the peak position is accompanied by an increase in intensity of the two peaks assigned to out-of-phase and in-phase anti-symmetric stretching vibrations of Si–O–Si groups, respectively.

Finally, the intensity of the extensive band structure between approximately 3600 and 3200 cm^{-1} related to stretching vibrations of the OH groups in Si–OH increases with rising O$_2$ content, having its maximum at the gas mixture ratio of 15 : 1 [32, 34]. For higher ratios, the intensity of this band structure is significantly decreasing again. A similar behavior of the peak intensity in relation to the gas mixture ratio can be identified for the peak at 926 cm^{-1} that is assignable to Si–OH stretching vibrations, too.

The considerably improving barrier properties of the SiO$_x$ films deposited at O$_2$: HMDSO gas mixture ratios higher than 10 : 1 are in accordance with the significant change in the FTIR spectra related to the ratios of 10 : 1 and 15 : 1, respectively, and particularly with the corresponding shift of the peak position of the in-phase anti-symmetric stretching vibrations of Si–O–Si groups in Figure 8.12: when increasing the O$_2$ content from 10 : 1 to 15 : 1, the FTIR analysis indicates a significant change of the stoichiometry of the SiO$_x$ films toward quartz-like layers in combination with a strong reduction of the Si–CH$_3$ functional groups. Additionally, the amount of Si–OH groups in the film can be reduced by increasing the O$_2$ content from 15 : 1 up to 40 : 1. It is known from literature that the presence of both CH$_3$ and OH groups is causal for reduced cross-linking in the SiO$_x$ film, showing minor barrier properties as a consequence[34]. A potential reason for the decreasing barrier properties toward O$_2$ of the films deposited at gas mixture ratios of 30 : 1 and even higher could be stress in the interface between the polymer substrate and the increasingly inorganic barrier coating, responsible for reduced adhesion and enhanced brittleness of the SiO$_x$ film[31, 40].

8.3.2.2 SiO$_x$ Barrier Films Deposited from O$_2$: HMDSN Gas Mixtures

Hexamethyldisilazane (HMDSN) was selected as a different silicon organic compound, replacing HMDSO as precursor in the working gas mixture. Comparable with the plasma deposition tests using HMDSO as precursor (see preceding Section 8.3.2.1), SiO$_x$ barrier films were plasma deposited from different O$_2$: HMDSN gas mixture ratios on PET foil samples. During plasma deposition, the PET foil samples were located in the middle of the vacuum vessel directly on top of the bottom plate (cf. Figure 8.6). The steady horizontal movement of the magnet configuration with the cycle time of approximately 60 seconds produced a homogeneous thickness of the SiO$_x$ films on the PET foil samples.

In order to be directly comparable with the results obtained for the O$_2$: HMDSO gas mixtures, the process parameters working pressure and mean microwave power were kept unchanged at $p = 6$–8 Pa and $P = 560$ W (in pulse mode, the pulsing frequency being set to $f = 1$ kHz and the duty cycle to 50%), respectively.

8.3.2.2.1 O$_2$ Permeation Measurements: Variation of the O$_2$: HMDSN Gas Mixture Ratio
Corresponding to the ECR plasma deposition of the SiO$_x$ barrier films from O$_2$: HMDSO gas mixtures, the O$_2$: HMDSN gas mixture ratio was varied in the range of 1 : 1–35 : 1. Furthermore, the thickness of the SiO$_x$ films on the PET foil samples taken for O$_2$ permeation measurement was again approximately 100 nm for all O$_2$: HMDSN gas mixture ratios applied. Therefore it was possible to

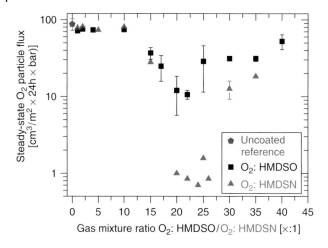

Figure 8.14 Steady-state O_2 particle flux as a function of the mixture ratio of O_2: HMDSO and O_2: HMDSN gas mixtures, respectively, obtained for the corresponding SiO_x barrier films deposited on PET foil by the ECR plasma source [48].

directly compare the OTR values referring to different O_2: HMDSO gas mixtures in Figure 8.9 with the OTR values measured for SiO_x films deposited from O_2: HMDSN gas mixtures (see Figure 8.14).

For gas mixtures in the range of 1 : 1 up to 10 : 1, almost no improvement of the barrier properties toward O_2 can be identified: the OTR results for both O_2: HMDSO and O_2: HMDSN gas mixtures in this range are still nearly equal to the OTR value of $87(\pm15)$ cm^3 $O_2/(m^2 \cdot 24$ h$\cdot$bar) of the uncoated PET foil samples. Comparable with the results obtained for the O_2: HMDSO gas mixtures, the OTR value is significantly reduced down to 28 cm^3 $O_2/(m^2 \cdot 24$ h$\cdot$bar) when increasing the O_2: HMDSN gas mixture ratio from 10 : 1 up to 15 : 1. Minimum OTR results are found for O_2: HMDSN gas mixtures in the range between 20 : 1 and 26 : 1. In comparison with the minimum OTR values of $12(\pm6)$ cm^3 $O_2/(m^2 \cdot 24$ h$\cdot$bar) and of $11(\pm1)$ cm^3 $O_2/(m^2 \cdot 24$ h$\cdot$bar) obtained for the O_2: HMDSO gas mixture ratios of 20 : 1 and 22 : 1, respectively (see Section 8.3.2.1.2), the OTR results measured for O_2: HMDSN gas mixtures in the range between 20 : 1 and 26 : 1 are even one order of magnitude lower: the minimum OTR value of 0.7 cm^3 $O_2/(m^2 \cdot 24$ h$\cdot$bar) was determined for the O_2: HMDSN gas mixture ratio of 24 : 1. For O_2: HMDSN gas mixture ratios of 30 : 1 and 35 : 1, the OTR values were increasing again, still being lower compared with the corresponding OTR values related to the O_2: HMDSO gas mixtures, but not as much as one order of magnitude.

In conclusion, similar behavior of the OTR results as a function of the O_2: HMDSO and of the O_2: HMDSN gas mixture ratio, respectively, could be identified. Nevertheless, the best performing SiO_x films deposited on PET foil samples from O_2: HMDSN gas mixtures provide OTR values approximately one order of magnitude lower compared with the minimum OTR results related to

O$_2$: HMDSO gas mixture ratios in the same range. FTIR analysis on best performing SiO$_x$ films deposited from O$_2$: HMDSO and O$_2$: HMDSN gas mixtures, respectively, at the same mixture ratio should indicate if the significantly lower OTR results related to O$_2$: HMDSN gas mixtures can be explained by a potentially different chemical composition of the corresponding SiO$_x$ films.

8.3.2.2.2   FTIR Analysis: Comparing Best Performing SiO$_x$ Barrier Films Deposited from O$_2$: HMDSO and from O$_2$: HMDSN Gas Mixtures

Equivalent to the FTIR analysis in Section 8.3.2.1.3, two pieces of aluminum foil (approximately 2 cm × 5 cm) were coated in parallel with the best performing SiO$_x$ films deposited from O$_2$: HMDSO and O$_2$: HMDSN gas mixtures, respectively, at the same mixture ratio. The thickness of the SiO$_x$ films on the pieces of aluminum foil turned out to be approximately 30–40 nm. The two coated pieces of aluminum foil were applied with their plasma-coated surfaces onto opposite sides of a KRS-5 crystal for FTIR analysis by use of a Bruker Vector 22 FTIR spectrometer in the attenuated total reflectance (ATR) mode.

The FTIR analysis of the best performing SiO$_x$ films deposited from the same gas mixture ratio of O$_2$: HMDSO and of O$_2$: HMDSN, respectively, should provide information to explain the OTR being approximately one order of magnitude lower for O$_2$: HMDSN ratios in the range between 20 : 1 and 26 : 1 compared with O$_2$: HMDSO ratios in the same range. Figure 8.15 shows the peaks of the different vibration bands at approximately the same wave numbers as were found in the FTIR spectra for higher O$_2$: HMDSO ratios in Figure 8.10. The peaks at 1228 cm^{-1} and at 809 cm^{-1} related to transverse optical out-of-phase anti-symmetric stretching vibrations and transverse optical bending vibrations of Si–O–Si groups, respectively, do not present any noticeable differences in peak intensity when

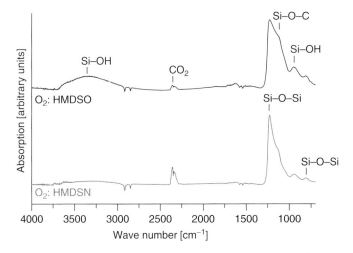

Figure 8.15   Infra-red spectra of the best performing SiO$_x$ barrier films plasma deposited from O$_2$: HMDSO and O$_2$: HMDSN gas mixtures at the same mixture ratio [48].

comparing the two FTIR spectra of the SiO_x films obtained from O_2: HMDSO and O_2: HMDSN, respectively [35, 39]. When looking at the shoulder at 1122 cm^{-1} and the peak at 947 cm^{-1} that are assignable to transverse optical stretching vibrations of Si–O–C and Si–OH groups, respectively, increased intensities in the FTIR spectrum of the SiO_x film deposited from O_2: HMDSO are perceptible [32]. However, the most obvious difference can be seen in the significantly higher intensity of the band structure between approximately 3700 and 2900 cm^{-1} in the FTIR spectrum of the SiO_x film deposited from O_2: HMDSO, correlated to transverse optical stretching vibrations of the OH groups in Si–OH [20, 41].

As these results clearly indicate higher concentrations of OH groups and carbon in the chemical composition of the SiO_x film deposited from O_2: HMDSO, similar conclusions can be drawn with respect to Section 8.3.2.1.3: the higher amount of OH groups and carbon in the SiO_x film deposited from O_2: HMDSO in comparison with the one deposited from O_2: HMDSN at the same gas mixture ratio, resulting in reduced cross-linking in the SiO_x film, might be the reason for the minor barrier properties toward O_2, as it is discussed in detail, for example, by Creatore *et al.* [34].

8.3.2.2.3 O_2 Permeation Measurements: Variation of the Film Thickness   ECR plasma deposition of SiO_x films on PET foil samples was performed at the O_2: HMDSN gas mixture ratio of 24 : 1 by variation of the coating thickness. Figure 8.16 shows that the OTR of the uncoated PET foil can be diminished by a factor of almost 30 already by plasma deposition of a SiO_x film of approximately 30 nm. An additional reduction of the OTR by a factor of four can be achieved when increasing the thickness of the SiO_x film to 90 nm. Finally, the OTR value of 0.4 cm^3 $O_2/(m^2 \cdot 24\,h \cdot bar)$ of the SiO_x film of approximately 215 nm after 480 seconds

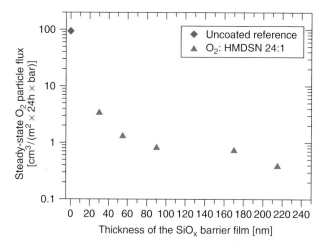

Figure 8.16 Steady-state O_2 particle flux as a function of the thickness of the SiO_x barrier films on PET foil samples, plasma deposited at the constant O_2: HMDSN gas mixture ratio of 24 : 1 by use of the ECR plasma source.

of plasma deposition time presents another decrease of the OTR by a factor of roughly two. The minimum OTR of 0.4 cm^3 O$_2$/(m^2·24 h·bar) corresponds very well with the results achieved by other research groups and with the present industrial standard [3, 4, 49].

8.3.3
SiO$_x$ Barrier Films on PP Foil

In parallel to the ECR plasma deposition of SiO$_x$ films from O$_2$: HMDSO gas mixtures on PET foil samples, corresponding plasma deposition tests were performed on PP foil samples 'Profol 0202/20' 95 µm in thickness by use of HMDSO as precursor, too. For these tests, both the ECR source presented in Figure 8.6 and a Duo-Plasmaline type of plasma source were used (see Section 2.2.1.3). The application of the Duo-Plasmaline type of plasma source should prove the feasibility of depositing efficient SiO$_x$ films from different low-pressure microwave plasma sources on PP as a representative of thermally sensitive polymer materials.

8.3.3.1 ECR Plasma Source: Comparing the Barrier Properties of SiO$_x$ Films Deposited on PP and on PET Foil by Variation of the O$_2$: HMDSO Gas Mixture Ratio

As the ECR plasma deposition tests on foils of PP and of PET were performed in parallel, almost identical process parameters were applied: HMDSO was used as precursor for plasma deposition of the single-layer SiO$_x$ films, the working pressure was kept in the range of $p = 6-8$ Pa, and the constant mean microwave power of approximately 560 W was applied in pulse mode (pulsing frequency $f = 1$ kHz, duty cycle 50%). Figure 8.6 shows the position of the PP foil samples: they were located in the middle of the vacuum vessel directly on top of the bottom plate. Just as in the ECR plasma deposition tests on PET foil, the PP foil samples were directly exposed to the plasma due to the steady horizontal movement of the magnet configuration underneath the bottom plate. The cycle time of the horizontal movement of the magnet configuration was approximately 60 seconds.

Based on the experience gained from the ECR plasma deposition tests on PET foil, the O$_2$: HMDSO gas mixture ratio was varied only in the range of 10 : 1 and 40 : 1. Different from the equivalent plasma deposition tests on PET foil, the thickness of the single-layer SiO$_x$ films was restricted up to a maximum of approximately 25–30 nm for all gas mixture ratios, as delamination of the SiO$_x$ film from the PP foil and subsequent formation of cracks in the coating could be observed from thicknesses of approximately 50 nm and higher.

Similar to the results obtained for ECR plasma deposition of SiO$_x$ films from O$_2$: HMDSO and O$_2$: HMDSN gas mixtures on PET foil samples (see Figures 8.9 and 8.14), the OTR value referring to the SiO$_x$ film deposited from the O$_2$: HMDSO gas mixture ratio of 10 : 1 on PP foil was even a bit higher than the OTR of 1131($\pm$37) cm^3 O$_2$/(m^2·24 h·bar) of the uncoated PP foil (see Figure 8.17). A significant decrease of the OTR was achieved by increasing the O$_2$: HMDSO gas mixture ratio from 10 : 1 up to 15 : 1. This behavior corresponds qualitatively

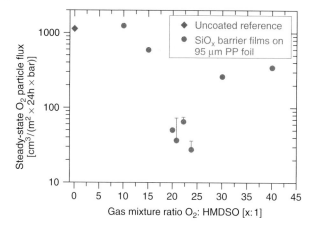

Figure 8.17 Steady-state O_2 particle flux as a function of the O_2: HMDSO gas mixture ratio, obtained for SiO_x barrier films deposited on PP foil by use of the ECR plasma source.

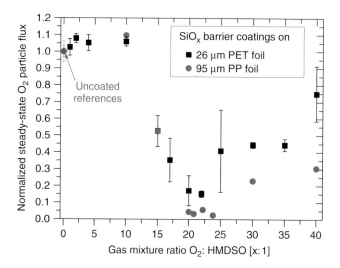

Figure 8.18 Normalized steady-state O_2 particle flux as a function of the O_2: HMDSO gas mixture ratio, obtained for SiO_x barrier films deposited by the ECR plasma source on PET and PP foil, respectively [31].

extremely well with the OTR results obtained for SiO_x films on PET foil deposited from ratios of 15 : 1 and smaller (Figure 8.18). This figure displays the OTR results as a function of the gas mixture ratio applied for ECR deposition on PET and PP foil, respectively. The values are normalized to the OTR value of the respective uncoated reference sample.

The minimum OTR results for the single-layer SiO_x films on PP foil samples were found for ratios in the range between 20 : 1 and 24 : 1, which agrees very well with the corresponding results on PET foil. The circles in Figure 8.17 refer to the minimum OTR values measured for the respective O_2: HMDSO gas mixture ratio, the error bars indicate the range of OTR results found for SiO_x films deposited repeatedly from the same O_2: HMDSO gas mixture ratio. The OTR result determined for the best performing SiO_x film on PP foil was 28 cm^3 $O_2/(m^2 \cdot 24\,h \cdot bar)$ for the ratio of 24 : 1. Compared with this OTR value, the OTR results for the films deposited on PP foil from the ratios of 30 : 1 and 40 : 1 turned out to be approximately one order of magnitude higher.

8.3.3.2   Duo-Plasmaline Plasma Source: SiO_x Barrier Films Deposited from O_2: HMDSN Gas Mixtures

The mode of operation of the Duo-Plasmaline plasma source is described in detail in Section 2.2.1.3. The set-up used for plasma deposition of SiO_x films on PP foil samples consists of two Duo-Plasmaline plasma sources arranged in parallel at the distance of 9 cm. The microwave power is applied equally to both sources from two opposite sides of the vacuum vessel via a power splitter. Homogeneous plasma treatment of samples on an area of up to approximately 30 cm × 30 cm was made possible by moving the substrate in the direction perpendicular to the two Duo-Plasmaline plasma sources. Compared with the sketch in Figure 8.19, the positions of the gas supply and the substrate holder were changed, that is, the substrate holder was arranged above and the gas supply below the two Duo-Plasmaline plasma sources. Therefore the substrate had to be fixed face down

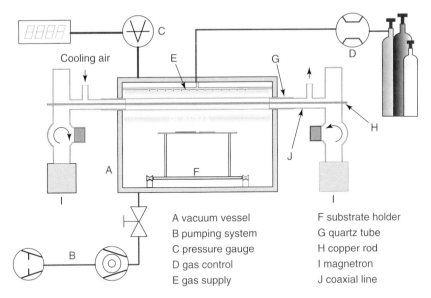

Cooling air

A vacuum vessel
B pumping system
C pressure gauge
D gas control
E gas supply

F substrate holder
G quartz tube
H copper rod
I magnetron
J coaxial line

Figure 8.19   Schematic of the experimental set-up of the Duo-Plasmaline type of plasma source for plasma deposition of 2D substrates. During plasma deposition of SiO_x barrier films on PP foil samples, the positions of the gas supply and the substrate holder were interchanged, that is, the substrate holder was arranged above and the gas supply below the two Duo-Plasmaline plasma sources [31].

onto the substrate holder, having the positive effect that no particles could fall on the surface of the substrate during plasma deposition, which would have caused defects in the plasma coating. When passing directly on top of the two Duo-Plasmaline plasma sources, the distance between the PP foil samples fastened to the substrate holder and the Duo-Plasmaline plasma sources was approximately 20 cm.

HMDSN was used as precursor in combination with O_2 as reactive gas to find out if the single-layer SiO_x films on PP foil can be significantly improved by plasma deposition with the Duo-Plasmaline type of plasma source, too. The admixture of argon to the O_2: HMDSN gas mixtures turned out to have a positive effect on the O_2 barrier properties. In a first step, the O_2: HMDSN gas mixture ratio was varied during deposition on PP foil at the working pressure of $p = 15$ Pa. The microwave power was kept constant at 720 W in continuous wave (cw) mode. In a second step, variation of the O_2: HMDSN gas mixture ratio was repeated at the elevated working pressure of $p = 30$ Pa. The thickness of the single-layer SiO_x films was kept in the range of approximately 25–40 nm so that the results would be comparable.

SiO_x films deposited from the O_2: HMDSN gas mixture ratios of 30 : 1 and of 40 : 1 at $p = 15$ Pa already show a slight decrease of the OTR by a factor of approximately 3.5 and 4.5, respectively, compared with the OTR of 1131($\pm$37) cm^3 O_2/(m^2·24 h·bar) of the uncoated PP reference (see Figure 8.20). The minimum OTR value of 38 cm^3 O_2/(m^2·24 h·bar) was obtained for the ratio of 47 : 1. When further increasing the O_2 content, the OTR is somewhat increasing again. This gradual deterioration of the barrier effect might be explained by the SiO_x films becoming more and more quartz-like, leading to a reduced adhesion to the PP foil samples and an enhanced brittleness that can be causal for incipient ablation of the barrier coating and subsequent formation of cracks in the SiO_x film.

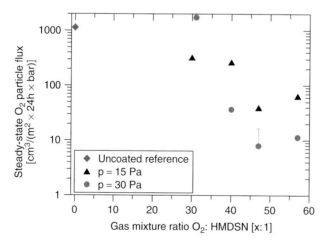

Figure 8.20 Steady-state O_2 particle flux as a function of the O_2: HMDSN gas mixture ratio with admixture of argon, obtained for SiO_x barrier films deposited on PP foil at the working pressure of 15 Pa and of 30 Pa, respectively, by use of the Duo-Plasmaline type of plasma source.

The measurement of the O$_2$ permeation for films deposited on PP foil at the higher working pressure of $p = 30$ Pa revealed noticeably lower OTR values for O$_2$: HMDSN ratios in the range of 40 : 1 and 57 : 1 (see Figure 8.20). Different from the OTR results obtained at $p = 15$ Pa, there was no significant increase of the OTR when the O$_2$ content was raised from 47 : 1 to 57 : 1, showing the OTR of 11 cm^3 O$_2$/(m^2·24 h·bar) for the ratio of 57 : 1 and 8 cm^3 O$_2$/(m^2·24 h·bar) for the best performing film deposited at the ratio of 47 : 1. After repetition of the plasma deposition test on PP foil at the same ratio of 47 : 1, the OTR value of 16 cm^3 O$_2$/(m^2·24 h·bar) was measured. This result represents a relatively good confirmation of the minimum OTR value of 8 cm^3 O$_2$/(m^2·24 h·bar) by taking into account that the second OTR value of 16 cm^3 O$_2$/(m^2·24 h·bar) could easily be caused by a higher number of minor defects in the SiO$_x$ film, maybe even due to a higher surface roughness of the second PP foil sample showing peaks and cavities that can hardly be covered by the SiO$_x$ film with a thickness of merely 25–30 nm.

8.3.4
ECR Plasma Deposition of SiO$_x$ Barrier Films on Polymer Trays Designed for Food Packaging

Injection-molded trays made of pure PP were used within ACTECO for plasma deposition of SiO$_x$ barrier films. These trays were divided into two compartments (see Figure 8.21). The outer diameter of the sealing rim on top was 18.0 cm, the inner diameter 16.5 cm, being tapered to approximately 16 cm on the bottom of the two compartments. The depth of the two compartments was 3.5 cm. According to the end-user requirements, SiO$_x$ films should be plasma deposited on the trays to achieve the target of 5 cm^3 O$_2$/(m^2·24 h·bar) OTR or even less. During plasma deposition, the sealing rim of the injection-molded trays had to be covered by a frame in order to maintain its sealability.

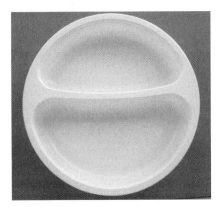

Figure 8.21 PP food tray sample as used for plasma deposition of SiO$_x$ barrier films.

8.3.4.1   ECR Plasma Deposition of SiO$_x$ Barrier Films without Directed Gas Supply and Customized Magnet Configuration: Variation of the Plasma Deposition Time and of the Distance between Sample and Plasma

Apart from the formation of less intense plasmas, ECR plasmas offer another decisive advantage compared with plasmas generated by the Duo-Plasmaline: the magnets can be easily arranged to form a customized ECR plasma, providing the homogeneous plasma deposition of 3D substrates in almost every shape, particularly with regard to the homogeneity of the coating thickness. Therefore the decision was taken to apply the ECR plasma source for plasma deposition of SiO$_x$ films on the injection-molded PP trays.

The PP trays were fixed to the substrate holder in the form of a mounting directly underneath their sealing rim, thus being located in the middle of the vacuum vessel at a height of approximately 110 mm above the bottom plate during plasma deposition (cf. Figure 8.22). Consequently, all inside and outside surfaces of the two compartments of the trays were directly exposed to the ECR plasma. The magnet configuration underneath the bottom plate was kept in a steady movement in order to ensure a uniform plasma treatment. Plasma deposition of the SiO$_x$ films was performed at the working pressure of $p = 6$–8 Pa and at the constant mean microwave power of approximately 560 W in pulse mode (pulsing frequency $f = 1$ kHz with the duty cycle of 50%).

The O$_2$: HMDSN gas mixture ratio was varied in the range of 29 : 1 up to 50 : 1. The optimum range of the plasma deposition time proved to be between 120 and 180 seconds.

The OTR results obtained by the ACTECO project partner Biophy during O$_2$ particle flux measurements performed on the larger compartment of the PP trays are presented in Figure 8.23. The Mocon O$_2$ transition rate test system OX-TRAN

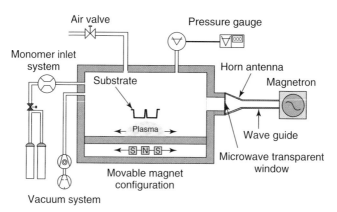

Figure 8.22   Schematic of the experimental set-up of the ECR plasma source, showing the position of the substrate in the middle of the vacuum vessel approximately 110 mm above the bottom plate during plasma deposition of the SiO$_x$ barrier film. The magnet configuration was moved during plasma deposition.

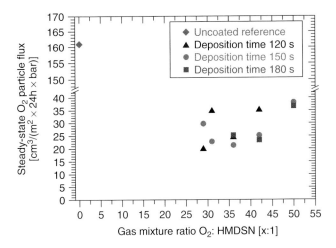

Figure 8.23 Steady-state O$_2$ particle flux as a function of the O$_2$: HMDSN gas mixture ratio, obtained for SiO$_x$ barrier films deposited on injection-molded trays made of pure PP by the ECR plasma source at the plasma deposition times of 120, 150, and 180 seconds, respectively.

Model 2/61 used was operated at the constant temperature of $T = 23\,°C$ and at approximately 50% relative humidity of the permeant air. For the plasma deposition time of 120 seconds, the minimum OTR value of 20 cm^3 O$_2$/(m^2·24 h·bar) was recorded for the O$_2$: HMDSN gas mixture ratio of 29 : 1. This corresponds to an improvement of the O$_2$ barrier properties by a factor of 8 compared with the uncoated trays that showed the OTR of 161 cm^3 O$_2$/(m^2·24 h·bar). The minimum thickness of the SiO$_x$ film was 10–15 nm on the inside bottom and a bit more than 15 nm on the outside directly underneath the sealing rim of the two compartments. This finding points to the fact that both the inside and the outside coating contributed to the barrier effect. When increasing the O$_2$ content, the plasma deposition rate is decreasing. Thus, when keeping the plasma deposition time fixed at 120 seconds, locally insufficient coverage of the polymer material by the SiO$_x$ film may occur, which can account for the minor O$_2$ barrier properties at higher O$_2$: HMDSN gas mixture ratios.

The SiO$_x$ films deposited on the PP trays at the plasma deposition time of 150 seconds showed a maximum barrier effect for the O$_2$: HMDSN gas mixture ratio of 36 : 1. The corresponding OTR was 21 cm^3 O$_2$/(m^2·24 h·bar). Again, the minimum thickness of the SiO$_x$ film of 10–20 nm on the inside bottom and of almost 30 nm directly underneath the sealing rim of the two compartments suggests that the barrier effect is due to film deposition on both the inside and outside of the PP tray. The maximum thickness of the SiO$_x$ film of 80 nm on the inside directly underneath the sealing rim and of more than 100 nm on the outside bottom of the two compartments proved to be already close to the threshold for delamination and subsequent crack formation in the SiO$_x$ film. Therefore, the gradual loss of barrier properties of the films deposited at lower O_2: HMDSN gas mixture ratios than 36 : 1 is most probably due to the maximum thickness of these barrier coatings already being beyond this threshold.

Finally, the minimum OTR result of 23 cm^3 O_2/(m^2·24 h·bar) for the plasma deposition time of 180 seconds was found at the O_2: HMDSN gas mixture ratio of 42 : 1. Almost no difference in film thickness could be determined between this film and the film deposited from the O_2: HMDSN gas mixture ratio of 36 : 1 at the deposition time of 150 seconds. The directly comparable results of the thickness measurements on the best performing SiO$_x$ films (i.e., the coatings deposited at 120 and at 180 seconds plasma deposition time), lead to the conclusion that the optimum coating thickness is a decisive factor regarding the quality of their barrier properties under these particular experimental conditions.

In a further step, the effect of the distance between the ECR plasma and the PP trays was investigated. At two different distances (110 and 165 mm), the O_2: HMDSN gas mixture ratio was varied for the constant plasma deposition time of 180 seconds. Note that the OTR results for the distance of 110 mm and the deposition time of 180 seconds in Figure 8.24 are taken from Figure 8.23. Comparing the OTR results for the two different distances in Figure 8.24 leaves a clear message: for O_2: HMDSN gas mixtures in the range of 36 : 1–50 : 1, the positioning of the trays at the larger distance of 165 mm above the bottom plate is favorable for improved barrier properties. The minimum OTR of 20 cm^3 O_2/(m^2·24 h·bar) was obtained for the O_2: HMDSN gas mixture ratio of 42 : 1 at the larger distance of 165 mm. Taking into consideration that in this case the film thickness was only marginally smaller compared with the corresponding film thickness for the smaller distance,

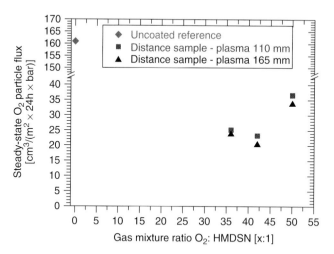

Figure 8.24 Steady-state O_2 particle flux as a function of the O_2: HMDSN gas mixture ratio, obtained for SiO$_x$ barrier films deposited on injection-molded trays made of pure PP by the ECR plasma source at different distances between the substrate and the bottom plate of approximately 110 and 165 mm, respectively. The plasma deposition time of 180 seconds was kept constant.

it can be concluded that the improved barrier properties are most probably due to the lower thermal impact on the injection-molded PP trays at larger distances.

8.3.4.2 Achieving Industrially Relevant Plasma Deposition Times by Directed Gas Supply and Customized Magnet Configuration

In the pre-evaluation tests by application of the experimental set-up of Figure 8.22, no set of process parameters, with the plasma deposition time being considerably less than 120 seconds, could be found that showed OTR results at least close to the minimum values presented in Figures 8.23 and 8.24 (see previous Section 8.3.4.1). For plasma deposition of the SiO$_x$ films on the injection-molded PP trays on industrial level in an inline or at least semi-batch process by treating several trays in parallel, the deposition time would have to be limited to a maximum of 60 seconds in order to comply with the present industrial specifications. Therefore, a new plasma deposition process had to be designed to noticeably increase the plasma deposition rate. In order to keep the OTR of the coated trays as low as possible by minimizing the density of coating defects particularly caused by mechanical stress in the interface between the barrier coating and the surface of the trays as well as in the SiO$_x$ film itself, the new plasma deposition process should additionally provide an improved homogeneity of the coating thickness on the entire surface of the tray, a higher mechanical flexibility of the films and a better film adhesion.

As a directed gas supply in combination with the ECR plasma being in close vicinity to the trays was supposed to produce both higher deposition rates and improved homogeneity of the thickness of the SiO$_x$ barrier films, a further concept for the experimental set-up was developed [48]. A new customized magnet configuration was designed that should provide for a considerably reduced thermal impact on the trays. The further concept for the experimental set-up presented in Figure 8.25 comprises an optimized gas supply for directed feed-in of the working gas O$_2$ and the precursor HMDSN into the ECR plasma produced by the new customized magnet configuration. The substrate holder was located approximately 60 mm underneath the new customized magnet configuration in the direction of the gas flow. Consequently, the trays were positioned in the middle of the vacuum vessel approximately 175 mm above the bottom plate. During plasma deposition, the magnet configuration directly underneath the bottom plate was kept fixed in its position in the turning point opposite to the horn antenna in order to reduce the effect of the ECR plasma right on top of the bottom plate to a minimum.

In order to achieve a better adhesion of the SiO$_x$ barrier films to the surface of the trays, tests were performed on plasma deposition of gradient layers: at the beginning of the plasma deposition, only HMDSN is fed into the plasma. The resulting coating has a polymer-like chemical structure due to a high content of carbon and hydrogen and is intended to provide a good adhesion to the surface of the polymer trays as well as to increase the mechanical flexibility of the entire gradient layer. Then O$_2$ is gradually added to the gas mixture. This procedure allows a continuous transition from the soft, polymer-like coating to a hard SiO$_x$ film necessary for considerably reducing the O$_2$ permeability.

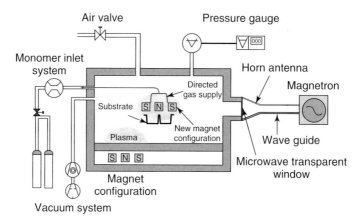

Figure 8.25 Schematic of the experimental set-up of the ECR plasma source, showing the position of the substrate during plasma deposition of SiO$_x$ gradient layers onto the inside of the PP trays, which are in the middle of the vacuum vessel directly underneath the directed gas supply and the new magnet configuration.

SiO$_x$ gradient layers from different O$_2$: HMDSN gas mixture ratios in the range of 40 : 1 up to 71 : 1 were plasma deposited on injection-molded trays made of pure PP at the plasma deposition times of 45 and 60 seconds, respectively, for analysis of the O$_2$ permeability. The mean microwave power of 1160 W in pulse mode (pulsing frequency $f = 1$ kHz, duty cycle 50%) as well as the working pressure of $p = 6$–8 Pa were applied during the plasma deposition processes. The O$_2$ particle flux measurements were performed on the larger compartment of the injection-molded trays made of pure PP by a Mocon O$_2$ transition rate test system OX-TRAN Model 2/61 at the constant temperature of 23 °C and at approximately 25% relative humidity of the permeant air.

The triangles and circles in Figure 8.26 indicate the gas mixture ratios applied at the end of the deposition processes of the SiO$_x$ gradient layers for deposition times of 45 and of 60 seconds, respectively. According to the results in Figure 8.26, the OTR decreases with lower content of O$_2$ in the gas mixture ratio. This would indicate a dominant effect of the coating thickness with regard to the improvement of the O$_2$ barrier properties, as the deposition rate increases when the O$_2$ content in the O$_2$: HMDSN gas mixture is reduced. Nevertheless, the barrier properties of the SiO$_x$ gradient layers can be significantly improved by reducing the deposition time from 60 to 45 seconds.

The minimum OTR of 13 cm^3 O$_2$/(m^2·24 h·bar) was obtained for the SiO$_x$ gradient layer deposited from the O$_2$: HMDSN gas mixture ratio of 40:1 at the plasma deposition time of 45 seconds. Thickness measurements showed only a few nanometers of coating thickness in maximum on the outer surface of the compartments for all the different sets of process parameters applied,

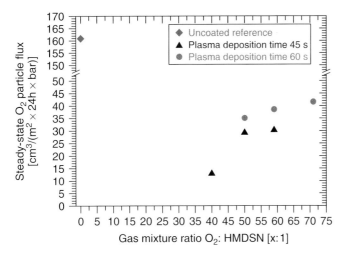

Figure 8.26 Steady-state O$_2$ particle flux as a function of the O$_2$: HMDSN gas mixture ratio, obtained for SiO$_x$ gradient layers deposited on injection-molded trays made of pure PP by the ECR plasma source, equipped with the new magnet configuration in between the substrate and the directed gas supply, at different plasma deposition times of 45 and of 60 seconds, respectively [48].

which confirmed the assumption that the barrier properties would be determined predominantly by the SiO$_x$ gradient layer inside the compartments of the trays. According to the thickness measurements on the SiO$_x$ gradient layer inside the two compartments of the tray deposited from the O$_2$: HMDSN gas mixture ratio of 40 : 1, the minimum thickness of this SiO$_x$ gradient layer on the bottom was approximately 30 nm and the maximum thickness directly underneath the sealing rim was nearly 65 nm. Due to the fact that the gas change from pure HMDSN to the final O$_2$: HMDSN gas mixture ratio of 40 : 1 was completed after approximately 30–35 seconds of the plasma deposition time, it can be estimated that the thickness of the hard SiO$_x$ film deposited during the remaining 10–15 seconds of the total plasma deposition time of 45 seconds did not exceed 10–20 nm. This leaves the conclusion that the mechanical flexibility provided by the polymer-like intermediate layer directly on the surface of the injection-molded trays made of pure PP is very important with regard to achieving excellent barrier properties toward O$_2$ permeability. The polymer-like intermediate layer seems to act as a kind of buffer between the surface of the trays and the hard SiO$_x$ surface layer. When comparing the OTR results in Figure 8.26 obtained for the different plasma deposition times of 45 and of 60 seconds at the O$_2$: HMDSN gas mixture ratios of 50 : 1 and 59 : 1, it can be assumed that the additional growth of the hard SiO$_x$ surface layer after 45 seconds of the plasma deposition time is detrimental with regard to the mechanical flexibility of the entire SiO$_x$ gradient layer, finally causing additional coating defects and an increase in the OTR.

8.4
Conclusions

A high percentage of food packaging materials is polymer-based, and their proportion in this field is steadily increasing. Excellent barrier properties toward a multitude of gases are an important criterion concerning the quality of polymer packaging. As packaging material produced from a single polymer cannot provide barrier properties toward most kinds of gases, it is very important to understand the diffusion process of gases through polymers in order to be able to design customized barriers to be added to the polymer. The fundamentals of diffusion, solubility, and permeability of polymers as well as corresponding diagnostic methods are described, and examples for barrier concepts are given. A very promising barrier concept concerning the permeability toward oxygen (O_2) is low-pressure microwave plasma deposition of silicon oxide (SiO_x) barrier films, performed by plasma sources of the Duo-Plasmaline type and based on the ECR principle, respectively.

SiO_x barrier films were plasma deposited on PET foil 26 μm in thickness by use of O_2 as reactive gas and HMDSO and HMDSN, respectively, as precursor in an ECR plasma set-up designed for the homogeneous plasma treatment of foil material. FTIR analysis of the chemical composition of the SiO_x barrier films performed in parallel with the measurement of the O_2 particle flux through the plasma-coated polymer foils showed that the content of carbon and the density of OH groups in the SiO_x barrier film had to be minimized and to be kept low, respectively, to obtain a considerable reduction of the permeability toward O_2. The OTR of ECR plasma-coated PET foils could be additionally reduced up to one order of magnitude by application of HMDSN as precursor instead of HMDSO, the minimum OTR achieved was 0.4 cm^3 $O_2/(m^2 \cdot 24\,h \cdot bar)$.

Plasma deposition experiments were also performed on PP foil 95 μm in thickness: the deposition of SiO_x barrier films from O_2: HMDSO gas mixtures by use of the ECR plasma set-up showed results qualitatively comparable with the plasma-coated PET foils. Efficient SiO_x barrier films on the thermally sensitive PP foil could be deposited even by use of the Duo-Plasmaline type of plasma source: the barrier properties toward O_2 of the uncoated PP foil could be improved by plasma deposition of SiO_x barrier films from O_2: HMDSN gas mixtures by a factor of approximately 120 at maximum, corresponding to the minimum OTR achieved of 8 cm^3 $O_2/(m^2 \cdot 24\,h \cdot bar)$.

The ECR plasma source proved to be advantageous with regard to the formation of a 3D plasma for customized deposition of SiO_x gradient layers on injection-molded trays made of pure PP. Starting with a polymer-like layer and ending in a hard SiO_x barrier film, these gradient layers provide good adhesion to the surface of the polymer material, high mechanical flexibility as well as excellent barrier properties concerning their permeability toward O_2. In first promising tests, the time for plasma deposition of the SiO_x gradient layers on the injection-molded trays made of pure PP could be reduced to 45 seconds, with the resulting OTR of 13 cm^3 $O_2/(m^2 \cdot 24\,h \cdot bar)$ being already close to the target of 5 cm^3 $O_2/(m^2 \cdot 24\,h \cdot bar)$.

Acknowledgements

The authors are indebted to their colleagues (in alphabetical order) Muhammad Iqbal Akbar, Jérôme Dutroncy, Dennis Kiesler, Martina Leins, Dr Andreas Schulz, Prof. Dr Uwe Schumacher, and Prof. Dr Ulrich Stroth for their contributions and for all their support.

References

1. Ashley, R.J. (1985) Permeability and plastics packaging, in *Polymer Permeability* (ed. J. Comyn), Elsevier Applied Science, London, pp. 269–308.

2. Fracassi, F., d'Agostino, R., Favia, P., and Van Sambeck, M. (1993) Thin film deposition in glow discharges fed with hexamethyldisilazane-oxygen mixtures. *Plasma Sources Sci. Technol.*, **2** (2), 106–111.

3. Lamendola, R. and d'Agostino, R. (1998) Process control of organosilicon plasmas for barrier film preparations. *Pure Appl. Chem.*, **70** (6), 1203–1208.

4. Lamendola, R., Favia, P., Palumbo, F., and d'Agostino, R. (1998) Plasma-modification of polymers: process control in PE-CVD of gas-barrier films and plasma-processes for immobilizing anti-thrombotic molecules. *Eur. Phys. J. Appl. Phys.*, **4**, 65–71.

5. Deilmann, M., Grabowski, M., Theiß, S., Bibinov, N., and Awakowicz, P. (2008) Permeation mechanisms of pulsed microwave plasma deposited silicon oxide films for food packaging applications. *J. Phys. D: Appl. Phys.*, **41** (13), 135207 (7 pp).

6. Vieth, W.R. (1991) *Diffusion in and through Polymers*, Oxford University Press, New York.

7. Comyn, J. (1985) Introduction to polymer permeability and the mathematics of diffusion, in *Polymer Permeability* (ed. J. Comyn), Elsevier Applied Science, London, pp. 1–10.

8. Rogers, C.E. (1985) Permeation of gases and vapours in polymers, in *Polymer Permeability* (ed. J. Comyn), Elsevier Applied Science, London, pp. 11–74.

9. Hopfenberg, H.B. (1974) *Permeability of Plastic Films and Coatings to Gases, Vapors, and Liquids. Polymer Science and Technology*, Vol. 6, Plenum Press, New York.

10. Müller-Plathe, F. (1994) Permeation of polymers – a computational approach. *Acta Polym.*, **45** (4), 259–293.

11. Huang, R.Y.M. and Rhim, J.-W. (1990) Theoretical estimations of diffusion coefficients. *J. Appl. Polym. Sci.*, **41** (3-4), 535–546.

12. Fels, M. and Huang, R.Y.M. (1970) Diffusion coefficients of liquids in polymer membranes by a desorption method. *J. Appl. Polym. Sci.*, **14** (3), 523–536.

13. Fujita, H. (1961) Diffusion in polymer-dilutent systems. *Fortschr. Hochpolymeren-Forsch.*, **3**, 1–47.

14. Crank, J. (1979) *The Mathematics of Diffusion*, Clarendon Press, Oxford.

15. Czeremuszkin, G., Latrèche, M., Wertheimer, M.R., and Sobrinho, A.S. (2001) Ultrathin silicon-compound barrier coatings for polymeric packaging materials: an industrial perspective. *Plasmas Polym.*, **6** (1-2), 107–120.

16. Creatore, M., Palumbo, F., d'Agostino, R., and Fayet, P. (2001) RF plasma deposition of SiO_2-like films: plasma phase diagnostics and gas barrier film properties optimisation. *Surf. Coat. Technol.*, **142-144**, 163–168.

17. Inagaki, N., Tasaka, S., and Nakajima, T. (2000) Preparation of oxygen gas barrier polypropylene films by deposition of SiO_x films plasma-polymerized from mixture of tetramethoxysilane and oxygen. *J. Appl. Polym. Sci.*, **78** (13), 2389–2397.

18. Lucovsky, G., Manitini, M.J., Srivastava, J.K., and Irene, E.A. (1987) Low-temperature growth of silicon dioxide films: a study of chemical bonding

by ellipsometry and infrared spectroscopy. *J. Vac. Sci. Technol. B*, **5** (2), 530–537.

19. Theil, J.A., Brace, J.G., and Knoll, R.W. (1994) Carbon content of silicon oxide films deposited by room temperature plasma enhanced chemical vapor deposition of hexamethyldisiloxane and oxygen. *J. Vac. Sci. Technol. A*, **12** (4), 1365–1370.

20. Walker, M., Baumgärtner, K.-M., Feichtinger, J., Kaiser, M., Schulz, A., and Räuchle, E. (2000) Silicon oxide films from the Plasmodul. *Vacuum*, **57** (4), 387–397.

21. Walther, M., Heming, M., and Spallek, M. (1996) Multilayer barrier coating system produced by plasma-impulse chemical vapor deposition (PICVD). *Surf. Coat. Technol.*, **80** (1-2), 200–202.

22. Walker, M. (2002) Plasma polymerized barrier films. *Recent Res. Dev. Appl. Polym. Sci.*, **1**, 281–312.

23. Friedrich, J.F., Wigan, L., Unger, W., Lippitz, A., Erdmann, J., Gorsler, H.-V., Prescher, D., and Wittrich, H. (1995) Barrier properties of plasma and chemically fluorinated polypropylene and polyethyleneterephthalate. *Surf. Coat. Technol.*, **74-75** (Pt 2), 910–918.

24. Walker, M., Baumgärtner, K.-M., Ruckh, M., Kaiser, M., Schock, H.W., and Räuchle, E. (1997) XPS and IR analysis of thin barrier films polymerized from C_2H_4/CHF_3 ECR-plasmas. *J. Appl. Polym. Sci.*, **64** (4), 717–722.

25. Feichtinger, J., Galm, R., Walker, M., Baumgärtner, K.-M., Schulz, A., Räuchle, E., and Schumacher, U. (2001) Plasma polymerized barrier films on membranes for direct methanol fuel cells. *Surf. Coat. Technol.*, **142-144**, 181–186.

26. Feichtinger, J., Kerres, J., Schulz, A., Walker, M., and Schumacher, U. (2002) Plasma modifications of membranes for PEM fuel cells. *J. New Mater. Electrochem. Syst.*, **5** (3), 155–162.

27. Köpnick, H., Schmidt, M., Brügging, W., Rüter, J., and Kaminsky, W. (2003) Polyesters, in *Ullmann's Encyclopedia of Industrial Chemistry*, 6th edn, Vol. 28, Wiley-VCH Verlag GmbH, Weinheim, pp. 75–102.

28. Whiteley, K.S., Heggs, T.G., Koch, H., Mawer, R.L., and Immel, W. (2003) Polyolefins, in *Ullmann's Encyclopedia of Industrial Chemistry*, 6th edn, Vol. 28, Wiley-VCH Verlag GmbH, Weinheim, pp. 393–495.

29. Song, J., Prox, M., Weber, A., and Ehrenstein, G.W. (1995) Self-reinforcement of polypropylene, in *Polypropylene: Structure, Blends and Composites* (ed. J. Karger-Kocsis), Chapman & Hall, London, pp. 273–294.

30. Walker, M., Meermann, F., Schneider, J., Bazzoun, K., Feichtinger, J., Schulz, A., Krüger, J., and Schumacher, U. (2005) Investigations of plasma polymerized barrier films on polymeric materials. *Surf. Coat. Technol.*, **200** (1-4), 947–952.

31. Schneider, J., Kiesler, D., Leins, M., Schulz, A., Walker, M., Schumacher, U., and Stroth, U. (2007) Development of plasma polymerised SiO_x barriers on polymer films for food packaging applications. *Plasma Process. Polym.*, **4** (S1), S155–S159.

32. Raynaud, P., Despax, B., Segui, Y., and Caquineau, H. (2005) FTIR plasma phase analysis of hexamethyldisiloxane discharge in microwave multipolar plasma at different electrical powers. *Plasma Process. Polym.*, **2** (1), 45–52.

33. Inoue, Y. and Takai, O. (1996) Spectroscopic studies on preparation of silicon oxide films by PECVD using organosilicon compounds. *Plasma Sources Sci. Technol.*, **5** (2), 339–343.

34. Creatore, M., Palumbo, F., and d'Agostino, R. (2002) Deposition of SiO_x films from hexamethyldisiloxane/oxygen radiofrequency glow discharges: process optimization by plasma diagnostics. *Plasmas Polym.*, **7** (3), 291–310.

35. Kim, H. and Jang, J. (1998) Infrared spectroscopic study of SiO_x film formation and decomposition of vinyl silane derivative by heat treatment. II. On copper surface. *J. Appl. Polym. Sci.*, **68** (5), 785–792.

36. Wrobel, A.M. (1987) Mechanism of plasma polymerization of

N-silyl-substituted cyclodisilazane: structure and properties of polymer film. *Plasma Chem. Plasma Process.*, **7** (4), 429–450.

37. Wrobel, A.M., Klemberg, J.E., Wertheimer, M.R., and Schreiber, H.P. (1981) Polymerization of organosilicones in microwave discharges. II. Heated substrates. *J. Macromol. Sci. A: Pure Appl. Chem.*, **15** (2), 197–213.

38. Agres, L., Ségui, Y., Delsol, R., and Raynaud, P. (1996) Oxygen barrier efficiency of hexamethyldisiloxane/oxygen plasma-deposited coating. *J. Appl. Polym. Sci.*, **61** (11), 2015–2022.

39. Gunde, M.K. (2000) Vibrational modes in amorphous silicon dioxide. *Phys. B: Condens. Matter*, **292** (3-4), 286–295.

40. Kim, H. and Jang, J. (1998) Infrared spectroscopic study of SiO_x film formation and decomposition of vinyl silane derivative by heat treatment. I. On KBr and gold surface. *J. Appl. Polym. Sci.*, **68** (5), 775–784.

41. Pai, P.G., Chao, S.S., Takagi, Y., and Lucovsky, G. (1986) Infrared spectroscopic study of SiO_x films produced by plasma enhanced chemical vapor deposition. *J. Vac. Sci. Technol. A*, **4** (3), 689–694.

42. Boyd, I.W. and Wilson, J.I.B. (1987) Structure of ultrathin silicon dioxide films. *Appl. Phys. Lett.*, **50** (6), 320–322.

43. Boyd, I.W. and Wilson, J.I.B. (1982) A study of thin silicon dioxide films using infrared absorption techniques. *J. Appl. Phys.*, **53** (6), 4166–4172.

44. Berreman, D.W. (1963) Infrared absorption at longitudinal optic frequency in cubic crystal films. *Phys. Rev.*, **130** (6), 2193–2198.

45. Röseler, A. (1990) *Infrared Spectroscopic Ellipsometry*, Akademie-Verlag, Berlin.

46. Harbecke, B., Heinz, B., and Grosse, P. (1985) Optical properties of thin films and the Berreman-effect. *Appl. Phys. A*, **38** (4), 263–267.

47. Häberle, E., Kopecki, J., Schulz, A., Walker, M., and Stroh, U. (2009) Deposition of barrier layers for thin film solar cells assisted by bipolar substrate biasing. *Plasma Process. Polym.*, **6** (S1), 282–286.

48. Schneider, J., Akbar, M.I., Dutroncy, J., Kiesler, D., Leins, M., Schulz, A., Walker, M., Schumacher, U., and Stroh, U. (2009) Silicon oxide barrier coatings deposited on polymer materials for applications in food packaging industry. *Plasma Process. Polym.*, **6** (S1), 700–704.

49. ALCAN Packaging, Products-High Barrier Films, Barriers-Comparison of Permeation Rates, *http://www. ceramis.com/en/produkte_high_ vergleich.thtml* (accessed on 28 December 2009).

9
Anti-wear Coatings for Food Processing

Maddalena Rostagno and Federico Cartasegna

9.1
Introduction

Anti-wear coatings are traditionally used in mechanics for several uses: automotive components, cutting tools for machining, tools and dies for primary and secondary transformations, and many other applications. This kind of materials, even if their development only started at the beginning of the 1970s, can be considered a quite established solution for increasing the performance of products or processes while keeping a high cost/benefit ratio. No re-design or equipment change is required, only a suitable substrate is needed, together with the willingness to test, under proper conditions, the newly coated tool.

The scenario described changes completely if the coatings use and application is analyzed in the framework of the food sector.

On the one hand, considering food processing, we cannot identify any novelties in the equipment for mass manufacturing industry: steel tools and dies are used for crushing, milling, and cutting food of different hardness (bones, meat, fruits) and consistency (milk, cheese, bakery). On the other hand, requirements are of course different for the health hazard connected to the improper use of substrates. Analyzing the actual legislation (at least in Europe), it can be seen that materials used for food processing are not standardized in a specific way. Inox steels, aluminum, and titanium alloys are the most common materials used and are officially recognized as safe and atoxic. No specific indications are given with respect to the use of alternative solutions: responsibility for the health and safety impact is left to the end-user, who must carry out all the necessary tests to show that the solution chosen is viable.

In this context, anti-wear coatings can be considered a brand new material if applied in the food processing tools and the ACTECO project (see Preface) has been the right place to exercise the development, characterization, and industrial testing.

The research activity started with an analysis of the state-of-the-art on anti-wear coatings in the food sector and analyzed the actual trends of the market. It was important to select the category of products suitable to be used in the food sector

Plasma Technology for Hyperfunctional Surfaces. Food, Biomedical and Textile Applications.
Edited by Hubert Rauscher, Massimo Perucca, and Guy Buyle
Copyright © 2010 WILEY-VCH Verlag GmbH & Co. KGaA, Weinheim
ISBN: 978-3-527-32654-9

and in this sense the actual standards and legislation sustained the choice of the physical vapor deposition (PVD) coatings, certifying them in this kind of application as totally atoxic and safe. This result allowed concentration on the investigation and development of the coatings for the specific case study selected: milling and cutting in the butchery sector. Consequently, the first priority was given to the functionalities required by this application field: improvement of the anti-wear properties in order to extend tool life and to reduce maintenance cost for the substitution of the worn tools. In this context, characterization techniques played a key role and it was of strategic importance to select the best characterization methodologies to monitor the structural and mechanical properties of the coatings capable to influence the anti-wear properties of the coatings during actual use. Coating design and development was carried out by Environment Park S.p.A. according to the industrial needs expressed by the Diad Group; both are ACTECO partners who actively collaborated in this four-year research program in the design of high-performing coatings that, in the end, allowed an increase in tool life duration of 300%, and thus proved the PVD coating potential for the food sector.

9.2
Recent Developments in PVD Coatings

The use of plasma surface treatment is actually growing in the food and food treatment industry, with benefits like improvement of process performance and efficiency, enabling new businesses or applications, process costs reduction or safety for products, people, and environment.

The wide range of plasma treatments applied allow achieving new materials and product properties comprising surface energy modification (e.g., activation for polymers), hydro-/oleophobicity or -philicity, deposition of biocompatible or barrier films, reduction of friction coefficient, corrosion resistance, and wear resistance.

Different plasma techniques are used to produce the above-mentioned surface treatments, but one that is widely diffused and applied is ceramic thin-film deposition for the enhancement of wear and corrosion resistance of metallic food processing components. Aside from the chemical and mechanical properties mentioned, the applied coatings are required to have proven properties of biocompatibility, following the present norm for the food contact.

Ceramic coatings dedicated to the development of thin films for wear resistance and chemical barrier properties are mostly based on titanium, which has a good biocompatibility. Some of the most popular ceramic coatings, like titanium nitride (TiN), which is widely used as anti-wear coating, are starting to be used with very good results for biomedical applications like hip prostheses, due to the high wear resistance and corrosion protection [1]. The fact that these coatings guarantee such high biocompatibility levels, suggests that the corrosion and migration standards for food contact will be satisfied.

Binary, ternary, and quaternary ceramic coatings are particularly suited for migration barrier function due to their chemical stability and micro- and nanostructure, which also have a proven high wear resistance.

In literature, comparative migration tests on steel surgical equipment are reported that show the effectiveness of TiN coatings as migration barrier for metals migration, in particular for Cr, Co, Ni, and Mo [2], see Table 9.1.

In particular new micro- and nanostructured ceramic coatings offer an advanced technological solution to obtain migration barriers for the heavy metals (Co, Cr, Mo) contained in the steels employed by the food industry.

A wide range of nitride coatings are actually in use the USA for food applications and are certified by the Food and Drugs Administration (FDA) [3] for specific food applications, while in the European Union (EU) specific certifications have to be required case by case, following present regulation.

In modern industry the improvement in surface character of components and production tools is often not a cost-effective or acceptable way of gaining real global improvements in environmental impact and manufacturing efficiency.

Since the introduction of new environmental protection measures in most European states and the USA, and since the onset of cost increases for waste disposal of cutting liquids and metal shavings in 1992–1998, several efforts have been made to substitute conventional cutting fluids or to develop techniques for dry machining (and in part for dry stamping). These were based on the development of new high temperature resistant and hard physical vapor deposition-chemical vapor deposition (PVD-CVD) coatings. For example considerable progress has been

Table 9.1 Data showing how the heavy metals migration potential is a general problem in the medical field; in particular, these analytical results show the effect on biomedical analysis alteration due to employment of metal stainless steel for medical sampling instruments.

Element	R^a	Element	R^a
Au	1.67	Mo	>3.3
Cd	>3.3	Ni	10.8
Co	>8	Sb	>3.3
Cr	12.5	Sc	2.5
Cs	2.9	Ta	3.2
Cu	>4	Tb	1
Fe	>2.5	Ti	>3.1
Hg	>2	V	>8
La	5	W	>3.8
Mn	125	Zn	>8

^{a}Ratio of the element quantity released by a standard stainless steel surgery cutter and a TiN coated one.

achieved with diamond and diamond-like coatings that are known to combine very low friction coefficients with ultra-high hardness (>80 GPa).

On the other hand, the use of diamond coatings on cutting tools is limited to non-ferrous materials, in fact steel alloys cannot be used as tool material coated by diamond or be machined by diamond because the steel alloys will be altered in microstructure by the diffusion of carbon species. Additionally, further restrictions on the substrate material are imposed by the catalytic effect of Fe, Co, and Ni, present in steel and hard metal. Therefore, alloys containing Fe, Ni, and Co and hard metals with a high Co content (higher that 6%) cannot be directly coated and machined with diamond. Alternatively, most ceramic hard PVD coatings such as TiN, Ti-Al-N, TiCN, and so on have high friction coefficients of the order of 0.5 (against steel) leading to excessive temperatures at the cutting edges of the tool.

The PVD deposition on cutting tools and dies with conventional binary coatings such as TiN, TiC and CrN and some ternary and quaternary coatings, respectively, such as Ti-Al-N, Ti-C-N, and Ti-Zr-C-N was employed by several tools producers, performing the deposition at temperatures below 400 °C in order not to induce an embrittlement of the substrate materials.

Additional progress for different applications is expected from new types of 'novel super-hard nanocomposite' coatings that combine high adhesion, high hardness, high temperature resistance with low friction coefficient, high temperature hardness, and a low thermal conductivity. The hard coatings being used presently include among others TiN, (TiAl)N, CrN, Al_2O_3, and multilayers. Their hardness is typically in the range of 20–30 GPa.

In the near future, an increase in demand for environmentally friendly, energy efficient, highly productive (high speed) precision machining is expected. Therefore, new coatings for dry and fast machining will be required and, consequently, new materials for coatings with better performance will be developed.

Novel super-hard nanocomposite materials have shown in several experimental examples super-hardness of ≥50 GPa reached by nitride based materials, such as nc-TiN/a-Si_3N_4, nc-W_2N/a-Si_3N_4, and nc-VN/a-Si_3N_4. The super-hardness is a consequence of that nanostructure: the strength and the hardness of materials are orders of magnitude smaller than the theoretical shear strength because of flaws, such as dislocation and microcracks which are present in any engineering material. In a nanocomposite, the multiplication and movement of dislocations is very efficiently blocked because of the small size of the nanocrystals and of the thin amorphous matrix. Nowadays, super-hardness has been achieved in many ternary and quaternary systems including (besides of the systems mentioned above) borides, carbides, or combinations of transition metal nitrides with soft metals and others. It is the universality of the design principle which allows one to tailor the properties of the coatings to the given application by an appropriate selection of a material combination with the desired properties. In such a way it is possible to design coating materials with high hardness, toughness, resistance against oxidation, and chemical reaction with the material to be machined, high adhesion, and so on.

9.3
Coatings Trends and Market Share

The antipollution policies already implemented, or being drawn up by the EU, raise awareness on environmental issues at industrial level to research and to develop of new products.

In particular, the areas of machining and molding are directly interested in these issues, since most operations are performed with the use of lubricants (pure oil and emulsions).

The total annual consumption of mineral lubricating oils is about 1 000 000 tonnes. The amount of oil allocated to the area of cutting fluids is equal to about 77 000 tonnes distributed as follows: 45 000 tonnes are used as whole oil (pure lubricant) and 32 000 tonnes are used as an emulsion (at 3%, 5% or 8% in water).

The annual amount of emulsions exhausted from the working area is over 1 500 000 tonnes, a relevant fact is that this amount exceeds the entire European production of lubricating oils. This fact urged the implementation by users, driven by the existing regulations, of a systematic policy on collection and disposal.

It must be taken into consideration that these figures do not include the cutting oils of vegetal origin, whose use is growing.

The use of high-speed and dry machining (that means high productivity and low environmental impact by eliminating the costs of disposal of waste lubricants) has steadily increased in the last ten years and much is already being done in research in order to innovate the existing production processes.

However, the intervention process, due to the high investment costs, is not always sustainable in the near future. It is, in fact, everyday experience for anyone involved in the sector, that the most effective low cost optimization concerns innovation in cutting tools.

The finishing of traditional substrates of cutting tools through PVD coatings enabled the possibility to reduce oil consumption through more efficient cutting processes including a significant improvement in terms of both increased lifetime and superior surface finishing (lower roughness, lower residual stresses).

Since the 1970s, the coating industry has been very dynamic, showing an enormous potential. In 2000 approximately 20–30% of product patents in Europe for cutting tools concerned the development of new coatings. Figure 9.1 clearly shows the vitality of this sector if compared with innovation in geometry, substrate, and other combinations.

The potential of such technological solutions is still not known worldwide, furthermore not all companies already using or internally producing coatings have the ability and knowledge to use coatings of complex composition. As example the aeronautical sector can be mentioned: application of new coatings is constrained by the lack of standards and benchmarking. Consequently, this field remains, rightly, very reluctant to introduce elements that may cause defects in manufactured parts, and thus, problems of reliability of machined components. In the automotive sector, coatings are widely used for cutting tools and for components in the valve-train

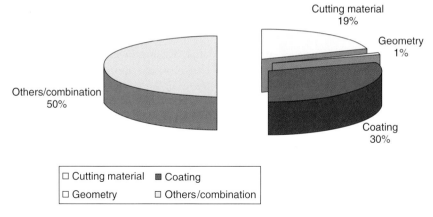

Figure 9.1 Distribution of main claims in patents for cutting tools in 2002 (Source: European Patent Office, Munich).

system (mainly in sport and luxury cars), but still a lot of development remains to extend the use of coatings also in components for mass-production cars.

The food sector is a specific niche in which national standards and regulations are compulsory for health, social, and ethical reasons. There are several standards related to the use of stainless steels and polymeric based materials, but few specifications are given for innovative materials such as coatings for which detailed indications for the end-users do not exist. This lack of normalization severely limits the diffusion of coatings used in the food sector, where traditional materials (steel and aluminum) are still considered the only reliable and safe solution.

9.4
Coatings Application in the Food Processing Sector

The corrosion of the machinery used for food processing is not only an economical cost (maintenance, substitution of corroded parts, production process interruptions), but above all a health and environmental cost. The corroded/eroded material is transferred into the food during the different processing phases and, consequently, it is systematically ingested by the final consumers, with concrete possibilities of accumulation and toxicology effects.

A typical example, at present under investigation in the USA, is the nickel release from stainless steels used in the bakery sector. It is suspected to be responsible for allergies and other pathologies, becoming more frequent even if the allowed nickel threshold has been continuously lowered.

Another typical example is in the cheese sector, where the processing of fresh cheese requires the use of cutters with very slippery surfaces, which means the lowest friction coefficient. The solution used at present are steels coated with fluorine-based coatings, having excellent anti-adhesion properties, but wear is very fast and the coating must be frequently replaced. The layers removed remain incorporated into the cheese and this can create potential allergenic risks in the long term.

In the food and biomedical sector growing caution is dedicated to the use of aluminum. Since the end of the last century, the aluminum toxicological and neuro-toxicological effects are under investigation: the hyper-aluminemia in humans has been detected as a possible cause of pathologies such as dialysis deficiency, pulmonary diseases, microcytic anemia, and others.

At present, the scientific community has strong suspicions that aluminum ingestion can be a possible cause of Parkinson's and Alzheimer's disease.

Recent studies have shown that if aluminum is absorbed through chemically stable and above all liophyle molecules, it is able to cross the hematoencephalic barrier and to penetrate into the neuronal cells.

Another important requirement for materials used in the food sector is the following: if abrasion or corrosion phenomena occur, the eroded material can have toxicological effects in the long term. In this sense it is also important that materials are chemically stable and inert in different working conditions (acid, salt, water, etc.).

For the above-mentioned reasons the food industry is a sector that opens great possibilities to thin-film coatings, because through their application it is possible to re-use traditional substrates such as 100Cr6 and stainless steels, thereby reducing health and social risks at present linked to these kind of materials.

PVD coatings present a unique combination of wear resistance, chemical and thermal stability, proved atoxicity (recognized by the US FDA) that can strongly revolutionize the food sector.

9.5
Coating Requirements in the Food Sector

The food processing sector covers a wide variety of applications from processing of human food (meat, bakery, liquids, fruits and vegetables, sweets, etc.) to feeders production for animals (zoo technique industry). As a consequence, the requirements of coatings in this field change considerably according to the specific addressed field. In the framework of the ACTECO project, the research activity has been focused on the development of coatings having high wear resistance (in terms of improved tool life and increased toughness) and a low friction coefficient. Exploitation and testing of the developed coatings has been mainly directed to the improvement of performances of milling and cutting processes for the meat industry in which steel tool abrasion is the main critical issue. FDA certification in relationship to corrosion resistance and atoxicity of PVD coatings has been considered sufficient to validate the use of the developed coatings for the selected application.

Wear resistance (see Section 9.5.1), is a property of primary importance for cutting tools: the capability of the tool to resist the abrasion deriving from the contact with the worked on material should be prolonged as much as possible in order to minimize the generation of erosion products that can act as additional counterparts at the interface and that can be detrimental for the tool life. Moreover, in the food sector eroded parts can generate a health hazard if ingested by the final consumers.

The friction coefficient (see also Section 9.5.2) is strictly dependent on the surface quality of the tool being considered. In this sense, the application of PVD coatings is essential when the substrate to be coated is a low-performing steel. Design and control of the friction coefficient are key issues to reduce abrasion at the interface and to increase tool life, even in the case in which low carbon steels are used.

The development of a coating for a specific application (see the case study analyzed in Section 9.7) requires an experimental study campaign in which reference coated samples and worn coated samples are pre-tested in controlled conditions and submitted to complete structural and mechanical characterization. The latter is important, because it allows coating designers and end-users to investigate and classify in detail the coating defects and their influence on tool life during use. As a consequence, the choice of the characterization techniques and of the properties to be monitored during the development and pre-testing should be carried out in advance for each specific application. In the framework of the ACTECO project, wear analysis and friction coefficient have been selected as priorities, and the best characterization methodologies identified to study these two properties (and to be able to predict and improve coating behavior) are described in Section 9.6.

A short overview is also given about the characterization of corrosion properties, describing the reference tests listed in the most recognized standards and regulations for the food sector. However, it remains up to the end-users to carry out, on a case-by-case basis, all the necessary test to grant a safe and atoxic processing.

9.5.1
Wear Resistance

The cutters used in the food industry are subject to abrasive wear, due to the cutting action on the food. The cutting edge of the tool is always kept very sharp. When a chromium plating treatment is carried out, it is only applied in the restricted area of the cutting edge in order to increase its hardness. Furthermore, the chromium layer cannot induce rounding of the cutting edge and must be deposited with precise control of the thickness.

The need to reduce abrasive wear in the cutters stems primarily from the need to increase tool durability and to reduce transfer (erosion) of the eroded material into the food.

As a consequence, the first main requirement for the coatings developed within ACTECO was to reduce friction between the cutter and the food, in order to reduce as much as possible the abrasion risk.

In order to reduce the friction, it is necessary to have coatings with the following characteristics:

- low friction coefficient
- medium-high hardness (not at the top in order to avoid drastic failures that could mean coating delamination due to unbalance between substrate and coating hardness)
- high toughness
- low thickness (2–3 µm).

9.5.2
Coefficient of Friction (COF)

The stress distribution during cutting depends mainly on the friction coefficient between the two surfaces in relative movement. *Friction* is defined as the resistance to relative sliding between two bodies in contact under a normal load. The coefficient of friction or COF, is determined by the ease with which two surfaces slide against each other. The (dimensionless) value of the COF is the ratio of the force required to slide the surfaces (F_t) and the force perpendicular to the surfaces (F_n):

$$[\text{COF}]f = \frac{F_t}{F_n} \tag{9.1}$$

A lower COF indicates that the surfaces are slicker – there is less resistance to the sliding motion.

Generally, the COF is considered independent from the contact nominal area, from F_n and from the sliding velocity v.

In the real world this is not completely true, since the friction coefficient depends on the sliding conditions and in particular on the physical and chemical conditions of the two surfaces in touch, that is on roughness, on the surface oxide conditions, and on the presence of absorbed layers.

9.6
Selection of Methodologies for Effective Characterization of Coatings for the Food Sector

Coating characterization is in every application field an indispensable step to reach the target of desired performance. In the case of coatings for the food processing tool sector, three kinds of characterization must be carried out before validating a solution:

- chemical and structural characterization (Table 9.2)
- tribological characterization (Table 9.3)
- atoxicity and corrosion characterization (Table 9.4).

The tables illustrate which kind of techniques are recommended for each characterization type.

Coating morphology and thickness are two fundamental properties to understand if the coating fulfils the targeted homogeneity and composition. Scanning electron

Table 9.2 Chemical and structural characterization.

Characterization technique	Properties
SEM + EDX	Coating morphology Coating thickness Qualitative chemical composition
Calotest + optical microscopy	Thickness Interlayers
XRD (glancing incidence)	Phase analysis
XPS	Analysis of chemical bonds and compounds Semiquantitative analysis of the coating chemical composition

Table 9.3 Mechanical characterization.

Characterization technique	Properties
Hardness	Coating adhesion Coating hardness
Pin-on-disk test	Friction coefficient Weight loss

Table 9.4 Atoxicity and corrosion characterization.

Characterization technique	Properties
Food compatibility test I: heavy metals release	Chemical stability
Food compatibility test II: oxidation	Oxidation resistance
Anticorrosion test: hot salt mist	Corrosion resistance

microscopy (SEM) and energy dispersive X-ray spectrometry (EDX) techniques can be applied to investigate coating morphology, to determine possible microstructural defects (inhomogeneities, pores), to measure coating thickness, and to perform qualitative and semi-quantitative chemical composition analysis.

Porosities and surface defects can promote bacterial retention and should be carefully investigated.

Coatings for food processing tools must show excellent wear resistance and hardness. Pin-on-disk test and Vickers microhardness can help to determine these important mechanical features.

On the other hand, these coatings cannot be brittle, otherwise wear/erosion products can remain in the treated foods, leading to a health hazard for the final consumers. This means that wear resistance and hardness must be accompanied by good toughness and adhesion. Both these properties can be evaluated in an indirect way through the Rockwell hardness test.

Atoxicity and corrosion resistance are also fundamental in order to avoid the release of materials from the coating into the food. The first property can easily be evaluated through specific food compatibility test (heavy metal release and oxidation), while the second can be investigated using the standard hot salt corrosion test.

Additional analysis such as X-ray photoelectron spectrosocopy (XPS) and X-ray diffraction (XRD) can be performed to investigate structural and chemical properties in more detail. The XPS technique reveals precise information about the chemical bonding of elements included in the coating composition (chemical compounds such as molecules are clearly detectable). Moreover, semi-quantitative chemical composition analysis is also possible.

XRD characterization (used in glancing incidence set-up) can be used to recognize all the crystalline phases contained in the coatings.

All this information builds up a consistent data sheet for a coating, remembering that the measured properties are strongly influenced by the selected substrate. In the following paragraphs the above-mentioned characterization techniques will be further described.

9.6.1
Chemical and Structural Characterization

9.6.1.1 Scanning Electron Microscopy (SEM)

SEM as detailed in Chapter 5 is a valuable tool for examining samples. The most immediate result of observation in the scanning electron microscope is that it displays the shape of the sample. It can also be used to determine the local composition, the crystal structure and orientation, and electrical and optical properties.

The SEM technique is widely used for coating characterization, because its high resolution allows the following features of coatings to be analyzed in detail:

- thickness
- microstructure
- defects (pores, cracks, in homogeneities, etc.)
- surface quality (droplets, finishing).

9.6.1.1.1 Application to Anti-wear Coatings for Food Processing Tools The previous properties can be detected by SEM analysis operating in two different ways:

- **Cross-section analysis of a coated samples**: As shown in Figure 9.2 this sample view detects the coating thickness and the coating microstructure. In the picture

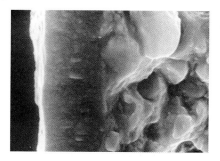

Figure 9.2 SEM analysis: cross-section section of a PVD coated sample (on the right-hand side of image it is possible to see the irregular matrix of substrate).

the coating presents a typical columnar structure and an average thickness of about 3 μm.

- **Front view of the coated sample**: Figure 9.3a shows the SEM analysis of a coated sample surface. It is possible to clearly detect droplets. These are small metallic particles (titanium particles in the given example) deriving from an insufficient reaction efficiency during the deposition. In fact a part of the evaporated metal from the target is not able to react with the inlet gas and condenses on the surface samples as metal element. EDX analysis confirms this hypothesis.

Droplets are not desirable, because these small particles lower the friction coefficient and the tribological properties of the coatings.

9.6.1.2 Energy Dispersive X-ray Spectrometry (EDX)

A different range of detectors can be used to complete SEM analysis, in fact the most interesting information arises from the collection and analysis of the signals obtained by the secondary electrons, the back-scattered electrons, and X-rays.

X-ray analysis allows us to obtain elementary information on the sample such as type of chemical elements, distributions and compositions. EDX is based on the fact that each chemical element is characterized by a specific X-ray emission spectrum. The EDX detector collects the emitted radiation and analyzes the energy content, calculating the correspondent wave length, that is typical for each chemical element. In this way it is possible to clearly detect all the chemical elements contained in the sample and, using proper standards, it is also possible to calculate a semi-quantitative chemical composition.

Typically, an EDX probe is used to obtain the following information:

- qualitative chemical analysis of the substrate, inter-layers and top-layers
- semi-quantitative analysis of the coatings
- local analysis of micro-impurities and microparticles (e.g., droplets).

9.6.1.2.1 Application to Anti-wear Coatings   Figure 9.3b reports the EDX analysis of droplets on a PVD coated steel sample. As can be observed, the EDX probe also detects the chemical composition of the substrate below.

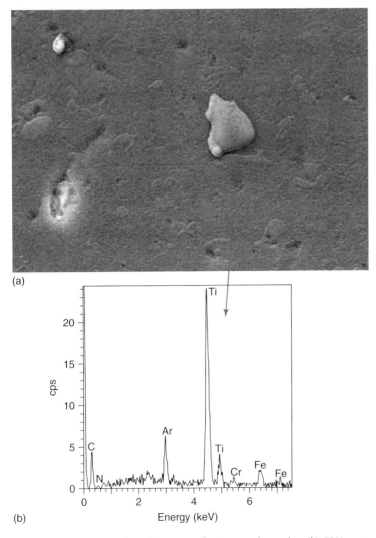

(a)

(b)

Figure 9.3 (a) SEM analysis: front view of PVD coated samples; (b) EDX spectrogram.

9.6.1.3 Calotest and Optical Microscopy (OM)

The Calotest is useful for a quick and precise measurement of the coating thickness and for the detection of the number and type of interlayers.

The coated sample is put against a rotating steel ball covered by a diamond paste. The friction between the ball and the surface of the coated sample generates an erosion mechanism that, when the trial parameters are properly calibrated, reaches the substrate of the sample. At this stage the test must be interrupted and the eroded sample is submitted to optical microscopy (OM) analysis. The image that appears is a circle in which all the interlayers are clearly visible. Using trigonometry it is possible to carry out, starting from the image, a quantitative measurement of the coating thickness. The sample can also be analyzed by EDX, in fact it is possible to make a local measurement in each layer that appears in the print.

The Calotest is typically used to obtain the following information:

- quantitative measurement of coating thickness (Calotest + OM)
- number of interlayers (Calotest + OM)
- qualitative analysis of interlayers chemical composition (Calotest + EDX).

9.6.1.3.1 Application to Anti-wear Coatings for Food Processing Tools A diagram explaining how it is possible to calculate the coating thickness starting from a Calotest image, is given in Figure 9.4. Figure 9.5 shows a typical image obtained by analyzing a Calotest image by OM,. which clearly reveals the number of interlayers. Figure 9.6 shows the EDX analysis carried out on the print taking into consideration each layer of the coating.

9.6.2
Mechanical Characterization

9.6.2.1 Hardness
Hardness is the first basic property to be measured in order to classify coatings in terms of mechanical resistance. *Hardness* is defined as the penetration resistance of a material, but it can give also indirect information about the coating wear resistance. In any case, hardness must be matched with a measurement of the coating toughness and of the friction coefficient in order to have a complete characterization of the wear behavior.

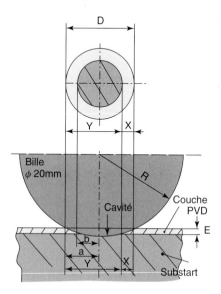

Figure 9.4 Calotest: thickness measurements.

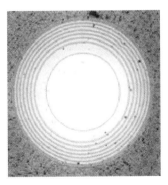

Figure 9.5 Calotest + optical microscopy: PVD coating interlayers.

Different hardness measurements techniques are available, depending on the shape and size of the used indenters.

In the case considered, two types of technique will be taken into consideration:

- Rockwell hardness (diamond steel sphere as indenter)
- Vickers microhardness (diamond pyramid as indenter).

Rockwell hardness is used to obtain a qualitative measurement of the coating adhesion to the substrate. In this case the observation by OM of the indented surface will also allow a qualitative indirect evaluation of the coating toughness. In fact, by analyzing the number of cracks surrounding the hardness track and comparing these data to proper established standards it is possible to classify the coating in term of adhesion and toughness. This method is also interesting for comparative studies among competitive coatings.

Vickers microhardness is used to measure the coating hardness, through standard measurements based on calculation of the length of the two print diagonals.

Load and indentation speed must be carefully calibrated in order to let the indentation print remain into the coating thickness, without reaching the substrate.

Rockwell hardness is used for a qualitative evaluation of the following features:

- coating adhesion (Rockwell test + OM)
- coating toughness (Rockwell test + OM).

The Vickers test is used for a quantitative measurement of the coating hardness.

9.6.2.1.1 Application to Anti-wear Coatings for Food Processing Tools   Figure 9.7 shows the Rockwell test result on two PVD coatings with similar characteristics. As can be seen, in Figure 9.7a the cracks surrounding the circle print are clearly visible. This is not the case for the coating shown in Figure 9.7b, which indicates a higher adhesion to the substrate than the previous one.

A typical indentation print obtained by Vickers microhardness is shown in Figure 9.8.

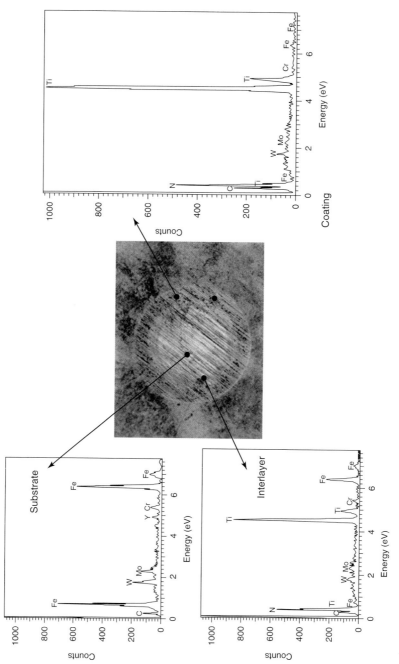

Figure 9.6 Calotest + EDX: local analysis of a PVD multilayer coating.

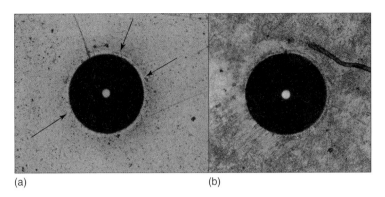

(a)   (b)

Figure 9.7 Rockwell adhesion test + optical microscopy:
(a) PVD coating with low adhesion; (b) PVD coating with
high adhesion.

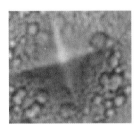

Figure 9.8 Vickers microhardness + optical microscopy:
PVD coating hardness indentation.

9.6.2.2 **Pin-on-disk**

Pin-on-disk testing is a commonly used technique for investigating sliding wear.
As the name implies, such apparatus consists essentially of a 'pin' in contact with a
rotating disk. Either the pin or the disk can be the test piece of interest. The contact
surface of the pin may be flat, spherical, or of any convenient geometry, including
that of actual wear components.

In a typical pin-on-disk experiment, the COF is continuously monitored as wear
occurs, and the material removed is determined by weighing and/or measuring
the profile of the resulting wear track. Changes in the COF are frequently indicative of a change in wear mechanism, although marked changes are often
seen during the early stages of wear tests as equilibrium conditions become
established.

The main variables which affect friction and wear are velocity and normal load.
In addition, specimen orientation can be important if retained wear debris affects
the wear rate.

Most commercial pin-on-disk testers use high loads (e.g., 100–1000 N) obtained
with a dead weight and large areas of contact.

The pin-on-disk test is used to determine the following characteristics:

- friction coefficient of the coatings
- weight loss of the coating.

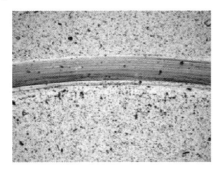

Figure 9.9 Pin-on-disk + optical microscopy: PVD coated sample track analysis.

9.6.2.2.1 Application to Anti-wear Coatings for Food Processing Tools A typical pin-on-disk track obtained on a PVD coated sample is reported in Figure 9.9.

9.6.3
Atoxicity and Corrosion Characterization

9.6.3.1 Food Compatibility: Heavy Metals Release
Evaluation of the release of heavy metals can be obtained with the migration test. The level of migration of the metals into the food should be determined for the worst foreseeable conditions, taking into account the natural content in the metal of the food stuffs itself. For the test a proper food simulant (corresponding to the food which has to be in contact with the tested component, for example, a fat simulant is olive oil) can be used, which has to stay in contact with the interested surface for 1 month at 25 °C [4].

The heavy metal release test is used to evaluate the coating compatibility in working conditions.

9.6.3.2 Food Compatibility: Oxidation Test
In the simple case of uniform corrosion, this reaction results in the formation of compounds of the metal (e.g., hydroxides) on the surface of the metal. The rate at which corrosion proceeds will depend in part on the composition of the aqueous medium: corrosion of iron in very pure water will be considerably slower than in water which contains, for example, acids or salts.

More complex corrosion patterns may occur, for example, 'pitting corrosion'. This occurs following 'attack' at discrete surface areas of the substrate that are susceptible because of, for example, surface imperfections or impurities in the metal. Pitting corrosion is generally seen as small, local, areas of corrosion.

9.6.3.3 Salt Spray Test
The salt spray test measures the corrosion resistance in a very corrosive environment. Sample materials are exposed to the aggressive vapor (sprayed solution of 50 g l^{-1} of NaCl in demineralized water at 35 °C) in a dedicated chamber. The sample surface is periodically observed by OM to determine the time (in hours)

before the formation of the first corrosion spot. The ISO standard reference that has been adopted is 9227.

9.7
Case Studies: Development and Characterization of Ceramic Coatings for Food Processing Applications

According to the general technical needs described in this chapter, this section describes a real coating development for a specific application for the food sector, starting from the technical needs analysis to the process scale-up for industrial production.

The activities presented were carried out by Environment Park S.p.A. in the context of ACTECO, according to the industrial needs expressed by Diad s.r.l. The goal was to increase the wear resistance of some components for the food processing industry like blades, knives, screw feeders, and hammers, to enhance product performance and internal processing efficiency and competitiveness. In particular, plasma PVD hard coatings were developed and industrialized for two types of tools: hammers for industrial butchery scraps treatments (meat, bones, etc.) and industrial blades for ham cutting. The next sections offer an example of the approach for the specific coating design and development for the industrial blades mentioned.

9.7.1
Relevant Substrates and Functionalities Required for Cutting Applications

As already mentioned, wear resistance is a key issue for all the materials used as substrate in tools involved in food treatment, because of the health and ethical issues involved (see Section 9.3).

Depending on the case study, different functionalities can be required, as described in Table 9.5, that summarizes for each application the state-of-the-art of materials and technological solutions. Food processing tools can be used in different operations: shearing, screwing, cutting, milling, and so on. In principle, each requires substrate materials having high wear resistance and toughness, even if for each specific application a dedicated solution can be envisaged. At the present moment, commonly-used substrates are stainless steels, aluminum alloys, or titanium (limited applications). They all are very efficient in granting high hygiene of the surfaces, easy cleaning and limited bacterial proliferation. These are very important properties, but do not solve the health risks deriving from long term ingestion of erosion products. PVD coatings could have the potential to break this barrier and to allow the development of an easy-to-clean and atoxic substrate surface capable of considerably reducing the release rate of particles and debris into the processed food. The US FDA has approved almost all the categories of PVD coatings, above all titanium based chemical compositions. This technological solution is capable not only of increasing functionalities of traditional substrates,

Table 9.5 Relevant substrates and functionalities required for food processing applications.

Applications in food processing tool sector	Anti-wear coatings for industrial shearing machines for food scraps (bones, meats, . . .)
	Anti-wear coatings for screw feeder machines for food scraps (bones, meats, . . .)
	Anti-wear coatings for industrial knives, Archimedean screw and blades (for meat, ham, . . .)
	Low friction coatings for inox dies (for spaghetti, pasta, . . .)
	Antibacterial, chemical resistance coatings for boxes, containers, reactor vessels
	Anti-adherent, Anti-wear coatings for industrial wine press
Substrate(s)/material(s) presently in use	100 Cr 6
	Inox steels (AISI 304)
	CP titanium
	Aluminum alloys
Functionalities required to the substrates presently in use	Corrosion resistance
	No porosities in the surfaces
	Easy cleaning of the surfaces (removal of bacterial species)
	Low bacterial retention after cleaning
State-of-the-art process(es) providing these functionalities	100Cr6: additional chromium plating treatment
	Fluorine-based coatings
	TiN PVD coatings (in very limited applications)
Functionalities/improvements required	Wear resistance
	Toughness
	Corrosion and erosion resistance (PVD coatings are certified for this use by the FDA).

but also of extending the lifetime of coated tools with a direct benefit on cost savings (reduction of used tools, maintenance cost, etc.).

9.7.2
Technical Analysis and Choice of the Proper Coating Chemistry and Technique

These components are traditionally made with chromium steels and stainless steels. Stainless steels have better corrosion resistance, but do not have the optimal mechanical properties such as hardness and wear resistance. For these reasons hard Chromium steels like 100Cr6 are preferred. 100Cr6 steel is typically employed for ball bearing production and presents high hardness after heat treatment and quenching.

As in many industrial sectors, also in the food treatment industry electrochemical chromium deposition is used on the cutting edge of blades to extend the lifetime and to protect the surface from oxidation. One of the main problems of this solution is the low wear resistance and coating adhesion on the steel blades, with potential release of chromium particles into the food.

The requirements set by the industrial partner are to maintain the actual production process and materials as well as to introduce a final coating which is able to enhance wear and oxidation resistance and which acts as an effective barrier to chromium migration (encapsulation effect).

The goal is to verify the application of ceramic coatings obtained by PVD technology to increase the component's wear resistance (and lifetime), to guarantee the barrier properties to chromium migration and to obtain good biocompatibility. If it fulfils these properties, PVD could be an effective alternative to the actual chromium galvanic coatings.

By the analysis of the technological needs and ceramic coating properties, the choice of TiN was made for a first attempt as these ceramic binary nitride coatings have optimal wear resistance, low friction coefficient, and high corrosion resistance [5]. Moreover, they guarantee an effective barrier against chromium migration into the foodstuff.

The main critical parameter for the TiN deposition process is the temperature. Actually, the heat treatment history of these components, made of 100Cr6 steel alloys, which have a quenching temperature around 180 °C, is not compatible with the typical TiN deposition temperature (which for PVD arc evaporation process is commonly around 400 °C). For this reason, it was necessary to study and develop an innovative TiN coating for the coating of steels with a low tempering temperature.

Coating parameters optimization was realized through iterative steps, in order to determine the restricted area in which converge of the different technological needs was reached.

Innovative ceramic coatings, resulting from recent research in the field of micro- and nanotechnologies, offer a relevant technological solution for the food-treatment industry due to the high mechanical properties and the chemical stability and biocompatibility. The ceramic coatings taken into consideration for the presented activity are realized with the PVD plasma process, obtained in a high vacuum chamber and with the innovative LARC (lateral arc rotating cathodes) technology (see Section 2.2.3).

Ceramic coatings are deposited on 100Cr6 steel samples and have the following properties:

- high hardness and wear resistance (microhardness up to 24 GPa)
- high adhesion (scratch test critical load 20 N)
- low friction coefficient (0.5)
- biocompatibility (coatings certified for orthodontic and biomedical applications and by the FDA for food applications)
- chemical and color stability (complete chemical stability under the maximum temperature of 600 °C).

The PVD LARC technology allows ceramic coatings to be obtained that minimize the presence of macroparticles and droplets present in traditional ceramic coatings. Figure 9.10 compares a traditional ceramic coating and one produced with LARC technology.

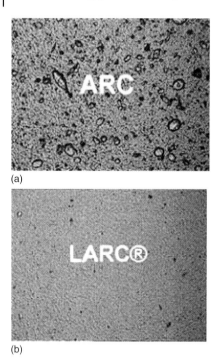

(a)

(b)

Figure 9.10 Optical microscopy comparison (at same magnification) of a (a) traditional ceramic coating (ARC) and (b) innovative coating developed with LARC technology. The uniformity and minimum presence of droplets of the LARC coating are obvious.

Droplets represent potential locations of ceramic coating cracks, with consequent migration of substrate components to the external environment. Crack sites also determine corrosion nucleation centers.

The high corrosion resistance of the TiN coatings developed has been demonstrated for the tapware sector (deposited on a rougher and more porous material) with more than 900 hours in salt spray test without a trace of corrosion [6]. These results can be extended to the food industry sector with improved properties, due to the more suited substrate for coating (steel compared with brass).

Moreover, the ceramic coatings for the tapware sector can be optimized with a microstructure dedicated for food application, thanks to the innovative LARC technology that allows coatings with variable stoichiometry for the deposition of gradient or multilayered structures.

The graduation of the lattice structure due to the different species concentration or stratification or to the nanoparticles concentration in the amorphous matrix, permits coatings that prevent identified species migration to be realized (Cr in this study). Interesting preliminary results were obtained with single, gradient and multilayered binary, ternary, and quaternary coatings [7] with high penetrating species like hydrogen. Based on chromium/hydrogen atomic radii ratio (exceeding 3.46),

Table 9.6 Assessed functional properties of the ceramic coating of this analysis (binary nitride titanium-based coatings).

Coating	Color	Microhardness (GPa)	Thickness (μm)	Friction coefficient	Max operating temperature (°C)
Titanium nitride (TiN)	Oro	24	3	0.55	600

a much higher barrier effect is expected compared with chromium penetration for the same coatings.

As a consequence, the considered ceramic coatings have an intrinsic wide applicability potential in the food industry. They present high chemical barriers properties, corrosion resistance, wear resistance, high toughness, and hardness up to 2450 μHV. The possibility of testing the food compatibility of quaternary-coatings would allow exploiting the even higher tribological and mechanical properties of such thin films.

Considering the technical features and the technological needs of the application, the choice for this study is the investigation of a binary nitride titanium-based system (or put more simple: TiN coating), whose technical properties are reported in Table 9.6.

In the USA, TiN coatings already exists for food applications and are certified by the FDA [3] for specific food applications, but they are obtained in different conditions and with a different technology. This study intends to validate TiN coatings (realized with the PVD LARC technology) with the current EU regulation, as well as to assess the applicability to provide technological solutions and improvements to the existing chromium plating.

9.7.3
Coating Development

The development of a new PVD ceramic coating dedicated to a specific application requires a very careful approach. In order to optimize the development time, resources and efficiency, it is most of the time a good choice to proceed with a first 'lab scale' phase which is carried out on dedicated samples.

In this case, the samples for the lab-scale tests are provided directly from the company, in order to reproduce the effective substrate surface finishing in terms of hardness, roughness, and morphology. To be precise, in this analysis only flat surfaces have been considered, which allow easy machining and uniform roughness control; for complex 3D surfaces it is more difficult to assure the same surface morphology at sub-micrometer level. Samples are prepared in 100Cr6 steel with a Rockwell hardness of 62HRC after quench and tempering (quench temperature 180 °C) with a surface finishing that simulate the effective (average)

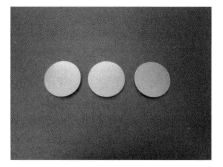

Figure 9.11 100Cr6 steel samples, before coating.

Table 9.7 100Cr6 sample properties.

Sample (shape)	Material	Dimension (mm)	Hardness	Roughness	
				R_a (μm)	R_z (μm)
Round (cylindrical) samples	100Cr6 (AISI E 52100)	31 × 4	60 + 4 (HRC)	1.2 ± 0.1	0.1 ± 0.01

industrial production conditions (roughness parameter Ra = 1.2 μm) (Figure 9.11 and Table 9.7).

The first step is the analysis of the compatibility of 100Cr6 steel with the typical PVD coating process conditions. The critical parameters are the melting temperature, the vapor pressure and the possible thermal history due to heat treatment of the metal substrates. In this specific case the critical parameter is only the tempering temperature of 100Cr6 steel. In fact, typical 'standard' TiN deposition temperature (400 °C) is not compatible with the substrate properties, because a heating above the tempering and/or annealing temperature would dramatically change the mechanical properties of the steel.

For this reason, it is necessary to study and develop a new 'low-temperature' TiN coating process for the coating of steels.

In particular, the coating process modifications involve:

- decrease of process maximum temperature from 400 °C down to 180 °C;
- reduction of ion etching power, because this process causes high local surface heating on substrates with consequent higher thermal stress.

This low-temperature coating process development is extremely critical because of two opposite technical needs: the low-temperature process required for maintaining the substrates hardness and the process temperature needed to obtain a ceramic coating with optimal mechanical properties, which is typically above 300 °C.

Before the coating, the substrates were cleaned with an ultra-sound cleaner that guarantees an optimal surface cleaning, applicable also on objects with complex geometry (3D surfaces).

After the cleaning phase, the samples are placed on the rotating carrousel of the PVD deposition chamber and the fully automated coating process can be started (see also Figure 2.13).

Rockwell C hardness sampling was performed before and after the coating process in order to verify the heating effect on 100Cr6 steel due to ion bombardment. Apparently, this bombardment had no effect on the substrate structure or on its mechanical properties.

With the OM analysis of the Rockwell indentation it was also possible to obtain information about ceramic coating adhesion to the substrate by the analysis of the indentation cracks geometry and propagation (results were compared with reference qualitative data from a database). This technique is now illustrated by comparing the results obtained on four different types of coating, referred to as *Coating 1* through 4. The results are presented in Table 9.8 and in Figure 9.12.

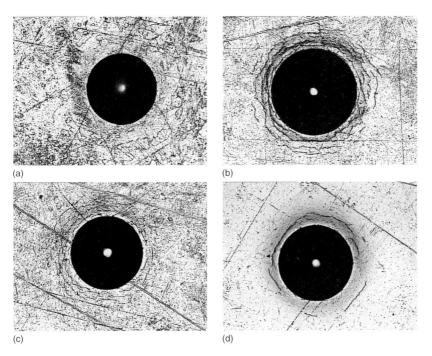

(a)

(b)

(c)

(d)

Figure 9.12 Four different types of coating. (a) Coating 1; (b) coating 2; (c) coating 3, and (d) coating 4.

Table 9.8 Overview of the properties obtained with the different deposition process parameters, illustrating the coating development on the chosen 100Cr6 steel substrate.

	Coating 1	Coating 2	Coating 3	Coating 4
Adhesion	Poor	Poor	Good	Good
Thickness (μm)	1.5	1.5	1.5	1.5
T_{max} ($^\circ$C)	170	185	204	197
HRC before	62	62	62	62
HRC after	62	62	60	62

The following sample analysis shows the results in the low-temperature coating development, from scarce adhesion, represented by wavy-like circumference cracks, to improved adhesion. The absence of radial cracks in the second type of coating represents elevated residual stress on the coating. Progressive decrease of the number and extension of crack patterns proved the development of coatings with less residual stresses and softer coatings with improved adhesion and good toughness. HRC before and after the coating process revealed the slight process impact on the third coating type, that showed better adhesion properties. The fourth coating type proved to be the best solution, providing best adhesion and unchanged substrate properties (see Figure 9.12d).

9.7.4
Case Study: PVD Coating of Saw Blades

The coating process scale up was carried out by Environment Park with the collaboration of the Swiss coating machines manufacturer Platit AG, which is equipped with a coating chamber machine for the large industrial scale production (π 300 R&D PVD LARC equipment).

The sample tools are saw blades used for cutting of food or packaging materials, their outer diameter is 350 mm. The tool surfaces had been polished by the manufacturer. The blades are made from standard 100Cr6 steel, that is, they have a good abrasion resistance but cannot endure temperatures significantly in excess of 200 $^\circ$C. Therefore, special recipes were created with mild process conditions. A long heating of at least 2 hours, normally 2 hours 30 minutes, was employed.

Two such low-temperature coating recipes, one for TiN (see previous section) and one for CrN (the development of which is not discussed here), were transferred to Platit and fine tuned to its equipment in terms of process parameter and thickness.

For the actual testing, the cleaning of the four saw blades was conveniently done in a Eurocold Thermovide single chamber cleaning equipment, using a special holder (Figure 9.13). The large blades were cleaned one by one, since the risk of

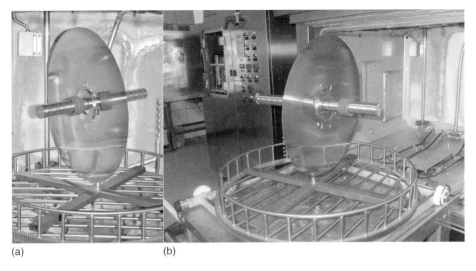

(a)   (b)

Figure 9.13 Cleaning set-up for the industrial blades.

(a)   (b)

Figure 9.14 Typical sample coating batch showing the dummy load and the position of the saw blades test pieces used for the Calotest and Rockwell measurements.

touching upon ultrasonic action or rinsing water jets was considered too big. Apart from this cleaning, the tools were not treated.

After the cleaning step, the samples were loaded into the test chamber, using the setup shown in Figure 9.14. A special fixture was welded, consisting of a base plate and a central stainless bar as centering pin. Steel rings were used as spacers, and stainless steel plates, previously modified to fit the central pin, as base and cover plates. The position of the test piece holder is marked in Figure 9.14a.

Samples coating characteristics are reported in Table 9.9.

Table 9.9 Characteristics of Platit sample coatings on Diad blades.

Tool material	Blade type	Coating type	T_{max} ($^\circ$C)	Coating thickness (μm)	HCR
100Cr6	Calibration pretest	CRN-LT200	201	2.00	66.6
100Cr6	Real blades	CRN-LT200	204	2.83	66.5
100Cr6	Calibration pretest	TIN-LT200	183	1.93	66.8
100Cr6	Real blades	TIN-LT200	182	2.86	66.5

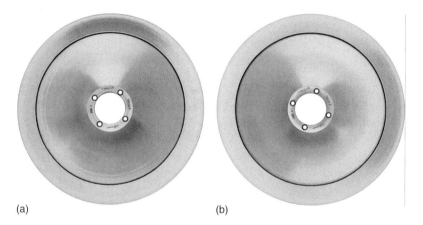

(a)   (b)

Figure 9.15 Coated 100Cr6 blades diameter 350 mm, (a) process TiN-LT200, (b) CrN-LT200.

Using a mobile 'pen-type' ballistic Rockwell tester, we found that the greater part of the 100Cr6 blades was soft. Only the circumference, that is, the outer rim of about 25 mm, had been hardened, probably by induction. There is a hardness gradient increasing toward the edge, at the closest point to the outer perimeter where measurement was still possible, 45 HRC was reached. The measurement scatter was fairly high.

After coating, values between 42 and 45 HRC were measured for CrN, for the TiN batch 43–45 HRC. A more detailed hardness study would require the destruction of the tool, which was not planned at this stage.

The coating appearance was shiny (Figure 9.15). The thickness was about 2.0 μm and the samples could be functionally used without restrictions.

The adhesion that was reached was very good for CrN, but only good for the TiN, despite the cathode cleaning step automatically carried out immediately before the metal ion etching. It is indeed a known fact that the best adhesion can be reached with Cr. The TiN adhesion should still be sufficient for regrinding (to recondition the blade cutting edge), especially as the cutting edge of the blade is much more exposed to ion bombardment compared to the test piece, and thus, more efficiently

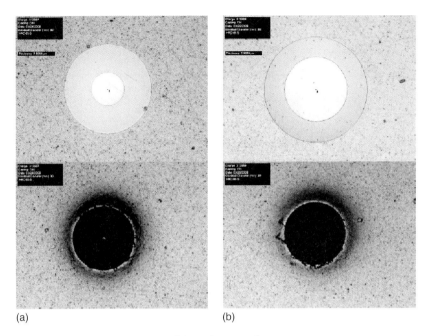

(a)   (b)

Figure 9.16 Optical micrographs of the Calotests and Rockwell imprints of the ultralow temperature (<200 °C) batches.

plasma etched. The coating surface is very smooth even though no post treatment was applied. A hand or machine polish may be used, if the final end-user requires a further improvement of the surface finish.

Despite the low cathode power used, the deposition time was no more than three hours for etching and nitride coating, which is approximately double the time of a standard coating process.

Summarized, the coating test was performed successfully, with a small compromise in the case of the low-temperature processes. The optical appearance of the coatings was smooth, and the cutting function should be excellent. Performance behavior of the 100Cr6 blades should be closely followed and eventually measures should be taken to further adjust the coating recipes to this very special case (Figure 9.16).

Of further importance is to learn more about the difference in regrindability between the CrN and the TiN coating.

9.7.5
Case Study: PVD Coating of Hammers for Food Treatment

Another industrial application was developed to increase wear and oxidation resistance of hammers for food scraps treatment. Hammers are mounted on mills and hosting different hammers, n.80 hammers are needed to operate a typical mill.

Figure 9.17 Comparison between worn and oxidized hammer (on the left) and a new one (on the right).

The interest of Diad is to eliminate the wear and oxidation problems on the surface of the components that cause continuous interruptions for maintenance causing long down times, which result in high decrement of the production.

As detailed in Section 9.6.1 andin Section 9.6.2, ceramic coated surfaces present high mechanical and chemical performance: high hardness (up to 3400 HV), wear resistance, low friction coefficient (down to 0.15), high chemical stability, corrosion resistance, biocompatibility, barrier properties to the migration of heavy metals, and penetrating ions.

Actually, the hammers (Figure 9.17) for meat and bones treatment were made of C20 steel, a material not suitable to the ceramic coating because it does not guarantee good adhesion. So, the first step was to select a new substrate material, the second step was to select a PVD coating able to enhance wear resistance and oxidation resistance.

Before being coated, the hammers are treated with a heat treatment in order to enhance the hardness of the substrate and to improve the adhesion of the coating.

Considering the technical needs of the application, it was decided to investigate both a binary nitride titanium-based system and a nitride chromium-based system.

The considered ceramic coatings have an intrinsic wide application potential in the food industry. They present high chemical barriers properties, corrosion resistance, wear resistance high toughness and hardness up to 2450 μHV.

The hammers were first cleaned with an ultra-sound cleaner that guarantees an optimal surface cleaning applicable also, on objects with complex geometry (3D-surfaces).

Figure 9.18 TiN coating of hammers.

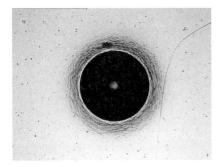

Figure 9.19 Rockwell indentation of PVD treated hammer on face-surface providing, the same HRC value with respect to uncoated hammer.

After the cleaning phase, the components were placed on the rotating carrousel of the PVD deposition chamber (Figure 9.18).

Due to the hammer material and heat treatment, a low-temperature coating process was selected (maximum temperature was about 200 °C). After the coating process the hardness of the hammers was verified to ensure it remained unaltered by the PVD treatment (result shown in Figure 9.19).

Large-scale testing of this solution led to on-site testing of hammers, which revealed an enhanced hammer lifetime of a factor of four, thus clearly by realizing the technological target of reducing maintenance frequency and the related production stops. The proven technical applicability of this solution led to the detailed economical sustainability assessment presented in Chapter 11.

9.8
Conclusions

In this chapter the particular case of ceramic coatings used for the food processing sector and the dedicated characterization methods for assessing the coating properties have been presented.

The actual problems connected to the use of electrochemical plating (in particular of chromium VI) on metallic surfaces for the food treatment industry has generated a growing interest for alternative solutions which are better in terms of effectiveness, efficiency, safety, and biocompatibility.

In this chapter two specific applications of PVD ceramic coatings have been presented for surface functionalization of cutting tools for food sector. Low-temperature PVD coating processes were developed to allow coating substrates of selected components.

In the first case study, saw blade substrates could be efficiently coated by a low-temperature deposition process, thus proving the applicability of inert and biocompatible PVD coating to substitute the chromium electroplating process. Tests also demonstrated the full compatibility of TiN PVD coating with the components material. The high adhesion, hardness, wear and corrosion resistance as well as the uniformity of the coating were validated.

In the second case study, PVC coating was applied to food processing hammers. The developed solution increased the wear resistance of the components, thus strongly minimizing the maintenance frequency and, consequently, significantly improving the productivity. Following the excellent technical results of this case study, the relative economical feasibility was assessed and found to be very promising too (see Chapter 11).

References

1. Liu, C. *et al.* (2003) *Surf. Coat. Technol.*, **163–164**, 597–604.
2. Carratù, B. and Aureli, P. (2007) Rapporti ISTISAN 03/45, Istituto Superiore di Sanità.
3. FDA (2009) Inventory of Effective Premarket Notifications for Food Contact Substances. *www.cfsan.fda.gov/~dms/opa-fcn.html.*
4. Council of Europe (2004) Guidelines on metals and alloys used as food contact materials. EC 1935/2004, *http://www.esb.ucp.pt/twt/embalagem/MyFiles/biblioteca/regulamentacao/MC/COE/COEmetais.pdf.*
5. Rylska, D., Sokołowski, J., and Klimek, L. (2003) Corrosion resistance investigations of prosthetic dental alloys coated by TiN. E-MRS Fall Meeting 2003, Symposium E.
6. CLEAN DECO (2005) LIFE00 ENV/IT/000213 Clean-Deco - Mid Term Report, Development of clean coating technology PVD for decorative application. Trattamenti Termici Ferioli & Gianotti S.p.A.
7. Clean NT Lab- Environment Park (2005) Migration of H-species through Nano-technological Coatings (internal report).

10
Physics and Chemistry of Nonthermal Plasma at Atmospheric Pressure Relevant to Surface Treatment

Yuri Akishev, Anatoly Napartovich, Michail Grushin, Nikolay Trushkin, Nikolay Dyatko, and Igor Kochetov

10.1
Introduction

The atmospheric pressure nonthermal plasma interaction with a surface is an extremely intricate object for detailed modeling. Complexity of plasma–surface interaction at conditions typical for polymer treatment consists in a wide variety of plasma and polymer surface parameters. In particular, the sort of a plasma-forming gas varies according to the purpose of polymer surface modification. Both the surface atomic profile and the chemical composition of a monomer vary for different polymers. Nowadays, we have a rudimentary knowledge about processes taking place at the boundary between plasma and polymer bulk.

The amount and sort of radicals approaching the surface can vary strongly with change of gas composition. In particular, small amounts of (gas) pollutants present in air or in technical gas admixtures, such as Ar or N_2, used for plasma discharge generation, can significantly influence the activation dose and can have a dramatic impact on the surface treatment efficiency. As will become evident from the following chapters, numerical modeling of such complex systems and its experimental validation demonstrate an opportunity to increase processing efficiency by a proper choice of the discharge gas composition. This not only helps new process engineering but also enhances the chemical and energetic efficiency of already investigated processes.

Generally, the plasma surface processing can be broken into the following stages:

1) plasma production;
2) transport of plasma activated gas to the polymer surface;
3) interaction of plasma products with polymer surface.

Plasma Technology for Hyperfunctional Surfaces. Food, Biomedical and Textile Applications.
Edited by Hubert Rauscher, Massimo Perucca, and Guy Buyle
Copyright © 2010 WILEY-VCH Verlag GmbH & Co. KGaA, Weinheim
ISBN: 978-3-527-32654-9

Consequently, a theoretical model of the whole technological process can be reduced to a combination of specific models for each of these stages.

Recently, the plasma physics scientific community has paid much attention to the development of atmospheric pressure nonthermal plasma sources, as discussed in Section 2.3.

Along with that progress in the discharge techniques, the theory of gas discharge formation and of plasma properties has been developed. The specific feature of nonthermal plasma produced by the dielectric barrier discharge (DBD), corona, and glow discharge is its high sensitivity to the sort of gas employed in plasma discharge. In contrast to thermal plasma, which can be completely characterized by a few parameters (gas pressure, temperature, dissipated power density), the number of parameters required for a full description of plasma properties is rather large. Typically, the electron energy distribution function (EEDF) is of a strongly nonequilibrium type with respect to other ionized and neutral species (see Chapter 1), while ions have nearly the same average energy as atoms and molecules. Free electrons effectively gain energy from the electric field, which further is dissipated in numerous reactive collisions (excitation of molecular vibrations, dissociation, ionization, and excitation of electronic states of atoms and molecules). For a correct description of nonthermal plasma, it is mandatory to treat the electron kinetics properly, as reported in Chapter 1.

The chemical kinetics of plasma involves many species produced in plasma including radicals and excited atoms and molecules. Above equilibrium concentrations of such species change the gas reactivity completely in comparison with the standard thermal reactivity. Moreover, it leads to the appearance of a number of transient compounds capable to react at nearly room temperature. Theoretical description of these processes is possible provided extensive knowledge about numerous rate coefficients is accumulated, which are generally derived from experimental data. Hence, electron kinetics and plasma chemistry have a fundamental value for theory.

Our main concern in this chapter is the physics and chemistry of atmospheric pressure nonthermal plasmas. This is because high-pressure nonthermal plasma allows elimination of expensive vacuum equipment and performance of continuous plasma processing that considerably enhances its incorporation into industrial production lines. As mentioned in Section 2.3 many kinds of gas discharges at atmospheric pressure have become more deeply involved in nonthermal plasma processing. Most of the atmospheric pressure discharges generate a nonequilibrium plasma due to the formation in the gas gap of large numbers of nonstationary microdischarges generating numerous current filaments (so-called streamers) of nano- or microsecond duration. As a result, highly reactive nonequilibrium conditions in cold gas can be provided at atmospheric pressure.

Two nonthermal plasma sources working at atmospheric pressure were used in our experiments for remote surface treatment. These sources are documented in Section 2.3.2.1.

10.2
Discharge Modeling

10.2.1
Full Kinetic Models and Reduced Model for Technological Plasma

The complexity of plasma description, in particular at high pressures, is associated with two points: (i) a vast amount of reactions taking place in the plasma, which are very specific for a given gas composition and (ii) non-uniformity of the plasma in the majority of discharge types. Evidently, taking into account both of these points presents a formidable challenge. Traditionally, an approach to modeling of realistic plasma devices consists of two phases. At a first step, plasma is considered to be uniform, and all transport processes are ignored. At this step, major attention is paid to the correct description of electron kinetics and reactions of excited particles and chemical radicals initiated by electrons. Because of a strong difference in masses of electrons and neutral molecules/atoms, electrons are strongly heated, even in a low electric field, and their kinetics is described by the EEDF. Molecular vibrations can easily be excited by electron collisions. Provided relaxation of vibrational energy is slow, vibrational degrees of freedom are also nonequilibrium. This requires description of molecular vibrations to be done in terms of a vibrational distribution function. Usually, rates of ion–molecular reactions are remarkably higher than rates of reactions of neutral species. It means that despite rather low number densities of charged particles, their role in the plasma evolution can be significant. The full kinetic model comprises self-consistent combination of electron and vibration kinetics in terms of respective distribution functions with chemical and ion–molecular kinetics in terms of reactive species concentrations. There are a few groups in the world using full kinetic models: Centro de Fizica dos Plasmas, Lisbon, Portugal; University of Bari, Italy; Centre de Physique Atomique de Toulouse, France; Institute of Physics, Belgrade, Serbia, and the SRC TRINITI, Troitsk, Russia.

Actually, for any specific system not all processes are equally important. The full kinetic model can be a good starting point for analyzing relative roles played by different species. An additional analysis of a specific system at specified conditions (gas temperature, pressure and composition, typical electric current density) allows one to make a significant simplification of a model. Such a reduced model containing a reasonably small amount of reactions can be employed in a second step for further analysis of effects of spatial non-uniformities and/or time variations. For example, it was shown [1], that to model an atmospheric pressure glow discharge in fast dry airflow one may neglect plasma density gradients and describe the evolution of the plasma as it is transported with the gas flow (plug-flow model). Such a plug-flow model was developed for atmospheric air plasma [1–5]. A good agreement between theoretical predictions and experimental measurements for fast airflow glow discharge was demonstrated.

The model incorporates kinetic equations for charged and neutral species including electrons, positive and negative ions, molecules, excited species, and radicals.

The production efficiency of the chemically active species is determined by the EEDF formed by electric field and collisions with atoms and molecules. The EEDF is found numerically by solving the electron Boltzmann equation (BE) (see Section 1.3). The influence of physical parameters such as gas temperature, pressure, and applied electric field on the EEDF shape can be described adequately by one dominant combined parameter controlling practically all important discharge characteristics: the reduced electric field strength E/N, where E is the electric field strength and N is the total gas density. The gas chemical composition is an important factor as well. For typical conditions of plasma treatment, E/N and the gas composition change in time. Therefore, the electron BE should be recalculated to provide the upgraded EEDF depending on new parameter values, when necessary.

Generally, discharges producing nonthermal plasmas at atmospheric pressure are inherently non-uniform and proceed in a form of a manifold of microdischarges (for DBDs) or as fast propagating streamers (for pulse coronas). However, the location of microdischarges (MDs) in DBDs is not fixed and varies from pulse to pulse, covering in average the entire electrode surface uniformly. Thus, it is reasonable to first employ a zero-dimensional model for simulation of evolution of a pulse repetitive single MD, and then to average out over the whole discharge space. Such an approach allows one to model plasma chemical processes in realistic gas mixtures. This approach was implemented successfully to generate nitrogen oxides in pulse-periodic DBDs [6].

In general, the evolution of MDs and streamers proceeds through the formation of self-organized transient structures that are controlled by correlations between the electric field and spatial charge distributions. All these factors contribute to the actual multi-dimensional modeling of discharges. Due to the complexity of space–time modeling, only the simplest kinetic models can be included. Microdischarge dynamics in DBDs of various geometries in oxygen has been numerically simulated by 3D fluid models [7]. Earlier theoretical studies have been reviewed [8], an approach to 2D MDs modeling has been reported [9]. Formation and evolution of streamers in pulsed coronas were analyzed with help of 3D axial-symmetry fluid models by several groups [10–15].

Typical diameters of MDs or streamers are $10–100\,\mu m$, and typical times of their existence are $10–1000\,ns$. Under such conditions, one has to take into consideration the effects of gas dynamics and dissipation processes like diffusion and thermal diffusion. M. Kushner [16] has made such modeling for MDs in a metal-dielectric-metal cylindrical device in Ar atmosphere. It was found that the properties of the MDs are nearly the same as those of classical glow discharges resulting in the sensitivity of the plasma parameters to variations of the secondary emission coefficients of the treated surface.

Actually, classic negative and positive corona discharges operate in a self-pulsation mode in a wide range of conditions. In the case of a negative corona, these are widely known as *regular Trichel pulses* [17], while positive coronas are described as current oscillations in a number of gases.

In air at atmospheric pressure, effects of electron and ion distributions non-locality as well as diffusion processes, could play a role only very close to the point electrode. It is known that the cathode layer parameters are insensitive to the kinetic model used. Therefore, the simplest fluid model can be taken where all the transport and kinetic coefficients are functions of the local value of E/N. To describe the pulse mode of the negative and positive point-to-plane coronas it is sufficient to solve the well-known continuity equations for electrons, positive ions, negative ions as well as the Poission equation, which are respectively Equations 10.1–10.4 :

$$\partial n_e/\partial t + \mathrm{div}\, n_e \mathbf{w}_e = (\nu_i - \nu_a)n_e + \nu_d n_n \tag{10.1}$$

$$\partial n_p/\partial t + \mathrm{div}\, n_p \mathbf{w}_p = \nu_i n_e \tag{10.2}$$

$$\partial n_n/\partial t + \mathrm{div}\, n_n \mathbf{w}_n = \nu_a n_e - \nu_d n_n \tag{10.3}$$

$$\mathrm{div}\mathbf{E} = e(n_p - n_e - n_n)/\varepsilon_0 \tag{10.4}$$

When the indexes e, p, and n refer to electrons, positive, and negative ions, respectively, n_p, n_e, and n_n are the positive ion, electron, and negative ion number densities, $\mathbf{w}_p$, $\mathbf{w}_e$, and $\mathbf{w}_n$ their drift velocities, ν_i, ν_a, and ν_d are the ionization, attachment, and detachment frequencies, e is the electronic charge, ε_0 is the permittivity of free space. The electron drift velocity and kinetic coefficients are to be determined from solving the electron BE, the ion drift velocities are calculated using the known ion mobilities. The current in the external circuit, I, is determined from Equation 10.5:

$$V = U_0 - RI \tag{10.5}$$

where V and U_0 are the discharge and power supply voltage, R is the ballast resistor.

This relatively simple system of differential equations with transport and kinetic coefficients calculated by an offline Boltzmann solver is rather hard for numerical simulations for typical geometries of electrodes and/or plasma objects. The challenge to numerical methods is a very strong variation in the size of the current channel as a function of time or distance. In a rigorous approach one has to solve 3D space and time dependent equations. Such an approach was realized for axial symmetry of discharges in [7, 10–15, 18]. Having knowledge from experiments about the structure of discharge, it is possible to reduce the dimensionality of the mathematical problem. Such an approach to modeling Trichel pulses in negative coronas and self-oscillations in positive coronas was realized [19–21]. A similar approach was employed to describe theoretically corona-to-glow transition taking place in fast flow atmospheric pressure discharge with current increase [22]. In all these cases, using the shape of the current channel, rather good agreement was achieved between predicted and observed discharge characteristics and parametric dependences of pulse repetition frequencies and average currents.

10.2.2
Electron Kinetics

Detailed knowledge of electron kinetics in nonthermal plasma is required for plasma chemistry modeling in technological gases (air-, N_2-, Ar- or He-based mixtures, and others). For a wide variety of discharge types (glow, silent, corona, pulse-periodic, continuous wave) the EEDF plays a key role in controlling the gas plasma parameters. In particular, the mean electron energy determines the excitation, dissociation, and ionization rates and the partition of the discharge energy input into different channels for the given gas mixture. Experimental study of the EEDF in atmospheric pressure plasma meets challenges associated with numerous fast transients in the plasma. Therefore, theoretical approaches are of particular importance. The most common theoretical approach to EEDF determination in a gas plasma is solving numerically the BE in the two-term approximation. This approach appears to be sufficiently simple, reliable and has been validated by comparison with more accurate models, for example, Monte Carlo calculations.

Electron motion in the gas under an applied electric field is determined by collisions with molecules and other electrons. Electron–electron collisions begin to play some role at a relatively high gas ionization degree, usually starting from values of 10^{-4}–10^{-3} for molecular gases. Here, gases under normal conditions are considered, that is, with the total molecule concentration more than 10^{25} m^{-3} (the same order of magnitude of Loschmidt's number, n_0, provided in Chapter 1). Typical electron densities in coronas and glow discharges are much lower than 10^{21} m^{-3} ($\sim 10^{-4} n_0$), hence the ionization degree is sufficiently low and electron–electron interactions may be neglected with a good accuracy.

The influence of physical parameters (such as gas temperature, pressure, applied electric field) on the EEDF shape can be described adequately by one main parameter, the reduced electric field strength E/N, which was already introduced before. Indeed, most of the calculated plasma characteristics for a fixed gas composition are a function of E/N only, without appreciable dependence on other physical parameters within their typical value range.

The most important components of air are O_2, N_2, CO_2, and H_2O. Their concentrations in the plasma discharge depend on many conditions and may vary in a wide range of values. Furthermore, numerous minor additives may be present in the gas, which do not exert strong influence on the shape of the EEDF. These admixtures may be strongly electronegative or chemically aggressive. To describe their role in the discharges it is sufficient to average cross-sections of interest with the EEDF calculated by neglecting these contaminants.

There is much data about electron collision cross-sections corresponding to different processes in the gases listed above and in some rare gases (Ar, He).

An important problem does exist with the formation of an electron scattering cross-section set for a given molecule. Actually, for many practical simulations of plasma devices, existing information about electron scattering cross-sections is excessive. For example, the cross-sections for electron scattering from N_2 molecules have been measured for the excitation of 17 electronic states [23] and the set recommended by Phelps *et al.* [24] includes excitation of 11 electronic states. Despite the wealth of information on specific cross-sections, molecule reactions rates in many of these excited states with other more complex molecules are not known, in particular for gas mixtures containing H_2O, CO_2, and O_2. In such a situation, one has two options for kinetic modeling. The first is to introduce voluntarily the quenching processes which are lacking; the second is to reduce the quantity of molecular states incorporated into the model. The second way is preferable when modeling electric discharge evolution in time and space. However, one has to take care of consistency of the BE analysis for the reduced number of collision processes accounted for in the model. Generally, it is necessary to create a reduced set of electron collision cross-sections for every species included into the kinetic model.

In order to guarantee that this reduced set of cross-sections is appropriate for kinetic modeling, this procedure should be followed. After formulation of the reduced kinetic model, the cross-sections for proper electron inelastic processes for every species are constructed from the known full set of cross-sections for this species. For example, excitation rates for a number of electronic levels of a molecule can be summed, and instead of a manifold of levels one effective lump level is introduced. Then, the procedure of adjustment of effective cross-sections should be applied similar to that developed earlier by A. V. Phelps [25]. Namely, transport coefficients (electron drift velocity, transverse and longitudinal diffusion coefficients) and kinetic coefficients (for ionization, attachment, excitation) are calculated by solving the BE. Comparing calculated transport and kinetic coefficients as functions of E/N with experimental ones, the differences between them are minimized by fitting amplitudes of effective cross-sections. The effective cross-sections found in this way differ from the cross-sections entering into the full set of cross-sections. The set of cross-sections generated with such methodology is traditionally called a *self-consistent set*. Change of any one of such cross-sections, made for example after appearance of new data in direct measurements, may result in a wrong description of the electron kinetics. Any change of cross-sections for processes included into the model should include as an obligatory step the procedure of fitting calculated transport and kinetic coefficients to the experimental data. The sets of electron collision processes accounted for in such reduced models of technological plasma for N_2, O_2, H_2O, CO_2 species are listed in Table 10.1. Numerical solving of the BE for the EEDF allows one to calculate the rate coefficients of the listed elementary electron-based processes as a function of E/N.

10.2.3
Plasma Chemistry

For the description of any particular plasma chemical system, one needs a lot of information about the mechanisms of chemical reactions and about rate coefficients for various conditions, which are characterized by gas pressure and temperature, excitation degree, and so on. The total number of different chemical species in a typical system is rather large. Reliable modeling of any particular system requires great efforts for the accumulation of data on processes in this system. Rather extensive data sets for modeling plasma chemistry in dry air can be found in Refs. [2, 30, 31]. Most detailed information about plasma chemistry in oxygen plasma can be found in Ref. [27]. Data for plasma chemistry at room temperature in polluted gases treated by beams of high-energy electrons are collected in Ref. [32].

Table 10.1   Reactions of electrons with molecules and reaction energy thresholds.

No	Process	Threshold (eV)	Comments	References
1	$O_2 + e \rightarrow O_2(vibr) + e$	0.193	v = 1,2,3,4	[26, 27]
2	$O_2 + e \rightarrow O_2(0.98) + e$	0.977	$O_2(a^1\Delta)$	[26, 27]
3	$O_2 + e \rightarrow O_2(1.63) + e$	1.627	$O_2(b^1\Sigma)$	[26, 27]
4	$O_2 + e \rightarrow O^- + O$	4.2	Dissociative attachment	[27, 28]
5	$O_2 + e \rightarrow O_2(4.5) + e$	4.48	$O_2(c^1\Sigma, C^3\Delta, A^3\Sigma)$	[27]
6	$O_2 + e \rightarrow O + O + e$	5.58	–	[27]
7	$O_2 + e \rightarrow O + O(^1D) + e$	8.4	–	[26, 27]
8	$O_2 + e \rightarrow O_2^+ + e + e$	12.1	–	[26, 27]
9	$O_2 + e \rightarrow O + O^+ + e + e$	19.5	–	[26, 27]
10	$O_2 + e + (O_2) \rightarrow O_2^- + (O_2)$	0.051	Three body attachment	[26]
11	$N_2 + e \rightarrow N_2(vibr) + e$	0.29	v = 1,2, . . . ,8	[2]
12	$N_2 + e \rightarrow N_2(E = 1) + e$	6.17	$N_2(A^3\Sigma)$	[2]
13	$N_2 + e \rightarrow N_2(esum) + e$	7.35	$N_2(B^3\Pi), N_2(C^3\Pi)$	[2]
14	$N_2 + e \rightarrow N + N + e$	9.0	–	[2]
15	$N_2 + e \rightarrow N_2^+ + e + e$	15.6	–	[2]
16	$H_2O + e \rightarrow H_2O(rot) + e$	0.040	–	[29]
17	$H_2O + e \rightarrow H_2O(vibr1) + e$	0.198	–	[29]
18	$H_2O + e \rightarrow H_2O(vibr2) + e$	0.453	–	[29]
19	$H_2O + e \rightarrow H + OH + e$	7.1	–	[29]
20	$H_2O + e \rightarrow H_2O^+ + e + e$	12.61	–	[29]
21	$H_2O + e \rightarrow OH + H^-$	5.7	–	[29]
22	$H_2O + e \rightarrow H_2 + O^-$	4.9	–	[29]
23	$CO_2 + e \rightarrow CO_2(vibr) + e$	0.083	–	[29]
24	$CO_2 + e \rightarrow CO + O + e$	6.0	–	[29]
25	$CO_2 + e \rightarrow CO_2^+ + e + e$	13.3	–	[29]
26	$CO_2 + e \rightarrow CO + O^-$	3.3	–	[29]

v/vibr. = vibrational level/ state; rot. = rotational state; in the comments column standard spectroscopic notation is used; bibliographic data sources are reported in the Chapter reference list.

Table 10.2 contains information about chemical reactions of neutral gas components appropriate for modeling of plasma polymer surface treatment in order to enrich the upper layers with oxygen. The enrichment of the polymer with oxygen results in improvement of its surface wettability.

10.2.4
Experimental UV, Optical, and Near Infra-red Emission Spectra

10.2.4.1   Air-based Discharges
The composition of active species generated by a specific plasma-jet source and their distribution along the plasma jet can be determined experimentally from

Table 10.2 Neutral species reaction rate coefficients are reported also taking into consideration absolute temperature dependence.

No	Process	Rate coefficient ($m^3 \ s^{-1}$, $m^6 \ s^{-1}$)	Reference
1	$O_2(4.5) + M \rightarrow O_2 + M$	2.3×10^{-20}	[33]
2	$O_2(4.5) + M \rightarrow O_2(a^1\Delta) + M$	1.86×10^{-19}	[33]
3	$O_2(4.5) + M \rightarrow O_2(b^1\Sigma) + M$	8.1×10^{-20}	[33]
4	$O + O + M \rightarrow O_2 + M$	$1.9 \times 10^{-42} \cdot T^{-1} \cdot \exp(-170/T)$	[33]
5	$O + O + M \rightarrow O_2(a^1\Delta) + M$	$1.3 \times 10^{-42} \cdot T^{-1} \cdot \exp(-170/T)$	[33]
6	$O + O + M \rightarrow O_2(b^1\Sigma) + M$	$6 \times 10^{-43} \cdot T^{-1} \cdot \exp(-170/T)$	[33]
7	$O + O + M \rightarrow O_2(4.5) + M$	1.2×10^{-46}	[33]
8	$O + O_2 + O_2 \rightarrow O_3 + O_2$	$8.6 \times 10^{-43} \cdot T^{-1.25}$	[34]
9	$O + O_2 + N_2 \rightarrow O_3 + N_2$	$5.6 \times 10^{-41} \cdot T^{-2}$	[35]
10	$O + O_3 \rightarrow O_2 + O_2$	$9.5 \times 10^{-18} \cdot \exp(-2300/T)$	[33, 35]
11	$O + O_3 \rightarrow O_2(a^1\Delta) + O_2$	$6.3 \times 10^{-18} \cdot \exp(-2300/T)$	[33, 35]
12	$O + O_3 \rightarrow O_2(b^1\Sigma) + O_2$	$3.2 \times 10^{-18} \cdot \exp(-2300/T)$	[33, 35]
13	$O + NO + M \rightarrow NO_2 + M$	$9.1 \times 10^{-40} \cdot T^{-1.6}$	[36]
14	$O + NO_2 \rightarrow NO + O_2$	$3.26 \times 10^{-18} \cdot T^{0.18}$	[30]
15	$O + NO_2 + M \rightarrow NO_3 + M$	$8.1 \times 10^{-39} \cdot T^{-2}$	[34]
16	$O + NO_3 \rightarrow NO_2 + O_2$	1.7×10^{-17}	[32]
		10^{-17}	[35]
17	$O(^1D) + N_2O \rightarrow N_2 + O_2$	4.4×10^{-17}	[36, 37]
18	$O(^1D) + N_2O \rightarrow NO + NO$	7.2×10^{-17}	[36]
19	$O(^1D) + N_2O \rightarrow O + N_2O$	10^{-18}	[34]
20	$O(^1D) + NO \rightarrow O_2 + N$	8.5×10^{-17}	
21	$O(^1D) + NO_2 \rightarrow O_2 + NO$	1.4×10^{-16}	[32]
22	$O(^1D) + O \rightarrow O + O$	7.5×10^{-17}	[38]
23	$O(^1D) + O_2 \rightarrow O + O_2(b^1\Sigma)$	$2.56 \times 10^{-17} \cdot \exp(67/T)$	[38]
24	$O(^1D) + O_2 \rightarrow O + O_2$	$6.4 \times 10^{-18} \cdot \exp(67/T)$	[38]
25	$O(^1D) + N_2 \rightarrow O + N_2$	$1.8 \times 10^{-17} \cdot \exp(107/T)$	[34]
26	$O(^1D) + O_3 \rightarrow O_2 + O + O$	1.2×10^{-16}	[33]
27	$O(^1D) + O_3 \rightarrow O_2 + O_2$	2.3×10^{-17}	[33]
28	$O(^1D) + O_3 \rightarrow O_2(a^1\Delta) + O_2$	1.5×10^{-17}	[33]
29	$O(^1D) + O_3 \rightarrow O_2(b^1\Sigma) + O_2$	7.7×10^{-18}	[33]
30	$O(^1D) + O_3 \rightarrow O_2(4.5) + O_2$	7.4×10^{-17}	[33]
31	$O_2(a^1\Delta) + O \rightarrow O_2 + O$	1.3×10^{-22}	[35]
32	$O_2(a^1\Delta) + N_2 \rightarrow O_2 + N_2$	1.4×10^{-25}	[34]
33	$O_2(a^1\Delta) + O_3 \rightarrow O_2 + O_2 + O$	$5 \times 10^{-17} \cdot \exp(-2830/T)$	[35]
34	$O_2(a^1\Delta) + O_2(a^1\Delta) \rightarrow O_2(b^1\Sigma) + O_2$	2.2×10^{-23}	[35]
35	$O_2(b^1\Sigma) + O \rightarrow O_2(a^1\Delta) + O$	8×10^{-20}	[38]
36	$O_2(b^1\Sigma) + O_2 \rightarrow O_2 + O_2$	4×10^{-23}	[36]
		1.5×10^{-22}	[38]
37	$O_2(b^1\Sigma) + N_2 \rightarrow O_2 + N_2$	2×10^{-21}	[35]
38	$O_2(b^1\Sigma) + O_2 \rightarrow O_2(a^1\Delta) + O_2$	4×10^{-23}	[38]
39	$O_2(b^1\Sigma) + N_2 \rightarrow O_2 + N_2$	2×10^{-21}	[35]

(continued overleaf)

Table 10.2 (*Continued*)

No	Process	Rate coefficient (m^3 s^{-1}, m^6 s^{-1})	Reference
40	$O_2(b^1\Sigma) + O_2 \rightarrow O_2(a^1\Delta) + O_2$	4×10^{-23}	[38]
41	$O_2(b^1\Sigma) + O_3 \rightarrow O_2 + O_2 + O$	1.5×10^{-17}	[38]
42	$O_2(b^1\Sigma) + O_3 \rightarrow O_2(a^1\Delta) + O_2 + O$	7×10^{-18}	[33]
43	$N_2(A^3\Sigma) + N_2(A^3\Sigma) \rightarrow N_2(\text{esum}) + N_2$	2×10^{-16}	
44	$N_2(A^3\Sigma) + O_2 \rightarrow N_2 + O + O$	2.29×10^{-18}	[39]
45	$N_2(A^3\Sigma) + O_2 \rightarrow N_2O + O$	4.6×10^{-21}	[39]
46	$N_2(A^3\Sigma) + N_2 \rightarrow N_2 + N_2$	3.7×10^{-22}	[37]
47	$N_2(A^3\Sigma) + NO \rightarrow N_2 + NO$	2.8×10^{-17}	[37]
		1.5×10^{-16}	[32]
48	$N_2(A^3\Sigma) + N_2O \rightarrow N_2 + N_2 + O$	1.4×10^{-17}	[37]
		8×10^{-17}	[32]
49	$N_2(A^3\Sigma) + NO_2 \rightarrow NO + O + N_2$	10^{-18}	[32, 37]
50	$N_2(\text{esum}) + M \rightarrow N_2(A^3\Sigma) + M$	10^{-19}	
51	$N_2(\text{esum}) + O_2 \rightarrow N_2 + O + O$	2×10^{-18}	[40]
52	$N + NO \rightarrow N_2 + O$	3.1×10^{-17}	[41]
53	$N + NO_2 \rightarrow N_2O + O$	1.4×10^{-18}	[34]
54	$N + NO_2 \rightarrow NO + NO$	6×10^{-19}	[37]
55	$N + O_2 \rightarrow NO + O$	$1.1 \times 10^{-20} \cdot T \cdot \exp(-3150/T)$	[35]
56	$N + O_2(a^1\Delta) \rightarrow NO + O$	10^{-22}	[34]
57	$N + O_3 \rightarrow NO + O_2$	10^{-21}	[34]
		5.7×10^{-19}	[35]
58	$N + N + M \rightarrow N_2 + M$	$8.3 \times 10^{-46} \cdot \exp(500/T)$	[35]
59	$N + O + M \rightarrow NO + M$	$1.8 \times 10^{-43} \cdot T^{-0.5}$	[35]
60	$N_2 + M \rightarrow N + N + M$	$6.1 \times 10^{-9} \cdot T^{1.6} \cdot \exp(-113200/T)$	[35]
61	$NO + M \rightarrow N + O + M$	$6.6 \times 10^{-10} \cdot T^{-1.5} \cdot \exp(-75500/T)$	[35]
62	$NO + O_3 \rightarrow NO_2 + O_2$	$9 \times 10^{-19} \cdot \exp(-1200/T)$	[35]
63	$NO + NO_3 \rightarrow NO_2 + NO_2$	2×10^{-17}	[34]
64	$NO + NO + O_2 \rightarrow NO_2 + NO_2$	$3.3 \times 10^{-51} \cdot \exp(526/T)$	[34]
65	$NO + NO_3 \rightarrow NO + NO + O_2$	$2.71 \times 10^{-17} \cdot T^{-0.23} \cdot \exp(-947/T)$	[32]
66	$NO_2 + M \rightarrow O + NO + M$	$1.8 \times 10^{-14} \cdot \exp(-33000/T)$	[35]
67	$NO_2 + NO_2 \rightarrow NO + NO + O_2$	$3.3 \times 10^{-18} \cdot \exp(-13540/T)$	[35]
68	$NO_2 + NO_3 + M \rightarrow N_2O_5 + M$	$5.3 \times 10^{-32} \cdot T^{-4.1}$	[34]
69	$NO_2 + O_3 \rightarrow NO_3 + O_2$	$1.2 \times 10^{-19} \cdot \exp(-2450/T)$	[32]
70	$NO_2 + O_3 \rightarrow NO + O_2 + O_2$	10^{-24}	[41]
71	$NO_2 + NO_3 \rightarrow NO_2 + NO + O_2$	$2.3 \times 10^{-19} \cdot \exp(-1600/T)$	[32]
72	$NO_3 + NO_3 \rightarrow NO_2 + NO_2 + O_2$	$7.5 \times 10^{-18} \cdot \exp(-3000/T)$	[32]
73	$N_2O_5 + M \rightarrow NO_2 + NO_3 + M$	$1.7 \times 10^2 \cdot T^{-4.4} \cdot \exp(-11080/T)$	[34]

exp stands for Neper exponentional function; *T* is a gas temperature (K); *M* stands for the second or third body in two or three molecule reactions; bibliographic data sources are reported in the Chapter reference list.

spectroscopic measurements (see also Chapter 4). Optical emission spectra of photons emitted by plasma jets of two different sources regarded as source No1 and source No2 (described in Section 2.3.2.1) were recorded with an Avantes spectrometer (AvaSpec-2048-FT-RM). Figure 10.1a shows the spectral lines of different species formed and excited in the plasma jet in ambient air. One can see that N and O atoms, NO and OH radicals, and especially molecular nitrogen

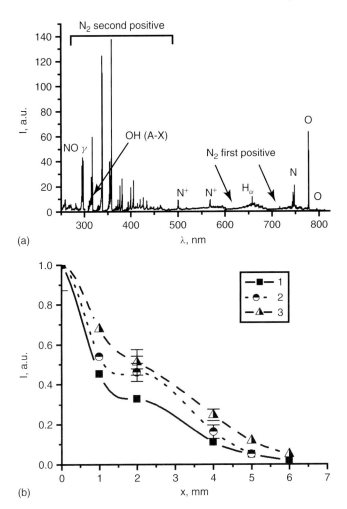

(a)

(b)

Figure 10.1 (a) Total optical spectrum in UV-Vis region of the radiation emitted from the plasma jet in airflow (plasma source No1). Gas flow velocity is 50 m s⁻¹. Discharge power is 35 W. (b) Longitudinal emission intensity distribution of active species in the same air plasma jet. Coordinate x = 0 is the outlet of the source.

The following contributions are considered: (1) $N_2(C^3\Pi_u, v' = 0 \rightarrow B^3\Pi g, v'' = 0)$ transition, $\lambda = 337.1$ nm, N_2 second positive system; (2) $O(^5S^0 \rightarrow {}^5P)$ transition, $\lambda = 777.2$ nm; (3) $OH(A^2\Sigma^+, v' = 0 \rightarrow X^2\Pi, v'' = 0)$ transition, $\lambda = 306$ nm. The radiation intensity is given in arbitrary units as a function of wavelength.

excited in $N_2(C)$ and $N_2(B)$ states give major input in the radiation from the air plasma jet. Decay in the light emission intensity of different active species along the plasma jet is shown in Figure 10.1b.

10.2.4.2 Nitrogen-based Discharges

Spectroscopic information on UV-Vis light emission from the middle of the discharge zone of the N_2-plasma source is shown in Figure 10.2a. The main contribution to emission from this region corresponds to 1^+ and 2^+ systems of nitrogen. Low intensity emission of CN is recorded only from the region close to the

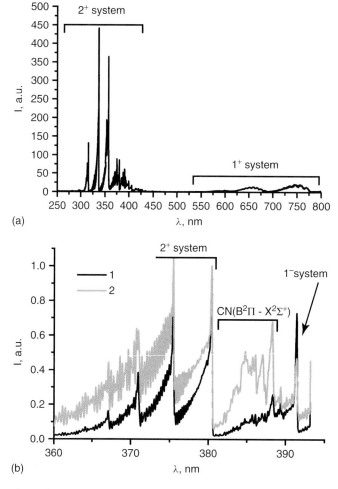

(a)

(b)

Figure 10.2 (a) UV-Vis spectrum of radiation emitted from the middle of the discharge zone (plasma source No2). Discharge power is 70 W. Gas flow rate of nitrogen is 440 cm³ s⁻¹. (b) Change in intensity of CN spectrum with location of the emitting discharge region inside plasma source No2. 1: region close to the inlet; 2: region close to the outlet.

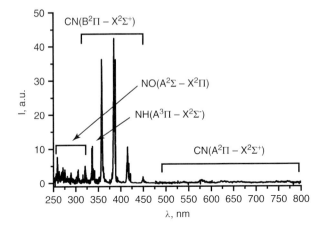

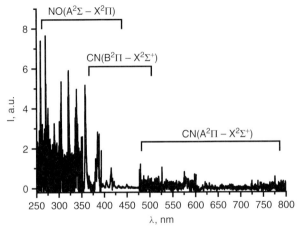

Figure 10.3 Afterglow UV-Vis spectrum of a radiation emitted from N_2-plasma jet (plasma source No2) at different distances away from the outlet of the plasma source nozzle. Discharge power is 70 W. Gas flow rate of nitrogen is 440 cm^3 s^{-1} (a) 1.5 mm away from nozzle and (b) 10 mm away from nozzle.

outlet of discharge zone (see Figure 10.2b). The afterglow UV-Vis spectrum emitted by the N_2-plasma jet at different distances away from the outlet of the plasma source nozzle are presented in Figure 10.3. We would like to draw attention to a surprising fact in the afterglow: despite the high purity of the nitrogen gas (99.999%) used as plasma-forming gas, the light emission in spectral lines does not correlate with emission from the excited states of nitrogen itself as is shown in Figure 10.2a. The very small impurities of radicals CN and NH give a main contribution to the light emission from the active N_2-plasma-jet afterglow at atmospheric pressure. It means that small impurities in the pure gas can influence strongly the efficiency of the plasma treatment.

As can be seen in Figure 10.3a, the light emission at short distances from the nozzle is caused, mainly, by radicals CN and NH while at the longer distances (10 mm or more) the essential contribution to the plasma emission is made by NO (γ-band) as shown in Figure 10.3b.

Under plasma processing, a narrow N_2-plasma jet strikes the sample, expands over the surface and occupies a large area. It is interesting therefore to know that a radial distribution of active species and gas temperature in the plasma over the treated zone will exist. Such experimental information is presented in Figure 10.4.

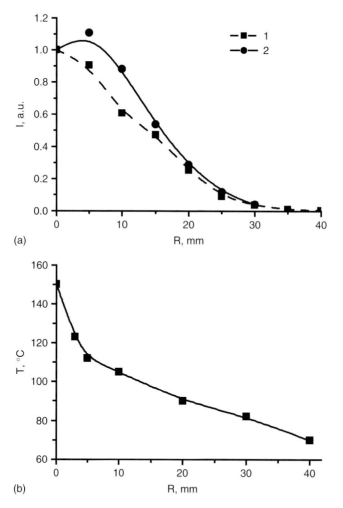

Figure 10.4 Radial distribution: (a) of light intensity of $CN(B^2\Sigma^+ \rightarrow X^2\Sigma^+)$ (1) and $NH(A^3\Pi \rightarrow X^2\Sigma^+)$ (2) emitted from N_2-plasma jet expanded over the surface to be treated; (b) of the gas temperature in the plasma jet expanded over the surface to be treated. The surface is placed 3 mm away from the nozzle of the plasma source. Discharge power is 70 W. Gas flow rate of nitrogen is 440 cm^3 s^{-1}.

So, due to the existence in pure nitrogen of trace gaseous admixtures (like CH_4, H_2O, O_2), the plasma composition can be changed drastically. Our experimental results presented above prove unambiguously that small gas additives play a crucial role in the generation of active species. Consequently, their interaction with the polymer surface has to be taken into account, both in theory and practice.

10.2.4.3 CF$_4$-based Discharges

Another subject of our investigations is CF_4 that is widely used as plasma-forming gas in the DBD systems for hydrophobic processing of fabrics and polymers. The purity of CF_4 used in the experiments is 99.5%, and air is the dominant impurity (about 0.5%). A survey spectrum of light emitted from DBD-plasma in gas mixture CF_4 (99.5% CF_4 + 0.5% air) for wavelength range from 240 to 800 nm is shown in Figure 10.5.

Emission spectra with higher resolution in two wavelength ranges 240–450 and 450–800 nm for two gas mixtures (99.5% CF_4 + 0.5% air) and (10% CF_4 and 90% nitrogen) are presented in Figure 10.6. One can see that CF_3 radicals form the main contribution to the light emission. It should be noted that the spectral lines of N_2-second positive system emitted from the gas mixture (99.5% CF_4 + 0.5% air) appear due to air admixture.

For the gas mixture (99.5% CF_4 + 0.5% air), a broad emission band because of CF_3 radicals is seen clearly in Figure 10.6b. Atomic spectral lines of the ionized oxygen observed in the spectrum appear probably due to the admixture of air in the working gas. The emission spectrum of pure nitrogen is shown in Figure 10.6b as superimposed curve for direct comparison.

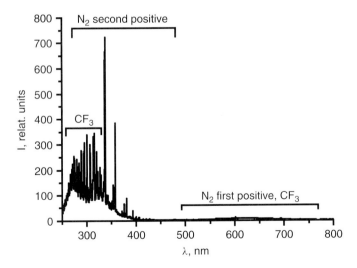

Figure 10.5 Survey of UV-Vis emission spectra from a DBD. Pressure of working gas $P = 1.2$ atm. Gas mixture is 99.5% CF_4 and 0.5% air.

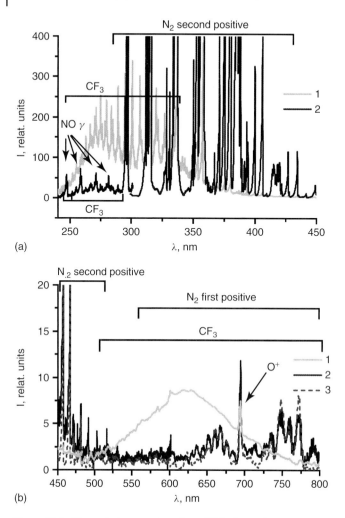

Figure 10.6 The emission spectrum from DBD-plasma in gas mixtures containing CF_4. Total gas pressure $P = 1.2$ atm. (1): 99.5% CF_4 and 0.5% air; (2): 10% CF_4 and 90% nitrogen; (3): pure nitrogen; (a) in the range 250–450 nm; (b) in the range 450–800 nm.

10.2.5
Influence of Impurities on Composition of Gas Activated by Nonthermal Plasma

One of the important parameters, which can serve to optimize the plasma surface treatment is a buffer (transportation) gas. A key criterion for the choice of buffer gas is the feasibility to realize a highly reduced energy input in nonthermal discharges. Besides, the energy consumed by the buffer gas should effectively be used for production of chemically active species and UV radiation. In experiments, many factors make it difficult to identify mechanisms for observed variations of discharge treatment efficiency of a specific surface. Theoretical modeling can significantly enhance analyses of experimental observations.

Here, an example is considered where the experimentally observed high efficiency of polypropylene (PP) and poly(ethylene terephthalate) (PET) surface treatment by a glow discharge in 'pure' nitrogen can be explained by the existence in the nitrogen gas of traces of other species at the ppm level [42]. Indeed, it was revealed experimentally (Figures 10.2 and 10.3) that the plasma emission spectrum of this type of discharge exhibits molecular systems of species, which can appear only from impurities.

Specification of the exact gas composition is usually given by the supplier. The typical level of contaminants is on the level of a few parts per million. In particular, in so-called 'pure' nitrogen there exist different admixtures like O_2, H_2O, CH_4, and other hydrocarbons. To evaluate their role, simulations were performed regarding the influence of these small additions (of O_2, H_2O, and a mixture of both components) for conditions close to our experiments with the microstreamer discharge.

It was found that, in the afterglow, atomic oxygen decays rather fast due to conversion to ozone. As a result, the atomic oxygen concentration in polluted nitrogen (10 ppm O_2) can be higher than in air. It is interesting to find a maximum in oxygen atom concentration in N_2 with variable concentration of O_2. Actually, this maximum depends on the moment when the measurement was done. Figure 10.7a shows the dependence of oxygen atom number density on concentration O_2 in the mixture $O_2 : N_2$ at 100 μs after start up of the electric discharge. There is the pronounced maximum of oxygen atom concentration 3.6×10^{22} m^{-3} at the concentration O_2 equal to about 0.5%. For comparison, the oxygen atom concentration is equal to 2.4×10^{20} m^{-3} and 1.7×10^{20} m^{-3} for 5 ppm O_2 initial concentration and 20% O_2 (synthetic air), respectively. Using a mixture of 0.5% O_2 in N_2, which is typical for so-called 'commercial purity nitrogen,' the oxygen atom density increases 200 times in comparison with dry air. In our calculation we used the electrical circuit with fixed parameters: discharge capacity, charging voltage on the discharge capacity and others. During the variation of the gas mixture in our simulation, the input discharge energy E_{in} changed in the range 600–750 J l^{-1}.

Figure 10.7b shows the dependence of the radical and active molecule number densities on the concentration of O_2 in the mixture $O_2 : N_2$ at 1 and 100 μs after starting the electric discharge. Note that, in contrast to Figure 10.7a,b uses the logarithmic scale. The first sampling time ($t = 1$ μs) corresponds in space to the end of discharge region. The second sampling time ($t = 100$ μs) corresponds approximately to the plasma transport time to reach the processed film surface. Hence, the different sampling times presented can also be interpreted as different spatial locations.

Figure 10.8a shows the dependence of the radical and active molecule number densities on the concentration of H_2O in the mixture of dry air with H_2O at the 1 and 100 μs after starting the electric discharge. The presence of a few percent of H_2O in the air decreases the oxygen atom density by a factor of 10 at 100 μs. The number density of OH radicals increases with increasing concentration of

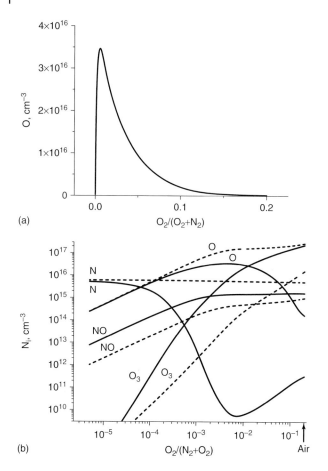

Figure 10.7 Dependence of the number density of oxygen atoms (a), radicals and active molecules (b) on the concentration of O_2 in the mixture $O_2:N_2$ at two different moments after start-up of the electric discharge. Solid line: after $100\,\mu s$; dashed line: after $1\,\mu s$. $P_0 = 1$ atm, $T_0 = 300$ K, $E_{in} = 600$ J l^{-1}.

H_2O. At the initial concentration of 5% H_2O, the OH radical number density at $t = 100\,\mu s$ is of the same order as the oxygen atom number density.

Simulations were also performed for nitrogen with additions of CH_4, O_2, and H_2O with equal concentrations 5 ppm. Our database on electron scattering cross-sections was extended by inclusion of a self-consistent set of cross-sections for CH_4 molecules. It is seen from Figure 10.8b that the admixture molecules are more destroyed in the afterglow than in the discharge (at $t = 0.5\,\mu s$. This is the typical situation for impurities, which are removed in reactions with active species produced in the discharge. The time-dependence of the density of a number of radicals produced in the discharge and in the afterglow is illustrated in Figure 10.9.

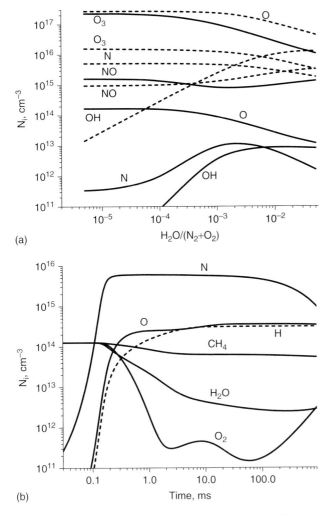

Figure 10.8 (a) The radical and active molecule number density as a function of H_2O in the mixture of dry air with H_2O at two moments after start-up of the electric discharge. (b) Atom and molecule number density as a function of time for the mixture of nitrogen and 5 ppm CH_4, 5 ppm O_2, and 5 ppm H_2O. Dashed line: 1 μs, solid line: 100 μs. Conditions are as for Figure 10.7.

To explain the discharge emission spectrum observed experimentally, information about how electronically excited molecules are formed is needed. Such information can be found in Table 10.3.

The amount and sort of radicals approaching the surface can vary strongly according to the gas composition. In particular, small admixtures (O_2, H_2O, C_xH_y) to pure gases, such as Ar or N_2, can significantly influence the activation dose. Our numerical data demonstrate an opportunity to increase processing efficiency by a proper choice of the components of work gas mixture.

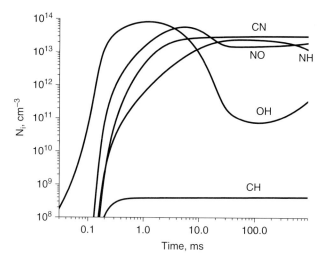

Figure 10.9 Radical number density as a function of time for the mixture of nitrogen and 5 ppm CH_4, 5 ppm O_2, and 5 ppm H_2O. Conditions are as for Figure 10.7.

Table 10.3 Reactions leading to electronically excited molecules and radicals are reported, with corresponding rate constants.

No	Reaction	Rate constant (m^3 s^{-1}, m^6 s^{-1})	Reference
1	$N_2(X, \Sigma, v > 12) + CN(X^2\Sigma) \rightarrow N_2(X, \Sigma) + CN(B^2\Sigma^+)$	1.2×10^{-17}	[43]
2	$N_2(A^3\Sigma) + CN(X^2\Sigma) \rightarrow N_2(X, \Sigma) + CN(B^2\Sigma^+)$	1.2×10^{-16}	[44]
3	$C + N + M \rightarrow CN(B^2\Sigma^+) + M$	9.4×10^{-45}	[44]
4	$N_2(A^3\Sigma) + OH(X^2\Sigma) \rightarrow N_2(X, \Sigma) + OH(A^2\Sigma)$	1×10^{-16}	[45]
5	$N_2(A^3\Sigma) + NO(X^2\Sigma) \rightarrow N_2(X, \Sigma) + NO(A^2\Sigma)$	6.9×10^{-17}	[46]

M stands for third body in three molecule reactions; bibliographic data sources are reported in the Chapter reference list.

10.3
Kinetic Model for Chemical Reactions on a Polypropylene Surface in Atmospheric Pressure Air Plasma

10.3.1
Description of Kinetic Model

10.3.1.1 Description of Chemical Reaction Modeling
Modeling of plasma surface interactions for conditions of air nonthermal plasma at atmospheric pressure is the most challenging part of the theoretical

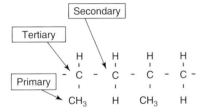

Figure 10.10 Three types of C atoms in polypropylene monomers.

approach to simulations of material processing by plasma. The model is based on the assumption that surface processing by plasma at atmospheric pressure is controlled by neutral chemistry reactions. The surface reaction mechanism for PP was taken from several references [47–51]. Here, we describe important features of this mechanism.

PP is a saturated hydrocarbon polymer with a carbon backbone containing hydrogen and methyl (CH_3) groups (see Figure 10.10). Gaseous particles approaching the surface of the perfect PP material meet only hydrogen atoms exposed to the outside. In general, the reactivity of the hydrogen atoms depends on the nature of the C atom to which they are attached. Carbon atoms in the PP monomer unit can be divided into three groups depending on the number of neighboring C atoms: primary, secondary, and tertiary C atoms (see Figure 10.10). According to the paper of R. Dorai and M. Kushner [47], the reactivity of hydrogen bound to these C atoms scales as: $H_{tert} > H_{sec} > H_{pri}$.

Table 10.4 Examples of surface radicals structure (symbol * designates chemical vacant site).

'In chain' tertiary alkyl radical	'End free' primary alkyl radical
$\begin{matrix} H & & H \\ | & & | \\ \sim C - \overset{*}{C} - C \sim \\ | & | & | \\ H & CH_3 & H \end{matrix}$	$\begin{matrix} H & H \\ | & | \\ * C - C \cdots \\ | & | \\ H & CH_3 \end{matrix}$
'In chain' secondary alkyl radical	'End free' secondary alkyl radical
$\begin{matrix} & H & H \\ & | & | \\ \sim \overset{*}{C} - C - C \sim \\ | & | & | \\ H & CH_3 & H \end{matrix}$	$\begin{matrix} H & H \\ | & | \\ \sim C - C * \\ | & | \\ H & CH_3 \end{matrix}$
'In chain' tertiary alkoxy radical	'End free' secondary alkoxy radical
$\begin{matrix} H & \overset{*}{O} & H \\ | & | & | \\ \sim C - C - C \sim \\ | & | & | \\ H & CH_3 & H \end{matrix}$	$\begin{matrix} H & H \\ | & | \\ \sim C - C - O* \\ | & | \\ H & CH_3 \end{matrix}$

Further we will use the term 'in chain' to designate radicals and groups formed in the polymer chain (not at the end of the chain) and the term 'end free' to designate radicals and groups formed at the end of polymer chain. 'End free' alkyl radicals appear due to the bond scission processes. Some of the radicals considered in the model are listed in Table 10.4.

For the description of reactions 'in chain' radicals and groups are designated using the tag 'PP'. For example: PP* is an in chain alkyl radical, PPO* is an in chain alkoxy radical, and so on. For the designation of 'end free' groups and radicals the tag 'R' is used. For example: R* is an end free alkyl radical, ROO* is an end free peroxy radical, and so on. When necessary to indicate explicitly the type of C atom (primary, secondary, or tertiary) with a given group or radical, then the terms 1PP, 2PP, 3PP, 1R, and 2R will be used. For example, 2PPO* is an in chain alkoxy radical based on a secondary C-atom. Note that the type of end free C-atoms can be only primary or secondary (see Table 10.4). The full list of processes taken into account in the model is presented in Table 10.5.

A brief description of the reaction mechanism for PP treatment in air plasma proposed in Refs. [47, 48] is as follows (Numbers refer to the reactions in Table 10.5). The modification of the PP surface starts with the abstraction of H atoms by O atoms and OH radicals and, respectively, production of alkyl radicals PP* (No. 1–2).

The alkyl radicals react with O and O_3 to form alkoxy radicals PPO*, with O_2 to produce peroxy radicals PPOO*, with OH to form alcohols PPOH and with H to regenerate PP monomer (No. 3–7).

Further reactions of the alcohol groups with O and OH radicals result in the abstraction of hydrogen atoms and formation of alkoxy radicals (No. 8–9).

The peroxy radicals can abstract hydrogen atoms from neighboring sites to produce hydroperoxide groups PPOOH and alkyl radicals. Similar processes take place for the alkoxy radicals, that is, alkoxy radicals abstract hydrogen atoms from neighboring sites to form alcohol groups and alkyl radicals (No. 10–11).

Besides, peroxy radicals can react with NO molecules to produce alkoxy radicals (No. 12).

There are also backbone scission processes due to the reactions of alkoxy radicals with PP backbone, scissions on tertiary carbon atoms lead to the formation of ketones $\sim(CH_3)C{=}O$ (No. 13)

and scissions on secondary carbon atoms result in aldehydes $\sim HC{=}O$ (No. 14).

Table 10.5 Processes included in the kinetic model for the plasma treatment of a PP surface in atmospheric pressure discharge in air.

No	Processes	Rate constant (m² s⁻¹), frequency (s⁻¹), or probability[a]	Comments, ref.
1	$PP + O \Rightarrow PP^* + OH$	$10^{-5}, 10^{-4}, 10^{-3}$	[47]
2	$PP + OH \Rightarrow PP^* + H_2O$	$0.0025, 0.05, 0.25$	[47]
3	$PP^* + O \Rightarrow PPO^*$	$0.01, 0.01, 0.1$	[47]
4	$PP^* + O_3 \Rightarrow PPO^* + O_2$	$0.5, 0.5, 1$	[47]
5	$PP^* + O_2 \Rightarrow PPOO^*$	$5 \times 10^{-4}, 2.3 \times 10^{-4}, 10^{-3}$	[47]
6	$PP^* + H \Rightarrow PP$	0.2, for all PP types	[47]
7	$PP^* + OH \Rightarrow PPOH$	0.2, for all PP types	[47]
8	$PPOH + O \Rightarrow PPO^* + OH$	7.5×10^{-4}, for all PP types	[47]
9	$PPOH + OH \Rightarrow PPO^* + H_2O$	8.2×10^{-3}, for all PP types	[47]
10	$PPOO^* + PP \Rightarrow PPOOH + PP^*$	$5.5 \times 10^{-20} \, (m^2 \, s^{-1})$, for all PP types	[47]
11	$PPO^* + PP \Rightarrow PPOH + PP^*$	$8 \times 10^{-18} \, (m^2 \, s^{-1})$, for all PP types	[47]
12	$PPOO^* + NO \Rightarrow PPO^* + NO_2$	0.02, for all PP types	[48]
13	$3PPO^* \Rightarrow 3PPO + 1R^* \, (3PPO - ketones)$	$500 \, (s^{-1})$	[47]
14	$2PPO^* \Rightarrow 2R^* + 2PPO \, (2PPO - Aldehydes)$	$10 \, (s^{-1})$	[47]
15	$2PPO + O \Rightarrow {}^*2PPO + OH$ $({}^*2PPO - carbonyl \, radicals)$	0.04	[47]
16	${}^*2PPO + OH \Rightarrow {}^*2PPO + H_2O$	0.4	[47]
17	${}^*2PPO + O \Rightarrow 2R + CO_2$	0.4	[47]
18	${}^*2PPO + OH \Rightarrow OH2PPO \, (OH2PPO - acids)$	0.12	[47]
19	$PPOO^* + HO_2 \Rightarrow PPOOH + O_2$	0.01, for all PP types	Estimated
20	$PPO^* + HO_2 \Rightarrow PPOH + O_2$	0.01, for all PP types	Estimated
21	$PPOO^* + OH \Rightarrow PPOH + O_2$	0.01, for all PP types	Estimated
22	$PPO^* + OH \Rightarrow PPOOH$	0.001, for all PP types	Estimated
23	$PPOO^* + PPOO^* \Rightarrow PPOOPP + O_2$	$10^{-19} \, (m^2 \, s^{-1})$, for all PP types	Estimated
24	$PPO^* + PPO^* \Rightarrow PPOOPP$	$10^{-19} \, (m^2 \, s^{-1})$, for all PP types	Estimated
25	$PPOO^* + PPO^* \Rightarrow PPOPP + O_2$	$10^{-19} \, (m^2 \, s^{-1})$, for all PP types	Estimated
26	$R^* + O \Rightarrow RO^*$	0.1, for all R types	[49]
27	$R^* + O_2 \Rightarrow ROO^*$	0.01, for all R types	[49]
28	$R^* + O_3 \Rightarrow RO^* + O_2$	0.001, for all R types	Estimated
29	$R^* + OH \Rightarrow ROH$	10^{-5}, for all R types	[49]
30	$RO^* + HO_2 \Rightarrow ROH + O_2$	0.01, for all R types	Estimated
31	$RO^* + OH \Rightarrow ROOH$	0.001, for all R types	Estimated
32	$ROO^* + HO_2 \Rightarrow ROOH + O_2$	0.01, for all R types	Estimated
33	$ROO^* + OH \Rightarrow ROH + O_2$	0.01, for all R types	Estimated
34	$ROO^* + NO \Rightarrow RO^* + NO_2$	0.02, for all R types	Estimated
35	$RO^* + PP \Rightarrow ROH + PP^*$	$10^{-20} \, (m^2 \, s^{-1})$, for all PP and R types	Estimated
36	$ROO^* + PP \Rightarrow ROOH + PP^*$	$10^{-20} \, (m^2 \, s^{-1})$, for all PP and R types	Estimated
37	$ROO^* + PPOO^* \Rightarrow ROOPP + O_2$	$10^{-19} \, (m^2 \, s^{-1})$, for all PP and R types	Estimated
38	$RO^* + PPOO^* \Rightarrow ROPP + O_2$	$10^{-19} \, (m^2 \, s^{-1})$, for all PP and R types	Estimated
39	$ROO^* + PPO^* \Rightarrow ROPP + O_2$	$10^{-19} \, (m^2 \, s^{-1})$, for all PP and R types	Estimated
40	$RO^* + PPO^* \Rightarrow ROOPP$	$10^{-19} \, (m^2 \, s^{-1})$, for all PP and R types	Estimated
41	$ROO^* + ROO^* \Rightarrow RO^* + RO^* + O_2$	$10^{-19} \, (m^2 \, s^{-1})$, for all R types	Estimated
42	$R^* + R^* \Rightarrow RR$	$10^{-19} \, (m^2 \, s^{-1})$, for all R types	[49]

[a]Three numbers correspond to primary, secondary, and tertiary PP radicals, respectively. Specific reaction paths are provided with corresponding rate constants, frequency, and reaction probabilities, according to available data in sources referenced in square brackets. 'Estimated' stands for data estimated by the authors.

In the kinetic model [47] there are no processes for the ketones degradation. For the aldehydes, there are processes for the production of carbonyl radicals ($\sim$C=O) by the abstraction of H atoms by O and OH radicals (No. 15–16).

$$
\begin{array}{c}
\text{H} \quad\;\; \text{O} \\[2pt]
| \qquad\;\; \| \\[2pt]
\sim\;\text{C}\;-\;\text{C} \;+\;\text{O}\;=>\; \sim\;\text{C}\;-\;\text{C}\;+\;\text{OH} \\[2pt]
| \qquad\;\; | \\[2pt]
\text{CH}_3 \quad\; \text{H} \qquad\qquad\qquad \text{CH}_3
\end{array}
$$

$$
\begin{array}{c}
\text{H} \quad\;\; \text{O} \\[2pt]
| \qquad\;\; \| \\[2pt]
\sim\;\text{C}\;-\;\text{C} \;+\;\text{OH}\;=>\; \sim\;\text{C}\;-\;\text{C}\;+\;\text{H}_2\text{O} \\[2pt]
| \qquad\;\; | \\[2pt]
\text{CH}_3 \quad\; \text{H} \qquad\qquad\qquad \text{CH}_3
\end{array}
$$

These carbonyl radicals further react with O atoms to produce CO_2 molecules and respective remainder radicals, that is, the model includes the processes resulting in etching of PP surface (No. 17).

$$
\begin{array}{c}
\text{H} \quad\;\; \text{O} \qquad\qquad \text{H} \\[2pt]
| \qquad\;\; \| \qquad\qquad | \\[2pt]
\sim\;\text{C}\;-\;\text{C} \;+\;\text{O}\;=>\; \sim\;\text{C}_{*}\;+\;CO_2 \\[2pt]
| \qquad\;\; * \qquad\qquad | \\[2pt]
\text{CH}_3 \qquad\qquad\qquad\;\; \text{CH}_3
\end{array}
$$

Carbonyl radicals can also react with OH to form acids $\sim$(OH)C=O (No. 18).

$$
\begin{array}{c}
\text{H} \quad\;\; \text{O} \qquad\qquad \text{H} \quad\;\; \text{O} \\[2pt]
| \qquad\;\; \| \qquad\qquad | \qquad\;\; \| \\[2pt]
\sim\;\text{C}\;-\;\text{C} \;+\;\text{OH}\;=>\; \sim\;\text{C}\;-\;\text{C} \\[2pt]
| \qquad\;\; * \qquad\qquad | \qquad\;\; | \\[2pt]
\text{CH}_3 \qquad\qquad\qquad\;\; \text{CH}_3 \quad\; \text{OH}
\end{array}
$$

Along with formation of ketones and aldehydes, backbone scission reactions lead also to the appearance of remainder radicals 1R* and 2R*. Processes with these radicals are not considered in [47].

Rate constants, frequencies and probabilities for the reactions discussed above were taken the same as in [47, 48]. In a previous paper from the authors [51], only these reactions were taken into account in the kinetic model. Here, a number of extra processes are included in the model in addition to the reaction mechanism presented in [47, 48, 51].

According to the results of discharge modeling, a noticeable amount of HO_2 radicals is formed in humid air plasma and these radicals can react with peroxy and alkoxy radicals to form hydroperoxides and alcohols, respectively [50]. These reactions are represented as No. 19 and 20, respectively. To estimate the probabilities of these reactions we have looked for analogous gas phase chemical reactions. It appeared that the rate constants for the gas phase reactions like

$$HO_2 + CH_3C(O)CH_2OO^* \Rightarrow CH_3C(O)CH_2OOH + O_2$$
$$HO_2 + (CH_3)3CCH_2OO^* \Rightarrow (CH_3)3CCH_2OOH + O_2$$

are about 10^{-17} m^3 s^{-1} [52], so the probability for reaction No. 19 is expected to be high. The value of this probability was estimated in the following manner [47].

It was assumed that a unit surface reaction probability corresponds to a gas phase kinetic rate constant ($\approx 10^{-16}$ m^3 s^{-1}). Then the surface reaction probability for the reaction No. 19 can be as high as 0.1. In our model the correspondent probability was taken 0.01 to fit the model results to the experimental data. Unfortunately, we have not found analogous gas phase chemical reactions for the process described as No. 20, so the correspondent probability was taken the same as for reaction No. 19. It should be noted that in Ref. [49] the probability for the reaction No. 19 is set to 10^{-5} by the authors without explanation. Following our results, this is probably an underestimation.

Termination reactions also occur between OH gas radical and alkoxy and peroxy surface radicals [50], they are listed as No. 21 and 22, respectively. The probabilities of these processes were taken as 0.01 and 0.001 respectively, again the choice is based on fitting the model results to the experimental data.

As it is pointed out in Ref. [50], the termination of alkoxy and peroxy radicals can occur by bimolecular recombination, reactions No. 23–25. The rate constants for all three processes were taken as 10^{-19} m^2 s^{-1}.

If the polymer peroxy and/or alkoxy radicals are in neighboring positions, they recombine to form stable cyclic peroxides or epoxides [50]. For example:

Reactions between peroxy and/or alkoxy radicals in different polymer chains cause crosslinking. Reactions between two peroxy radicals and between peroxy radicals and HO$_2$ gas radicals can also lead to formation of C=O groups [50]. At present, these processes are not included in the kinetic model.

Next, we focus on the processes with the participation of 'end free' radicals that are included in the model. These reactions are similar to those for 'in chain' radicals. The following processes are taken into account (Numbers refer to the reactions in Table 10.5):

- Formation of alkoxy and peroxy radicals in reactions with active plasma species (No. 26–28)

- Formation of alcohols and hydroperoxides in reactions with active plasma species (No. 29–33)
- Decay of 'end free' peroxy radicals in reactions with NO molecules (No. 34)
- Extraction of H atoms from polymer surface by 'end free' alkoxy and peroxy radicals (No. 35–36)
- Recombination reactions between 'end free' and 'in chain' alkoxy and peroxy radicals lead to crosslinking (No. 37–40)
- Recombination reactions between 'end free' peroxy radicals are also included in the model (No. 41).

The corresponding probabilities and rate constants were partially taken from [50] and partially were assigned by ourselves by analogy with 'in chain' radical reactions. It was assumed that the typical value of the rate constant is 10^{-19} m^2 s^{-1}.

Finally, alkyl radicals can react with each other to crosslink. We included in the model crosslink processes for the 'end free' alkyl radicals (No. 42) with an estimated rate of 10^{-19} m^2 s^{-1} [49].

Note that the role of these processes under the considered conditions (atmospheric pressure air discharge) is minor and the density of crosslinked alkyl radicals is small, since the rate of the alkyl radicals decay due to reactions with active plasma species is essentially higher than the rate of crosslinking.

10.3.1.2 Description of Surface Concentration Modeling

The evolution of the surface densities of different functional groups (including H-sites) with treatment time was calculated by numerical solution of the balance equations (Equation 10.6) [51]:

$$\frac{dN_i}{dt} = Q^i_{prod} - Q_{loss}, i = 1, 2, 3 \ldots M \tag{10.6}$$

where N_i [m^{-2}] is the surface concentration of i-th functional group and M is the number of functional groups considered in the model. The terms in the right hand side of Equation 10.6 describe the total rates of production and loss of the i-th functional group assigned through the reaction rate constants and reaction probabilities.

The kinetic model described includes reactions of three different types: (i) reactions between plasma species and surface species (including etching); (ii) reactions between different surface species; and (iii) chain backbone scission reactions due to interaction of surface species with polymer backbone. The explicit expressions for the corresponding terms on the right hand side of Equation 10.6 depend on the type of reaction. Let us illustrate the procedure by particular examples.

10.3.1.2.1 Abstraction of H Atoms from H-sites by OH Radicals

$$\text{H-site (surface)} + \text{OH (gas)} = \text{alkyl radical (surface)} + \text{H}_2\text{O (gas)} \tag{10.7}$$

Partial production rate of alkyl radicals (equal to the decay rate of H-sites) in reaction No. 10.7 is expressed as (Equation 10.8):

$$Q_{\text{prod}}(\text{alkyl rad}) = Q_{\text{loss}}(\text{H-sites}) = J_{\text{OH}} \times \gamma \times N_{\text{H}}/K \qquad (10.8)$$

where N_{H} is the surface concentration of H-sites, K is the total surface density of H-sites on the untreated polymer surface, γ is the reaction probability per unit area, and J_{OH} is the flux of OH radicals to the polymer surface, which is estimated as (Equation 10.9):

$$J_{\text{OH}} = n_{\text{OH}} \times v_{\text{OH}}/4 \qquad (10.9)$$

where n_{OH}· is the concentration of OH radicals in the plasma near the surface and v_{OH} is their thermal velocity.

10.3.1.2.2 Abstraction of H Atoms from H-sites by Alkoxy Radicals

$$\text{Alkoxy radical(surface)} + \text{H-site(surface)} = \text{Alcohol group (surface)}$$
$$+\text{Alkyl radical (surface)} \quad (10.10)$$

Partial production rates of alcohol groups and alkyl radicals (and the decay rates of H-sites and alkoxy radicals) for this process are calculated as (Equation 10.11):

$$Q_{\text{prod}}(\text{alcohol}) = Q_{\text{prod}}(\text{alkyl rad}) = Q_{\text{loss}}(\text{alkoxy rad})$$
$$= Q_{\text{loss}}(\text{H-sites}) = N_{\text{alkoxy}} \times k \times N_{\text{H}} \qquad (10.11)$$

where N_{alkoxy} is the surface concentration of alkoxy radicals and k is the rate constant of the reaction represented by Equation 10.10

10.3.1.2.3 Chain Backbone Scission Due to Interaction of Alkoxy Radicals with the Polymer Backbone

$$\text{Alkoxy radical (surface)} + \text{Backbone} = \text{Aldehyde (surface)}$$
$$+\text{Backbone scission radical} \quad (10.12)$$

Partial production rates of aldehydes and backbone scission radicals (and decay rate of alkoxy radicals) for this process are calculated as (Equation 10.13):

$$Q_{\text{prod}}(\text{aldehyde}) = Q_{\text{prod}}(\text{scission}) = Q_{\text{loss}}(\text{alkoxy rad}) = N_{\text{Alkoxy}} \times \tau^{-1} \quad (10.13)$$

where τ is the characteristic reaction time.

10.3.2
Results of Modeling and Comparison with Experimental Data

Conditions under consideration are similar to those in Ref. [51]. The initial surface densities of all the functional groups (except H-sites) were taken to be zero. To specify the initial surface concentration of H-sites the surface density of C atoms (the number of C atoms in PP monolayer per cm^2) was first estimated. It was done by using the known density of the PP polymer ($\approx$910 kg m^{-3}) and assuming that a thickness of a PP monolayer is 0.5 nm. According to these estimations the surface density of C atoms is about 2×10^{19} m^{-2}. Accordingly, the initial surface density of H-sites was chosen 2×10^{19} m^{-2}, whereas the relative surface densities of primary, secondary and tertiary sites were taken 0.25/0.5/0.25, respectively [47].

The degree of the surface modification depends on the concentrations of active species in the gas phase and on the time of processing. In our calculations concentrations of active species (O, OH, HO$_2$, H, O$_2$, and O$_3$) in the gas phase were considered as time-independent parameters.

Gas phase species concentrations were calculated using the plasma kinetic model described in Refs. 1, 51: [O] = 8×10^{18} m^{-3}, [O$_2$] = 5×10^{24} m^{-3}, [O$_3$] = 1.8×10^{20} m^{-3}, [OH] = 1.8×10^{18} m^{-3}, [H] = 4×10^{14} m^{-3}, [NO] = 1.3×10^{19} m^{-3}, [HO$_2$] = 1.1×10^{19} m^{-3}. Gas temperature and pressure were 300 K and 1 atmosphere, respectively.

For the simulations, it was assumed that the PP film is treated during 3 seconds, followed by storage in ambient air and temperature. Calculations were performed from 0 to 10 seconds to include the post discharge period. Results of the calculations are shown in Figures 10.11–10.14.

Let us briefly discuss the general features of the time variation of the surface concentrations of the different groups and radicals during the actual and post treatment period. The decay of the H-sites density under the considered conditions is mainly due to the abstraction of H atoms by OH radicals. The characteristic decay time for 3PP and for 2PP during treatment time (3 seconds) is shorter than the one for 1PP, and concentrations of these sites at the end of the time interval become very small with respect to their initial values (see Figure 10.11a). The density of 1PP is decreased after the end of treatment, this is due to the abstraction of H atoms by alkoxy and peroxy radicals formed during treatment.

The abstraction of H atoms from H-sites leads to the formation of 'in chain' alkyl radicals, surface concentrations of which get maximum ($\sim$10^{15} m^{-2}) at about 10^{-3} s, that is, at the onset of the plasma discharge (Figure 10.11b). With time, these radicals are transformed by the reactions with O$_2$ and O$_3$ to form peroxy and alkoxy radicals. As the free sites 2PP and 3PP on the PP surface are depleted, the densities of the alkyl radicals decrease. The maximum surface density of the 'end free' alkyl radicals (which appear due to the bond scission processes) is about $\sim$10^{14} m^{-2}, and is reached about 0.5 seconds after the start of the treatment. These radicals are also transformed by the reactions with O$_2$ and O$_3$ to form 'end free' peroxy and alkoxy radicals. After the end of the treatment, the main source of alkyl radical production (i.e., the abstraction of H atoms by OH radicals and formation of 'in chain' peroxy radicals, which are responsible for the bond scission processes) is diminished and the surface concentration of these radicals is decreased. During the post treatment period, 'in chain' alkyl radicals are still formed due to the abstraction of H atoms by alkoxy and peroxy radicals stored during treatment. 'End free' alkyl radicals are also still formed, since there is some amount of 'in chain' alkoxy radicals, which produce bond scissions.

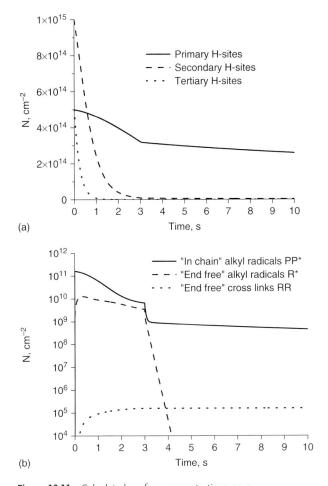

Figure 10.11 Calculated surface concentrations as a function of time: (a) of H-sites; (b) of alkyl radicals and crosslinks. The actual treatment time is 3 seconds.

Surface concentrations of alkoxy and peroxy radicals (see Figure 10.12a) are not very high (in total $\sim 3 \times 10^{18}$ m^{-2} at t = 3 s). Under discharge conditions the decay rates of these radicals are rather high, because of reactions with OH and HO$_2$ plasma radicals. After the treatment, the decay is provided by recombination reactions. Concentrations of aldehydes, acids and carbonyl radicals are small and the concentration of ketones is about $\sim 3 \times 10^{18}$ m^{-2} (see Figure 10.12b).

Figure 10.13a shows the calculated surface concentration of the hydroperoxides and the alcohols that are introduced into the PP surface. For both species, the concentration of 'in chain' groups is higher than the one of 'end free' groups. The total concentration of the products of recombination (see Figure 10.13b) is about 1.9×10^{18} m^{-2} at t = 10 s. ROOPP and ROPP groups (in total $\sim 0.98 \times 10^{18}$ m^{-2}) as well as some part of PPOOPP and PPOPP groups are the crosslinks. Note that

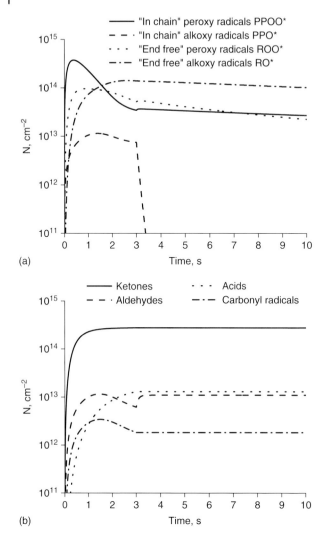

Figure 10.12 Calculated surface concentrations as a function of time: (a) of alkoxy and peroxy radicals; (b) of different groups and radicals with double bonded (=O) oxygen atoms. The actual treatment time is 3 seconds.

the concentration of crosslinked 'end free' alkyl radicals is very small and is not shown in the figures.

The total concentration of O atoms incorporated onto PP surface is about 3.0 × 10^{19} m^{-2}, the concentration of the 'in chain' O atoms is almost twice the concentration of the 'end free' atoms (see Figure 10.14a).

According to our calculations (see Figure 10.14b) at the end of the plasma treatment ($t = 3$ s) the number of C atoms etched from 1 m^2 is about 2.5 × 10^{18}, that is, about 13% of the upper PP layer is etched. Moreover, a large number of bond

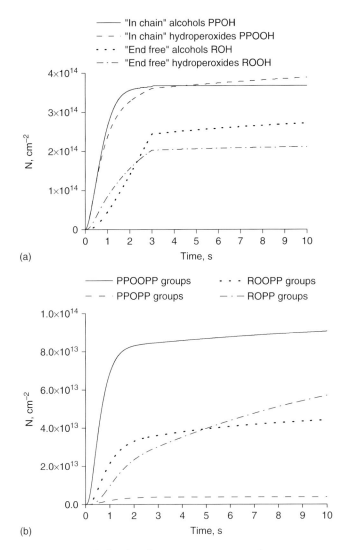

Figure 10.13 Calculated surface concentrations as a function of time: (a) of alcohols and hydroperoxides; (b) of products of alkoxy and peroxy radicals recombination. The actual treatment time is 3 seconds.

scissions appeared. From the standpoint of experiments it means that the upper layer of PP polymer contains a lot of low molecular weight moieties. In fact, such a situation corresponds to over-treatment, since these moieties can easily be removed from the surface, for example, by mechanical means. In general, the treatment time should be long enough to provide the necessary degree of oxidation but it should be restricted to avoid noticeable destruction of the upper layer of polymer.

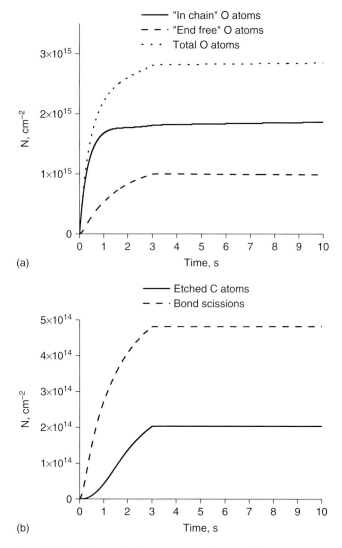

Figure 10.14 Calculated surface concentrations as a function of time; (a) of O atoms; (b) of bond scissions and number of etched C atoms. The actual treatment time is 3 seconds.

It follows from our studies that under the considered conditions, the optimized treatment time should be about 1 second.

Now let us compare results of calculations with available experimental data [51]. It follows from Figure 10.14a that the characteristic time for the surface concentration of O atoms introduced into the PP surface is about 1–2 seconds. This value is in a reasonable agreement with that estimated from the results of contact angle measurements as a function of treatment time [51].

We can also compare the calculated and measured O/C ratios. According to X-ray photo electron spectroscopy (XPS) analysis under experimental conditions [51], O/C ratio (at $t = 3$ s) is about 0.25. The characteristic depth from where the signal was collected was about $L_0 \sim 7.5$ nm, this depth corresponds to 15 PP atomic layers.

It is possible to deduce from the simulation results the value of the O/C ratio, which is also found from XPS analyses, provided that the concentration of O atoms introduced in the surface of the PP polymer is known [51]. To do this, we have assumed that the XPS signal intensity from C atoms located at different depths exponentially decreases with depth with a characteristic length λ. Then, the signal from i-th layer ($i = 0, 1, 2, \ldots$; $i = 0$ denotes the upper layer) of C atoms can be calculated as (Equation 10.14):

$$S_{Ci} = k_X \times k_C \times [C_1] \times \exp[(-\xi/\lambda)i] \qquad (10.14)$$

where $[C_1] = 2 \times 10^{19}$ m^{-2} is the surface density of C atoms in a layer, ξ is the thickness of the PP monolayer (0.5 nm), k_X is a numerical factor proportional to the power of the X-ray source, and k_C is a numerical factor proportional to the probability of the electron emission from C atoms.

The total signal from C atoms is calculated as (Equation 10.15)

$$S_C = \sum_{i=0}^{\infty} S_{c,i} = k_X \cdot k_C \cdot [C_1] \cdot \sum_{i=0}^{\infty} \exp[(-\xi/\lambda) \cdot i] \qquad (10.15)$$

In our model O atoms are introduced only in the upper layer of PP polymer. For this reason, the XPS signal intensity from O atoms can be estimated as (Equation 10.16)

$$S_O = k_X \times k_O \times [O] \qquad (10.16)$$

where $[O]$ is the surface concentration of O atoms and k_O is a numerical factor proportional to the probability of the electron emission from O atoms.

The O/C ratio measured in experiments is the ratio of signals from O and C atoms normalized to corresponding probabilities of the electron emission. Consequently, the O/C ratio is calculated as (Equation 10.17)

$$\frac{O}{C} = \frac{S_O/k_O}{S_C/k_C} = \frac{[O]}{[C_1]} \cdot \frac{1}{\displaystyle\sum_{i=0}^{\infty} \exp[(-\xi/\lambda) \cdot i]} \qquad (10.17)$$

To estimate the value of λ let us note that in experiments the depth of analysis was about $L_0 = 7.5$ nm [47], that is, the signal was collected from about 15 PP atomic layers. The contribution from deeper layers is neglected. Therefore, it is reasonably to assume that $\lambda = L_0/3 = 2.5$ nm, since in this case 15 PP atomic layers provide about 95% of the total signal calculated from Equation 10.15

For the given values of λ and ξ (and taking into account only 15 PP atomic layers) Equation 10.17 can be rewritten as Equation 10.18:

$$\frac{O}{C} = \frac{S_O/k_O}{S_C/k_C} = \frac{[O]}{[C_1]} \cdot \frac{1}{\sum\limits_{i=0}^{14} \exp(-i/5)} \approx 0{,}19 \cdot \frac{[O]}{[C_1]} \tag{10.18}$$

According to our calculations, at $t = 3$ s the total surface concentration of O atoms is about 3×10^{19} m^{-2}. The corresponding O/C ratio estimated by formula 10.18 is about 0.29, which is in agreement with our experimental data as O/C = 0.25 was reported [51].

10.4
Conclusions

Nonthermal plasma produced by gas discharges at atmospheric pressure is an attractive tool for polymer surface treatments, for example, for increasing or decreasing the surface wettability. Despite numerous successful demonstrations of polymer surface improvement by this technology the quantitative description of processes taking place on the plasma–surface boundary is still far from being satisfactory. To achieve a better understanding of these processes, one has to develop an extended kinetic model of nonthermal gas discharges in various environments as well as a kinetic model for interface chemical processes, thereby taking into account the fluxes of radicals and of excited species on the surface. The state-of-the-art regarding theoretical modeling is described and the typical challenges highlighted. The modeling approach is illustrated via a case study which related to PP surface treatment by nonthermal plasma at atmospheric pressure. The validity of the model was demonstrated via the O/C ratio of the treated PP surface. This parameter offers a good opportunity as it can be deduced from the simulation results and it is also experimentally available via XPS measurements. A satisfactory agreement between the simulated and experimental value could be obtained.

Experimental validated numerical models also help understanding the dynamics of complex systems such as plasmas and plasma–surface processing, along with providing useful indications for further process engineering and optimization. This approach also helps enhancing the research and development efficiency by dramatically decreasing the number of experimental trials. This in turn avoids an experimental total heuristic approach, which often forces to explore the whole process parameters domain.

Acknowledgement

The authors thank the Russian Foundation for Basic Research RFBR (No. 08-02-0061a) for supporting this contribution part.

References

1. Akishev, Yu.S., Deryugin, A.A., Elkin, N.N., Kochetov, I.V., Napartovich, A.P., and Trushkin, N.I. (1994) Calculation of air glow discharge spatial

structure. *Plasma Phys. Rep.*, **20** (5), 437–441.

2. Akishev, Y.S., Deryugin, A.A., Karal'nik, V.B., Kochetov, I.V., Napartovich, A.P., and Trushkin, N.I. (1994) Numerical simulation and experimental study of a atmospheric-pressure direct-current glow discharge. *Plasma Phys. Rep.*, **20** (6), 511–524.

3. Akishev, Y.S., Deryugin, A.A., Kochetov, I.V., Napartovich, A.P., and Trushkin, N.I. (1994) Generation efficiency of chemically-active particles in self-maintained discharge. *Plasma Phys. Rep.*, **20** (6), 525–532.

4. Napartovich, A.P., Akishev, Y.S., Deryugin, A.A., Kochetov, I.V., and Trushkin, N.I. (1993) Glow discharge with fast gas flow for flue gas processing, in *Non-Thermal Plasma Techniques for Pollution Control*, NATO ASI Series, Vol. G 34, Part B (eds B.M.Penetrante and S.E. Schultheis), Springer-Verlag, pp. 355–370.

5. Akishev, Y.S., Deryugin, A.A., Kochetov, I.V., Napartovich, A.P., and Trushkin, N.I. (1993) DC glow discharge in air flow at atmospheric pressure in connection with waste gases treatment. *J. Phys. D: Appl. Phys.*, **26** (10), 1630–1637.

6. Stefanović, I., Bibinov, N.K., Deryugin, A.A., Vinogradov, I.P., Napartovich, A.P., and Wiesemann, K. (2001) Kinetics of ozone and nitric oxides in dielectric barrier discharges in O_2/NO_x and $N_2/O_2/NO_x$ mixtures. *Plasma Sources Sci. Technol.*, **10** (3), 406–416.

7. Gibalov, V.I. and Pietsch, G.J. (2000) The development of dielectric barrier discharges in gas gaps and on surfaces. *J. Phys. D: Appl. Phys.*, **33** (20), 2618–2636.

8. Eliasson, B. and Kogelschatz, U. (1991) Modeling and applications of silent discharge plasmas. *IEEE Trans. Plasma Sci.*, **19** (2), 309–323.

9. Steinle, G., Neundorf, D., Hiller, W., and Petralla, M. (1999) Two-dimensional simulation of filaments in barrier discharges. *J. Phys. D: Appl. Phys.*, **32** (12), 1350–1356.

10. Babaeva, N.Y. and Naidis, G.V. (2000) Modelling of streamer propagation, in *Electrical Discharges for Environmental Purposes: Fundamentals and Applications* (ed. E.M. van Veldhuizen), Nova Science Publishers, Inc., New York, pp. 21–48.

11. Kulikovsky, A.A. (1994) The structure of streamers in N2. I. fast method of space-charge dominated plasma simulation. *J. Phys. D: Appl. Phys.*, **27** (12), 2556–2563.

12. Kulikovsky, A.A. (1994) The structure of streamers in N_2. II. Two- dimensional simulation. *J. Phys. D: Appl. Phys.*, **27** (12), 2564–2569.

13. Kulikovsky, A.A. (1997) Positive streamer between parallel plate electrodes in atmospheric pressure air. *J. Phys. D: Appl. Phys.*, **30** (3), 441–450.

14. Kulikovsky, A.A. (1997) The mechanism of positive streamer acceleration and expansion in air in a strong external field. *J. Phys. D: Appl. Phys.*, **30** (10), 1515–1522.

15. Pancheshnyi, S.V., Starikovskaya, S.M., and Starikovskii, A.Yu. (2001) Role of photoionization processes in propagation of cathode-directed streamer. *J. Phys. D: Appl. Phys.*, **34** (1), 105–115.

16. Kushner, M.J. (2005) Modelling of microdischarge devices: plasma and gas dynamics. *J. Phys. D: Appl. Phys.*, **38** (11), 1633–1643.

17. Trichel, G.W. (1938) The mechanism of the negative point to plane corona near onset. *Phys. Rev.*, **54** (12), 1078–1084.

18. Akishev, Y.S., Kochetov, I.V., Loboiko, A.I., and Napartovich, A.P. (2002) Numerical simulations of Trichel pulses in a negative corona in air. *Plasma Phys. Rep.*, **28** (12), 1049–1059.

19. Napartovich, A.P., Akishev, Y.S., Deryugin, A.A., Kochetov, I.V., Pan'kin, M.V., and Trushkin, N.I. (1997) A numerical simulation of Trichel-pulse formation in a negative corona. *J. Phys. D: Appl. Phys.*, **30** (19), 2726–2736.

20. Akishev, Y.S., Grushin, M.E., Kochetov, I.V., Napartovich, A.P., and Trushkin, N.I. (1999) An establishment of regular Trichel pulses in a negative corona in air. *Plasma Phys. Rep.*, **25** (11), 922–927.

21. Akishev, Y.S., Grushin, M.E., Deryugin, A.A., Napartovich, A.P., Pan'kin, M.V., and Trushkin, N.I. (1999) Self-oscillations of a positive corona

in nitrogen. *J. Phys. D: Appl. Phys.*, **32** (18), 2399–2409.

22. Akishev, Yu.S., Grushin, M.E., Kochetov, I.V., Napartovich, A.P., Pan'kin, M.V., and Trushkin, N.I. (2000) Transition of a multipin negative corona in atmospheric air to a glow discharge. *Plasma Phys. Rep.*, **26** (2), 157–163.

23. Itikava, Y., Hayashi, M., Ichimura, A., Onda, K., Sakimoto, K., Takanayagi, K., Nakamura, M., Nishimura, H., and Takanayanagim, T. (1986) Cross-sections for collisions of electrons and photons with nitrogen molecules. *J. Phys. Chem. Ref. Data*, **15** (3), 985–1010.

24. Phelps, A.V. and Pitchford, L.C. (1985), 26th report. JILA Information Center Report, Boulder.

25. Pack, J.L., Voshall, R.E., and Phelps, A.V. (1962) Drift velocities of slow electrons in krypton, xenon, deuterium, carbon monoxide, carbon dioxide, water vapor, nitrous oxide, and ammonia. *Phys. Rev.*, **127** (6), 2084–2089.

26. Eliasson, B. and Kogelschatz, U. (1986) Basic Data for Modeling of Electric Discharge in Gases: Oxygen, Asea Brown Boweri Forschungszntrum CH-5405, Baden, KLR 86-11C.

27. Ionin, A.A., Kochetov, I.V., Napartovich, A.P., and Yuryshev, N.N. (2007) Physics and engineering of singlet delta oxygen production in low-temperature plasma. *J. Phys. D: Appl. Phys.*, **40** (2), R25–R61.

28. Rapp, D. and Englander-Golden, P. (1965) Total cross sections for ionization and attachment in gases by electron impact. I. Positive ionization. *J. Chem. Phys.*, **43** (12), 1464–1479.

29. Yousfi, M., Azzi, N., Segur, P., Gallimberti I., and Stangherlin I. (1987) *Electron–molecule Collision Cross Sections and Electron Swarm Parameters in Some Atmospheric Gases (N_2, O_2, CO_2 and H_2O)*, Centre de Physique Atomique de Toulouse and Instituto di Elettrotecnica ed Elettronical Universita di Padova, Toulouse, Padova.

30. Kossyi, I.A., Kostinskii, A.Y., Matveev, A.A., and Silakov, V.P. (1992) Kinetic scheme of the nonequilibrium discharge in nitrogen-oxygen mixtures. *Plasma Sources Sci. Technol.*, **1** (3), 207–220.

31. Capitelli, M., Ferreira, C.M., Gordiets, B.F., and Osipov, A.I. (2000) *Plasma Kinetics in Atmospheric Gases*, Springer-Verlag, Berlin.

32. Maetzing, H. (1991) Chemical kinetics of flue gas cleaning by irradiation with electrons, in *Advances in Chemical Physics*, Vol. 80 (eds I.Prigogine and S.A. Rice), John Wiley & Sons, Inc., pp. 315–402.

33. Zakharov, A.I., Klopovskii, K.S., Osipov, A.P., Popov, A.M., Popovicheva, O.B., Rakhimova, T.V., Samorodov, V.A., and Sokolov, A.P. (1988) Kinetics of processes excited by a self-sustaining volume discharge in oxygen. *Plasma Phys. Rep.*, **14** (3), 191–195.

34. Mukkavilli, S., Lee, C.K., Varghese, K., and Tavlarides, L.L. (1988) Modelling of the electrostatic corona discharge reactor. *IEEE Trans. Plasma Sci.*, **16** (6), 652–660.

35. McEwan, M.J. and Phillips, L.F. (1975) *Chemistry of the Atmosphere*, Edward Arnold Ltd, London.

36. Atkinson, R., Baulch, D.L., Cox, R.A., Hampson, R.F., Kerr, J.A., and Troe, J. (1989) Evaluated kinetic and photochemical data for atmospheric chemistry: supplement III – IUPAC subcommittee on gas kinetic data evaluation for atmospheric chemistry. *J. Phys. Chem. Ref. Data*, **18** (2), 881–1097.

37. Person, J.C. and Ham, D.O. (1988) Removal of SO_2 and NOx from stack gases by electron beam irradiation. *Radiat. Phys. Chem.*, **31** (1-3), 1–8.

38. Feoktistov, V.A., Lopaev, D.V., Klopovsky, K.S., Popov, A.M., Popovicheva, O.B., Rakhimov, A.T., and Rakhimova, T.V. (1993) Low pressure RF discharge in electronegative gases for plasma processing. *J. Nucl. Mater.*, **200** (3), 309–314.

39. Fraser, M.E. and Piper, L.G. (1989) Product branching ratios from the $N_2(A^3\sum u+) + O_2$ interaction. *J. Phys. Chem.*, **93** (3), 1107–1111.p.

40. Slovetskii, D.I. (1980) *Mechanisms of Chemical Reactions in Nonequilibrium Plasma*, Nauka, Moscow.

41. Gentile, A.C. and Kushner, M.J. (1995) Reaction chemistry and optimization of plasma remediation of N_xO_y from

gas streams. *J. Appl. Phys.*, **78** (3), 2074–2085.

42. De Geyter, N., Morent, R., Leys, C., Gengembre, L., and Payen, E. (2007) Treatment of polymer films with a dielectric barrier discharge in air, helium and argon at medium pressure. *Surf. Coat. Technol.*, **201** (16-17), 7066–7075.

43. Hubenák, J. and Krcma, F. (2000) Determination of a hydrocarbon concentration in the N_2 dc flowing afterglow and its application. *J. Phys. D: Appl. Phys.*, **33** (23), 3121–3128.

44. Pintassilgo, C.D., Cernogora, G., and Loureiro, J. (2001) Spectroscopy study and modelling of an afterglow created by a low-pressure pulsed discharge in N_2-CH_4. *Plasma Sources Sci. Technol.*, **10** (2), 147–161.

45. Dilecce, G., Simek, M., and De Benedictis, S. (2001) The $N_2(A^3\sum u+)$ energy transfer to $OH(A^2\sum +)$ in low-pressure pulsed RF discharges. *J. Phys. D: Appl. Phys.*, **34** (12), 1799–1806.

46. Fresnet, F., Baravian, G., Magne, L., Pasquiers, S., Postel, C., Puech, V., and Rousseau, A. (2002) Influence of water on NO removal by pulsed discharge in $N_2/H_2O/NO$ mixtures. *Plasma Sources Sci. Technol.*, **11** (2), 152–160.

47. Dorai, R. and Kushner, M.J. (2003) A model for plasma modification of polypropylene using atmospheric pressure discharges. *J. Phys. D: Appl. Phys.*, **36** (6), 666–685.

48. Dorai, R. and Kushner, M.J. (2002) Plasma surface modification of polymers, *http://www.ee.ualberta.ca/icops2002* (accessed on 30 January 2009).

49. Bhoj, A.N. and Kushner, M.J. (2007) Continuous processing of polymers in repetitively pulsed atmospheric pressure discharges with moving surfaces and gas flow. *J. Phys. D: Appl. Phys.*, **40** (22), 6953–6968.

50. Rabek, J.F. (1995) *Polymer Photodegradation: Mechanisms and Experimental Methods*, 1st edn, Chapman and Hall, London.

51. Akishev, Yu., Grushin, M., Dyatko, N., Kochetov, I., Napartovich, A., Trushkin, N., Tran, Minh Duc., and Descours, S. (2008) Studies on cold plasma-polymer surface interaction by example of PP- and PET-films. *J. Phys. D: Appl. Phys.*, **41** (23), 1–13.

52. ANSI 17-2Q98. (1998) *Standard Reference Database*, American National Standards Institute, New York. *http://kinetics.nist.gov/kinetics/* (accessed on 9 February 2009).

Part III
Economical, Ecological, and Safety Aspects

Plasma Technology for Hyperfunctional Surfaces. Food, Biomedical and Textile Applications.
Edited by Hubert Rauscher, Massimo Perucca, and Guy Buyle
Copyright © 2010 WILEY-VCH Verlag GmbH & Co. KGaA, Weinheim
ISBN: 978-3-527-32654-9

11
Economic Aspects

Elisa Aimo Boot

11.1
Market Analysis: an Overview

11.1.1
Textile Market Analysis

11.1.1.1 General

The European textile industry continues to be an important European manufacturing industry in overall employment, despite fairly flat or slowly growing demands and falling employment. Its importance resides in the fact that this industry represents a complete production chain which extends from the raw product to the supply of larger consumers and intermediate markets. The European textiles and clothing industry is a major player in world trade, the first in textile exports and the third in clothing [1].

The textile and clothing industries in Europe, North America, and some other developed countries are facing some big challenges today. Therefore, the shift to high added value functionalized (technical) textiles is considered essential for their sustainable growth. The growing environmental and energy-saving concerns will also lead to the gradual replacement of many traditional wet chemistry-based textile processing methods. The textile sector is highly dependent upon traditional manufacturing processes based upon wet chemistry and the application of heat. As is well known, wet and heat-based processing is costly and very environmentally burdensome, consuming, and wasting profligate quantities of chemicals and raw materials as well as huge amounts of energy and water as explained in Chapter 12. The environmental consequences of the millions of tonnes of oxidative and toxic by-products currently generated by the European industry are well known and the financial burden on the industry because of this is ever increasing. Economically and environmentally, reduction in water consumption is becoming a real necessity because of the continuously increasing costs of supplying water and purifying effluents and, also, because of the high energy costs incurred by the after-treatment drying. The shift to plasma treatment should enable the move to dry, low temperature processing which will be based on small quantities

Table 11.1 Worldwide consumption of technical textiles [2].

	2000	2005	2010	% Annual growth	
	Ktons			05/2000 (%)	10/2005 (%)
America	5078	5811	6857	3	3
Europe	4096	4665	5448	3	3
Asia	6961	8482	10 602	4	5
Others	556	623	724	2	3
Total	16 691	19 581	23 631	3	4

of reactive precursors. Consequently, the exhaust products will be limited and either harmless (water vapor, CO_2) or treatable by solid scrubbers that are easily manipulated. Plasma technology, when developed at a commercially viable level has a strong potential to offer in an attractive way the achievement of a new type of functionalizing in textiles.

11.1.1.2 Technical Textiles

The technical textile worldwide market is increasing with an annual growth of 3–5%. Technical textiles are characterized by the requirement for specific functions as communication, medical, protection against heat and cold, lightness, and so on. This market has a high potential, and Europe represents a major share, as shown in Table 11.1

The technical non-wovens market grows at an annual rate of 5–6% (Figure 11.1), more rapidly than traditional textiles such as wovens or knitted fabrics. Whereas most of European textile production is threatened by low-cost countries, European production still represents 30% of non-woven textiles. Non-wovens represented about 25% of technical textiles in 2005.

11.1.1.3 Hydrophobic and Oleophobic Textile Market

With specific reference to the topics treated in Chapter 6 and in Chapter 12 concerning assessment of the environmental impact, we provide an overview on the hydrophobic and oleophobic textile market, which at the present development stage are the functionalities that appear to be the most feasible and sustainable application of plasma technology.

For example, hydrophobic and oleophobic properties for professional and consumer clothing and furniture upholstery address a potentially large global market. The major needs are to increase the durability of the existing treatments and to replace the wet chemical processes which are large consumers of chemicals, energy and water.

The studies within ACTECO (see Preface) estimated the potential market of plasma technologies at about €50 million for hydro- and oleophobic treatment and over €100 million for hydrophilic treatment in 2013 (Figure 11.2)

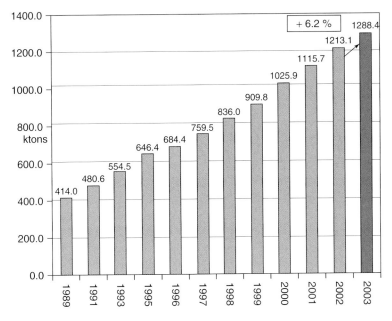

Figure 11.1   Production of technical non-wovens in Europe [3].

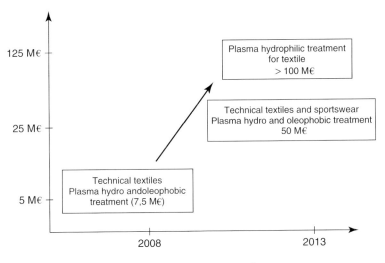

Figure 11.2   Potential market of plasma technologies €50 million for hydro- and oleophobic treatment.

11.1.2
Biomedical Market Perspective

Medical technology covers a very wide range of products: aids for the disabled, active implantable devices, anesthetic/respiratory devices, dental devices, electromedical, hospital equipment (hardware), imaging, *in vitro* diagnostic devices, ophthalmic and optical devices, passive implantable devices, single use (disposable) devices, and surgical (re-usable) instruments.

The medical technology industry is an important economic player in Europe, with some 11 000 enterprises (80% of which are small and medium enterprises (SMEs) employing about 435 000 people (Germany: 110 000, France: 40 000, Switzerland: 40 000, UK:60 000) most of whom are highly qualified, and market sales of €72.6 billion (nearly of 33% of the world market share)[4].

The medical technology market has many niche product lines, each representing a modest market share. Each product requires specific manufacturing or distribution skills with the result that many SMEs concentrate on a single product line or on a specific geographical area.

The biomedical applications can be divided into:

• medical devices: catheters, filters, and syringes
• implants and biomaterials: stents, prostheses, contact lenses, abdominal meshes
• medical technical textiles: wound dressing.

We now give more detailed market data regarding the latter two groups.

For the first group of 'implants and biomaterials' the world market is estimated around €25 billion with an annual growth of 5–7%, one third of this market belongs to the USA as shown in Table 11.2. Orthopedic biomaterials represent €8 billion with an annual growth of 7%. It includes implants like stents, prostheses, or scaffolds for tissue re-engineering [5].

The group of 'medical textiles' includes hospital laundry (35% for medical, personnel, and patient clothes, 60% for sheets, 5% for others), care devices for

Table 11.2 Global market of main implants.

Device	Number/year in USA	Number/year in the world	Biomaterial
Pacemaker	430 000	1 290 000	Polyurethane
Renal dialyzer	16 000 000	48 000 000	Cellulose
Dental implant	300 000	900 000	Titanium
Breast implant	192 000	576 000	Silicone
Stent (cardiovascular)	>1 000 000	>3 000 000	Stainless steel, NiTi, Co–Cr
Heart valve	200 000	600 000	Pig valve, PyC, Ti, Co–Cr
Catheter	200 000 000	600 000 000	Silicone, Teflon
Hip and knee prosthesis	500 000	1 500 000	Titanium, Co–Cr, PE
Vascular graft	250 000	750 000	PTFE, PET
Contact lens	30 000 000	90 000 000	Silicone acrylate
Intraocular lens	2 700 000	8 100 000	PMMA

PE: polyethylene; PTFE: polytetrafluoroethylene; PMMA: poly(methyl methacrylate);
PET: poly(ethylene terephthalate)

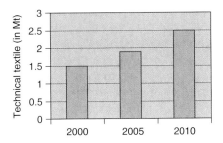

Figure 11.3 Evolution (in Mt) of the technical textiles in the medical market segment. (Source: David Rigby Associates, http://www.davidrigbyassociates.co.uk/)

bandages, (bio-)textiles for implants. The medical sector accounts for 10–15% of technical textiles, which represents worldwide around 2 Mt (see Figure 11.3). In particular, antibacterial textiles are developing in the medical field, because it brings hygienic benefits, even though the development of resistant bacteria is an important issue here.

The application of plasma technology in the biomedical sector may contribute, for example, to the reduction of chronic inflammatory response, that is, to a decreased occurrence of nosocomial infections. This clearly leads to an improved quality of life for the operated patients via the dramatic reduction in the incidence of post-operative pain. Moreover, this has also important social and economic benefits, both directly (reduced health care costs) and indirectly (faster average return to normal activities after a common surgical practice if nosocomial infections can be largely prevented).

11.1.3
Food Packaging Market Potential

Food applications represent a $3.3 billion market for barrier packaging film, with an overall $9 billion global market for flexible packaging that incorporates barrier films. About 80% relates to perishable food, the other 20% to dry foods. Annual growth of barrier packaging films in the USA is forecast to be nearly 5% [6].

High-performance barriers are used in the food industry for:

- dry food such as snack foods, baked snacks, dry mixes, pet foods, and coffee
- perishable food such as fresh meats, processed meats, cheese, and dairy, prepared foods.

Today alimentary packaging films are already treated by plasma to give them hydrophilic properties. This treatment allows these films to be printed with water-based inks, which is otherwise impossible.

For packaging vegetables, mist deposition on the film (referred to as *fogging*) is a marketing issue. A plasma treatment with anti-fogging properties would be very interesting. The consequent economic impact would be very high: for example, current multilayered barrier polypropylene trays and foils are expensive (high energy and raw materials costs) and slow to produce. Moreover, they give rise to high environmental impacts and are heavy relative to the weight of the food content.

Innovative plasma barrier nanocoatings onto the basic trays would imply a reduction of more than 50% in raw material and energy use, thereby also reducing the weight by more than 50%. These effects generate an increased user appeal as they lower the final product cost via reduced shipping costs and reduced environmental taxes on packaging.

A further example of what surface treatment can achieve in this industry is the concept of active food packaging where a targeted trapping of microorganisms or of food conserving chemicals (biocides, anti-oxidation, etc.) within the packaging will increase food shelf-life, reduce waste, and deliver health and safety enhancement for the final consumers.

11.2
Case Study: Up-Scaling of the Plasma Treatment of Hammers for Meat Milling

In this section we present an example of the application of physical vapor deposition (PVD) technology (see Chapter 9) in industry. In particular, we explain the study regarding the economic impact of introducing a PVD coating solution for the production of hammers for food processing. The target was to increase wear resistance of the hammers to enhance product performance and improve the internal processing efficiency and competitiveness, as explained in Chapter 9.

Next to the obligatory reference scenario, two solutions are considered: the first considers outsourcing of the coating activities, the second assumes in-house coating. The latter opportunity allows to enhance the companies' technological know-how and provides a clear added value to the production processes and the products. This solution allows improvement in the processing efficiency by optimizing process cost. Both solutions permit the improvement of the production quality as they enable implementation of the technical specifications for the coating: providing barrier effects against the release of allergenic substances and biocompatibility with food processing equipment.

The whole PVD deposition process is automatically controlled. All deposition processes are optimized for production efficiency: they are pre-programmed and fully automatic in operation. So the PVD technology does not require specific competences and can be easily integrated in the manufacturing process. Indeed, the employment of one dedicated operator is sufficient for complete system management, although his constant or extended engagement during the deposition process is not needed. The operator is only required to load the components onto the carousels, activate the washing cycle, transfer the components in the PVD reactor, start the coating process, and unload the components from the reactor. Further, the operator is needed in case of an unexpected interruption of the coating cycle. The deposition systems have an automatic control and, therefore, they can be monitored from a remote location without the direct presence of an operator.

It is important to stress the point that each suggested solution is already industrially up-scalable as the plasma processing units are commercially available. The developed process recipes (see Chapter 9) are coded in digital format compatible with the industrial units control software. Thus, direct transfer of the developed recipes to large scale reactors of the same PVD unit is possible.

The sustainability analysis considers the following three scenarios:

1) **Reference scenario**: this scenario considers the actual costs for the hammer procurement and its in-process durability as a benchmark.

2) **Outsourcing**: this scenario considers hammer manufacturing using top level materials adequate to receive a top level dedicated ceramics coating. The coating process is assumed to be carried out by a specialized coating center at reference market price.

3) **In-house**: this solution considers, as in the previous scenario, top level materials with a top level ceramics coating. However, in this case the coating process is assumed to be carried out within the food processing company, that is, in-house surface treatment of the hammers before mounting them in a meat mill is considered. This scenario is based on the number of hammers yearly consumed, which is based on the yearly throughput of the food processing company. The use of one PVD system operated by a single operator and using one washing system is considered as it complies with the current production scheme of the company. As work load, two shifts per day are assumed on the basis of 252 working days. The amortization flat rate related to the PVD system has been distributed over a five-year time period. Other financial solutions need to be considered specifically for each case.

11.2.1
Analysis of the Reference Scenario

For the reference scenario, it is assumed that the hammers are manufactured from C40 steel.

We consider a typical mill which hosts 80 hammers and which is operating for 252 effective working days per year. Given a hammer lifetime of 12 working days, the estimated resulting average hammer consumption per processing unit is 1680 hammers per year. This amounts to an estimated cost of about €94 000 per year per mill, assuming a hammer unit price of €56. The number of maintenance operations per year is about 20, leading to a considerable amount of process downtime. In this conservative analysis, we do not consider economic losses due to these process interruptions, although they are thought of as being one of the primary sources of low processing efficiency.

11.2.2
Analysis of Scenario 2 – Outsourcing

For scenario 2, the hammers are manufactured from K110 (AISI D2) steel, a material with improved mechanical properties compared with C40 steel. The average estimated cost of a single K110 hammer is about 40% higher compared with the average cost of a single C40 hammer (the type considered in the reference scenario).

Each mill hosts 80 hammers. As indicated by similar tests in severe operating conditions, the duration of coated hammers is 41 days for full time operation (two shifts per day), which is equal to 656 hours lifetime. Considering 252 effective working days per year, the estimated average hammer consumption per processing unit is 492 hammers per year amounting to an estimated cost of about €86 000 per year per mill for a hammer unit price of €175.

The unit price per hammer mentioned consists of the basic price for the hammer (about €56 + 40% = about €78) and the coating cost (about €97).

This latter value assumes that the hammers are coated by a specialized coating center; the cost is estimated on the basis of detailed historical data. So the total cost for a single coated hammer is roughly three times as high as for the non-coated hammers considered in the reference scenario. It has to be pointed out that this analysis does not consider logistic and delivery cost, which may also have a relevant influence in the economic evolution. On the other hand, possible discounts based on number of coated pieces are not taken into consideration, as they depend basically on commercial agreements, which may change from coating center to coating center. But to be conservative with the figures, we only include the cost of the production methods themselves.

Summarized, comparing the reference scenario and the outsourcing scenario, it follows that the latter solution, which considers hammer manufacturing using top level materials adequate for being coated with top level dedicated ceramics coating (developed in a coating center), shows to be an economically profitable solution because the annual saving exceeds 10%, and does not depend on the number of mills used.

11.2.3
Analysis of Scenario 3 – In-house

This scenario considers identical hammers as for scenario 2: K110 (AISI D2) steel hammers, coated with a top level ceramics coating. Consequently, the lifetime of the hammers is identical to the one of scenario 2 (41 days or 656 hours). But, here the coating is assumed to be deposited in-house, that is, within the food processing company itself.

In this scenario, the coating is performed by using a medium-high capacity reactor. The average cost of the PVD coating per hammer is not fixed, but depends on the number of mills used and, eventually, on the total number of hammers to be coated per year.

If we consider 252 effective working days per year, the estimated average hammer consumption per processing unit is 492 hammers per year per mill. This amounts to an estimated cost of about €238 per year per mill for a hammer unit price of about €484, based on about €78 for the basic hammer and €406 for the coating procedure. The deduction of the price for the coating procedure is explained in the next paragraph.

Table 11.3 The cost of coating one hammer in function of the number of mills, including the operation cost and the investment cost for the PVD equipment, assuming an amortization period of five years.

Number of mills	Coating cost per hammer (€)
1	406
2	203
3	135
4	101
5	81
6	67
7	58
8	50
9	45
10	40
11	36
12	33
13	31
14	29
15	27

Table 11.4 Initial investment cost for solution with one PVD coating system.

Item	Estimated costs (k€)
PVD unit (200 lt. capacity)	600
Cleaning unit	200
Total	800

11.2.4
Investment and Operating Cost

Now, we explain the calculation of the price for the coating procedure. This coating cost takes into account the initial investment to purchase the PVD unit including the cleaning machining, and the operating costs. The amortization is spread over five years. As the investment cost is included, the average coating cost rapidly drops if the number of mills increases, as reported in Table 11.3. The reason is of course that the amortization cost is distributed over more hammers. Table 11.4 summarizes the preliminary estimate of the items of the initial investment, assuming the procurement of 1 PVD coating system.

After the investment, one of the most relevant parts of the cost assessment regards the energy consumption and substitution of the metallic cathodes, the precursor gas consumption being negligible.

Table 11.5 Maximum annual operation cost for the PVD and washing systems, assuming a work load of full reactor saturation (corresponding to 26 hammers per batch) basing on two shifts per day (per six batches per day) per 252 operational days per year.

Item	Annual cost (k€)
Operators on PVD system	12.0
Electric energy	10.5
Consumables and maintenance	31.5
Total	**54.0**

It should be noticed that the PVD process does not produce any by-products that need to be treated. This is quite significantly different from traditional wet processes such as Cr electroplating where process water needs an accurate post treatment before being emitted into the environment.

The operator working on the PVD system does not need any particular technical or scientific skill; an operational training course is sufficient. In this cost assessment the annual cost of the labor related to this type of work has been considered.

Operating costs regarding the use of the considered system (with medium-high capacity, in which it is possible to lodge 26 hammers per coating process), costs of energy consumption and materials have been coherently evaluated on the basis of the maximum variable cost (corresponding to full PVD reactor saturation) with the proposed technical solutions (one PVD job coating unit supported by one washing unit), taking into account ordinary and extraordinary maintenance. This evaluation has been done on the basis of detailed historical series of use and in reference to updated information on spare parts. The prospect in Table 11.5 summarizes the operating costs used for the realization of the current analysis.

Solutions dependent on the actual hammer consumption and hammers coating (as a function of mills number) reduce the operating (variable) cost proportionally to effective equipment use and operator allocation to coating activity.

11.2.5
Comparative Analysis of All Three Scenarios

The analysis of the third scenario (in-house coating) was focused on the determination of the unit cost for deposition (cost for coated components), as a function of the productive strategies and amortization of the systems. Therefore, the initial investment cost and operating costs were evaluated (see Tables 11.4 and 11.5).

Based on one operating mill, the in-house option does not prove profitable with respect to the reference scenario because the coating cost is too expensive with respect to the uncoated standard solution. However, if more mills are used

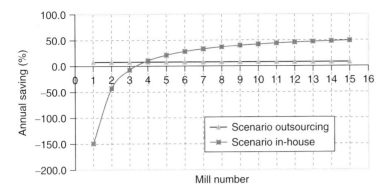

Figure 11.4 Annual savings (in %) within the first five years (during amortization) as a function of the number of mills.

the annual percentage saving rapidly increases, also beating the first innovative solution, that is, hammers coated by external coating center.

Actually, the in-house solution annual savings becomes positive with respect to the reference scenario (non-coated hammers) when at least four mills are used. Considering a production scenario with 14 mills the saving amounts to almost 50% for the in-house solution (Figure 11.4) corresponding to absolute values exceeding €623 000 per year.

From the sixth year on there is no longer the cost of amortization and the annual saving for the internal coating solution with respect to the reference solution amounts about 8% for one mill. Considering 14 mills the saving rate is about 60% for the in-house solution, corresponding to an absolute value exceeding €783 000.

11.2.6
Final Considerations

Different production scenarios have been considered. The comparative analysis, carried out to identify the best production solution for the company, regarded the production of hammers for food processing. The study suggests that the PVD solution is suitable in terms of economic benefits.

The first proposed solution, outsourcing of the coating of the hammers is profitable as it guarantees an annual saving, even if only one mill is considered.

The second solution, in-house coating, is technically feasible and proves to be economically competitive, provided some minimum requirements are fulfilled: assuming an amortization of the PVD equipment over five years, at least four mills need to be used. If this requirement is met, the in-house coating solution has the following benefits:

- complete control of the value chain and production costs;
- further enhancement of the technological know-how of the company's productive process;
- economic advantages concerning specific production;

- efficiency and production solutions;
- possibility of using PVD technology for other internal company applications (for example, coating of mechanical tools used);
- good automation level and process flexibility.

Note that this analysis does not consider the savings due to the dramatic increase of the production throughput, resulting from strongly reduced operation downtimes and related maintenance costs. Indeed, even for the conservative assumption of the increase in the hammers' average lifetime with a factor of 3.5 (from 12 to 41 days), this increased lifetime leads to a proportionally decreasing number of operation stops, which leads to reduced production and maintenance costs. The computed savings increase with the number of in-production mills considered.

To conclude, the presented solutions appear to be not only economicly sustainable but also convenient and of strategic importance for the food processing itself.

References

1. Euratex (2004) European Technology for the Future of Textile and Clothing – A Vision for 2020, December 2004.
2. David Rigby Associates (2002) World Market forecast to 2010.
3. European Disposables and Non-wovens Association, EDANA Releases 2006 Statistics, (2007), *http://www.nonwovens-industry.com/*
4. Wilkinson, J. (2009) An Introduction to Medical Technology industry, March 2009 *http://www.eucomed.org/.*
5. ASM International (2004)Medical Device Materials II: Materials and Processes for Medical Devices, St. Paul, MN, USA, August 25-27, 2004.
6. Kline Group (2005) High-Performance Barrier Packaging Films USA 2005: Business Analysis and Opportunities, *http://www.klinegroup.com/reports/y381a. asp.*

12
Environment and Safety

Massimo Perucca and Gabriela Benveniste

12.1
Introduction to LCA

Life cycle assessment (LCA) is a methodological tool that applies life cycle thinking in a quantitative way to environmental analysis of activities related to processes or products (goods and services) to assess their environmental burden. A central characteristic of LCA is the holistic focus on products or processes and their functions, considering upstream and downstream activities. LCA has been applied to study processes, products, and commodities as in the automotive, packaging, or energy sectors and sometimes it has become standard practice to assess the sustainability of different options [1–3]. LCA has also been introduced in the eco-design perspective of products, which comprises the selection of the best technically and economically sustainable solutions complying with the design functionalities, as well as including the lowest environmental impact choices of design and materials to be employed.

The life cycle concept understands that LCA of a product includes all the production processes and services associated with the product through its life cycle (Figure 12.1), from the extraction of raw materials through production of the materials which are used to manufacture the product, over the use of the product, to its recycling and/or ultimate disposal of (some of) its constituents. Such a complete life cycle is also often named 'cradle to grave.' Transportation, storage, retail, and other activities among the life cycle stages are included, where relevant. Hence the life cycle of a product is identical to the complete supply-chain of the product plus its use and end-of-life treatment.

In an LCA study, for each single process step, the use of resources, raw materials, parts and products, energy carriers, electricity, and so on are inventoried and documented as 'Inputs' while emissions to air, water, and land as well as waste and by-products are recorded on the output side and regarded as 'Outputs.' For products coming from the technological upstream processes considered as inputs of the system investigated under LCA methodology their 'environmental history' has also to be included in the calculations by taking into account their indirect upstream activities. For waste products, the subsequent treatment processes have

Plasma Technology for Hyperfunctional Surfaces. Food, Biomedical and Textile Applications.
Edited by Hubert Rauscher, Massimo Perucca, and Guy Buyle
Copyright © 2010 WILEY-VCH Verlag GmbH & Co. KGaA, Weinheim
ISBN: 978-3-527-32654-9

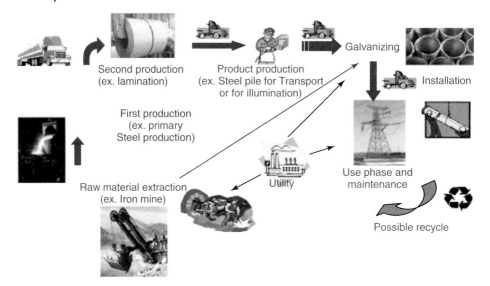

Second production
(ex. lamination)

Product production
(ex. Steel pile for Transport
or for illumination)

Galvanizing

Installation

First production
(ex. primary
Steel production)

Raw material extraction
(ex. Iron mine)

Utility

Use phase and
maintenance

Possible recycle

Figure 12.1 Illustration of a general product life cycle.

to be included accordingly. Indeed, the total balance of inputs from, and outputs to the environment (thought of as being external to the defined system) affects the environmental burden and is the basis for a later analysis and assessment of the environmental effects related to the investigated product or process.

Therefore, an LCA study describes and analyzes, in a quantitative way, all the important environmental aspects of a product system or technology. The study is the result of an iterative procedure of scope definition and data collection,

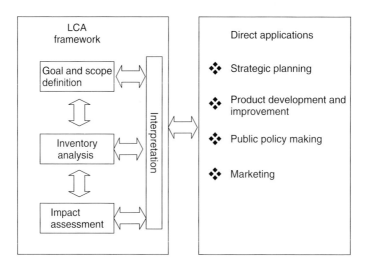

Figure 12.2 LCA framework and applications according to ISO 14040.

followed by analysis and interpretation. It is today's practice that an LCA should, as a minimum requirement, follow the ISO 14040 standards (see Figure 12.2). Performing a specific LCA at a detailed level may require comprehensive data collection on site and from suppliers and related human resources. Therefore, in order to keep the analysis feasible and realistic, it is convenient to strive at making an LCA as simple as the goal of the LCA allows.

In accordance with the standard ISO 14040, the LCA consists of four interrelated phases as presented in Figure 12.2.

According to ISO 14040, the general framework of the assessment phase is composed of several mandatory elements that convert inventory results into environmental indicators. For this analysis the following impact categories are considered (such characterization factors are also recognized by the system for environmental product declarations – EPDs [4, 5]): global warming potential (GWP), acidification potential (AP), photochemical ozone creation potential (POCP), and eutrophication potential (EP).

It is interesting to stress the point that LCA is not just intended to stand alone or assess a more or less absolute and abstract impact on the environment, but has to be considered a dedicated tool to perform comparative analysis of solutions. To such an extent LCA can be included into decision making in various fields such as:

- product development and improvement
- process and service operation
- strategic planning
- technological impact assessment
- public policy making
- marketing.

LCA is used by industry and authorities for various purposes. Growing environmental awareness among consumers and other important stakeholders have led many industries to consider the environmental performance of the life cycle of their products. With the goal of producing greener, more environmentally friendly products, LCA is used to:

- support methodologies or tools aimed at developing greener products, such as design for environment (DfE) or design for recycling (DfR);
- compare different design options during product development;
- identify the most important environmental problems (hot spots) in the life-cycle of one's product (system analysis) and of competitors' products (benchmarking);
- document improvements in the environmental performance of products.
- select amongst suppliers in a green supply chain management (e.g., for institutions);
- communicate the environmental performance of products or services, through the use of environmental labels and product declarations (awareness growth and marketing).

At production sites and organizations level LCA helps to focus on the relevant processes and to avoid shifts between life cycle stages by:

- quantifying indirect effects which occur outside the production site but are caused by the demand of products and services on site;
- benchmarking sites to find optimization potentials.

Environmental management schemes like EMAS (Eco-Management and Audit Scheme, IEMA) or ISO 14001, as well as sustainability and environmental reporting use information derived from LCA as well.

LCA also provides support to public policy applications. The life cycle thinking approach is promoted in policy making by, for example, the integrated product policy (IPP) strategy. IPP is a voluntary approach and seeks to minimize the environmental effect of a product by looking at all phases of a product's life cycle and taking action where it is most effective.

Implementation of the IPP is attained with a variety of tools, including measures such as economic instruments, substance bans, voluntary agreements, environmental labeling, and product design guidelines. For example:

- Environmental labels indicate the environmental performance of the product or service, based on multiple criteria over the entire life.
- EPD are a communication format for quantified LCA information using predetermined parameters based on independently verified rules for the product category.
- Green procurement guiding purchasers, both public and private, in taking environmental considerations into account in procurement processes.
- Waste management strategies, such as take-back responsibility for certain product types (e.g., cars and items of electronics and household appliances), and making manufacturers liable to take their products back after end-of-use, thus motivating them to design and construct the products with their disposal in mind.
- Green taxation of products which reflects the environmental costs (including external factors) that the product inflicts on society throughout its life cycle.
- Ecodesign supports product developers in reducing the environmental impact of a product early in the process of product development (design phase).

Authorities can also use the holistic assessment principle of LCA in the environmental assessment of major societal action plans, of legislation or more specifically of different ways of providing services like transportation, electricity generation, or waste treatment.

12.2
Environmental Impact of Traditional Surface Processing: the Reason for Developing Innovative Solutions Supported by Dedicated LCA

Surfaces processing and treatment encompasses a wide range of industrial productive sectors. Indeed just to recall a few of the many examples that can be considered, the performances of major industrial applications in the field of health, food, textile, and environment depend very strongly on the physical-chemical properties of surfaces. For instance, the functional properties of catheters, contact lenses, vascular prostheses ... and so on, are strongly dependant on the ability of the device surface to repel proteins. Similarly, the operating performance of heat exchangers in the food processing industry depends on their capability to avoid biofilm formation. Along the same line of ideas, the properties of textile surfaces are a major parameter for finishing operations including dyeing and for the final properties of clothes or technical textiles such as hydro-phobic/philic oleo-phobic/philic. Packaging of ready-to-consume food involves large volumes of plastic materials whose surface may be optimized for barrier properties. Food processing involves use of rough technologies which still have a wide improvement margin, for example, processing components can be surface treated against wear, corrosion, and/or release of heavy metals [6]. Clearly, this also guarantees improved consumer safety and health.

The quest for efficient technological solutions is not only focused on efficiency and competitiveness but also has a strong relation to sustainability and environmental preservation. On this account, at the development stage of innovative treatments the assessment of reductions of energy and water consumption used during textile processing, employment of more homogeneous and recyclable plastic materials for packaging, and the use of more efficient hyperfunctional surfaces for biomedical applications, become crucial.

Just to recall one of the traditional industrial sectors, the textile industry as we know it today is a 300-year-old industry. Generally speaking, the large number of basic industrial operations presently run for sizing/de-sizing, dyeing, and finishing have been the very same for the last three centuries. Until now, all the technological developments have been focused into the automation and refinement of the existing operative principles, aiming at improving results on quality and productivity, always to perform the very same operations, only with better and larger production capacity. Dyeing and finishing fabrics is, still today, a series of wetting and drying operations, as it always has been, with a huge consumption of water, energy, investment in space, and heavy equipment, with very large consumption of chemical products (detergents, dyes, resins, etc). The effective incorporation of products in the final fabrics is only a small percentage. Thus large quantities of waste and ecologically unfriendly effluents are the result. Hence, the textile industry strongly needs radical technological improvements at different process stages, for instance, to:

- find new solutions for sizing/de-sizing operation, without any loss of quality and productivity at weaving, decreasing or eliminating the consumption of water, energy, and starch;
- improve the limited efficiency of machines, efficient use of energies (electrical, thermal), and to decrease the use of chemical products. For example, regular dyeing of 1 kg of any textile based on natural fiber yarns, by reactive dye, requires about 120 kg of water, most of it at the average temperature of 50 °C.
- improve the actual poor efficiency on the quality/irregularities on the final product;
- provide novel efficient processes to enhance dyeability, dye up-take, and fastness;
- provide eco-efficient treatments to attain super hydro-/oleo-phobic or super hydrophilic textiles for clothing, upholstery, and technical textiles applications;

- strongly decrease the high aggression to the environment stemming directly from the industrial production or, afterwards, from the final product because of the use of harmful compounds. For example, for chemical cleaning using more eco-friendly detergents or more efficient dry cleaning can be considered.

Surface molecular re-engineering applied to the textile industry, during the preparation, dyeing, and finishing of the fibers and fabrics, thus incorporating new functions directly aimed at the final product application, might be a new path to follow. The interest in developing new technologies in an industrial textile company is not only motivated by the market but also by the need to innovate into the direction of new and better products, produced in shorter time, further functionalized at a better cost and with higher flexibility, complying with the most strict principles of quality and made with the protection of the natural environment [7].

In order to motivate the need of devising new sustainable surface treatment solutions we consider one of the many examples that could be taken into account, and we suggest an evaluation exercise which consists in the assessment of water and energy consumptions due to traditional dyeing, printing, and finishing processes for textiles. For this evaluation, and to keep ourselves on the safe side, we assume conservative figures: an overall average consumption of 50 l of water to process roughly 1 kg of textile is assumed. This amount of water is heated up to 50–70 °C for the selected textile treatments, therefore setting an energetic equivalent of about 3 kWh consumed to process 1 kg of textile. This technical reference input is generally considered valid for a reference period of about 20 years, since technology has not changed dramatically in this time span. Now, if we consider the European and worldwide textile production in one reference year, we may figure out what was the estimated overall amount of water and energy used for the selected textile processes. For instance if we refer to the year 2002, which was still representative for standard European textile manufacture and at the same time not too far back in the past, we get astonishing results. In 2002 the world production of synthetic filament yarn amounted to about 19 million tonnes, of which about 0.8 million tonnes in Europe. Further, the world production of cotton, wool, and man-made fibers amounted to about 35 million tonnes that year, of which about 5.5 million tonnes in Europe. Getting on with our simple exercise we find that for synthetic fibers 940 billion liters of water and 55 billion kWh have been consumed worldwide, of which 131 billion liters and 8 billion kWh in Europe. At the same pace for cotton and wool fibers 1.75 trillion liters of water and 102 billion kWh have been consumed, of which 275 billion liters of water and 16 billion kWh in Europe. Since these absolute figures are so large that one misses the gage, some comparisons to figure out the extent of the resources used are convenient. The total energy consumed in the world (in 2002) just for the specific textile processing was equivalent to the total industrial energy (textile, mechanical industry, chemical industry, etc) consumed in the same year in Germany, while the water consumption corresponded to the total water (domestic and industrial) consumed in Czech Republic in 2002. The energy used for dyeing and finishing in Europe (2002) is equivalent to 60% of the total textile energy employed in Europe and equivalent to 10% of the total industrial energy consumed in Germany in the same year, while the water consumption corresponded to 1% of the total domestic water used in Europe in 2002 (data source: Eurostat 2002) [8]. These comparisons provide an example of the environmental emergency related to the industrial use of natural resources and forces mankind to strive for sustainable development of innovative eco-efficient solutions. This is the framework in which LCA comparative analysis is considered a powerful tool to devise new low environmental impact treatments.

Although LCA is an assessed and coded procedure, each LCA analysis is applied to a specific production process or product, and represents a case study. In particular, pertaining to surface processing and treatment to get enhanced products and component performances, the LCA approach is definitely not a trivial question and needs the development of a specific methodology. This is what has been done in the analysis of specific cases in the framework of the ACTECO project (see Preface) and what we want to refer to in the next paragraphs as a detailed example of environmental sustainability analysis of plasma processes in comparison to traditional treatments in order to assess the eco-efficiency of plasma technology.

12.3
LCA Applied to Plasma Surface Processing: Case Studies

The LCA main objective in the following considered case studies is to compare surface functionalization technologies as an alternative to conventional ones and to analyze how the origin and type of energy and input materials employed in each process affect the overall evaluation of environmental burden. While analyzing the surface treatment process, we provide fundamental comments on the LCA procedure as well as specify the dedicated methodology to analyze surface processing. The assumptions and the computational choices made, as well as the main characteristics of the reported LCA case studies may be summarized as follows.

- Most of the data used during the model implementation are primary, that means that they have been collected on site by using ad hoc questionnaires. Secondary data, obtained by databases, previous analysis, or published reports, have been used with respect to the production and delivery of energy carriers (electricity, natural gas, etc.) and to the production and delivery of all raw materials entering the production plants.
- Mass and energy balances have been calculated following the general principles of ISO 14040 standard.
- The comparison between all the technologies with different analyzed energy mixes has to be considered as a first order approximation result.
- The software Boustead Model 5 [9] is used as calculation model and as the main source of secondary data.

All the case studies presented have been developed in the framework of the ACTECO project with the principal objective of comparing the actual and future surface functionalization technologies for textile, biomedical, and food packaging/processing products, with specific reference to their environmental life-cycle burden.

Some representative case studies may be selected as a reference basis to show how a dedicated LCA methodology can be devised and applied to the specific investigation of surface functionalization. In particular, for textile applications, we consider treatments to obtain hydrophobic and oleophobic properties, while for food processing we address to coating methods to provide anti-corrosion properties on metal surfaces. In particular, among the most representative case histories we report:

- deposition of ceramic coatings by physical vapor deposition (PVD) ion-plating;
- SiO_x deposition by radio frequency plasma treatment;
- chromium electroplating;
- comparative LCA analysis between traditional and plasma processes for poly(ethylene terephthalate) (PET) textile substrates to obtain oleophobic properties;
- comparative LCA analysis between traditional and plasma processes for PET+cotton textile substrates to obtain hydrophobic properties;
- comparative LCA analysis between traditional and plasma processes for PET+cotton textile substrates to obtain oleophobic properties.

12.3.1
Scope, Functional Unit, and System Boundaries

The first phase of LCA analysis carried out for these specific applications is characterized by the definition of the functional unit (FU), the system boundaries and the data categories of the study.

The functional unit is defined as the quantified performance of the product system for use as a reference unit in a LCA study. The system boundaries are the interface between a product system and the environment or other product systems. Data categories are aggregations of input data from inventory collection according to the specific environmental impact. A well known impact category is the climate change (formerly 'global warming') potential. The most well-known emission in this category is carbon dioxide. All emissions which produce a potential contribution to the greenhouse effect are assigned to this category. This way, the number of data is considerably reduced and the results can be better interpreted by referring directly to the environmental impacts. Since the inventory data are related to the quantitative reference of the product or process, this relation also exists in the life cycle impact assessment. The impact assessment results and also data obtained from the inventory analysis can be used jointly for the (later) interpretation phase of the LCA study [10].

For each process considered the most suitable FU has been chosen to which all the input and output data are referred.

Because the products employed to obtain anti-corrosion properties vary in shape and quality, it was decided to express the results in terms of $1\,m^2 \times 1\,\mu m$ of coated surface for PVD and SiO$_x$ since this FU represents the reference unit that gives the same performance of the treated surface. For chromium electroplating the reference unit is $1\,m^2 \times 3\,\mu m$, which is the unit that is equivalent in terms of performance for this treatment to the unit that has been used for the other two treatments.

Regarding the fabrics functionalization, the FU chosen is 1 kg of treated material. Due to the fact that the two different substrates (PET and PET+cotton) cannot be compared in terms of surface morphology and chemical-physical characteristics, it was decided to carry out a specific analysis for each substrate, using as a reference an invariant quantity: mass. Choosing a mass unit as FU ensures that whatever the process is, one always considers the same quantity of material, eliminating the possible shrinking, or enlargement of the surface area after the treatment. Note that the mass difference before and after the surface treatment can be considered as negligible.

The boundaries of the considered systems include all phases from raw material extraction to the production and coating of a generic product. Product life durability and end of life have not been taken into account in this analysis

The analysis does not take into consideration:

- the production and the transport of the substrate materials;
- the production and the end-of-life of the surface treatment plant;
- the end-of-life of the coated products.

As a general rule, the system boundaries applied for these processes correspond to Figure 12.3.

Concerning textile applications, the boundaries comprehend pre-treatment and finishing plasma or traditional processes, as shown in Figure 12.4. Both plasma and traditional treatments consist of two-step processes: a cleaning pre-treatment and the functionalization treatment itself. Cleaning treatments consist of a de-sizing (for cotton containing fabrics or PET fabrics), and are necessary to remove undesired additives that will make the treatment process difficult. At the

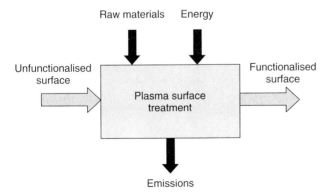

Figure 12.3 Representation of system boundaries related to surface functionalization process.

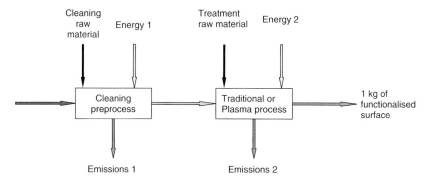

Figure 12.4 System boundaries for specific textile case studies.

moment, the cleaning process that has been analyzed within this project consists of traditional wet washings, using water and specific surfactant-detergent chemicals.

In general one of the main difficulties that technicians and scientists come across while carrying out LCA analysis is data availability, accuracy, and precision. Another hampering issue in LCA is confidentiality of the data employed in the analysis. Indeed materials and chemical composition are sometimes restricted. In this case and in order to go forward with the analysis one has to set specific hypotheses about general chemical compositions whose formulation is available. These reference compounds are assumed to have equivalent performances with respect to the reference ones. This assumption has also been made in some case studies considered here.

12.3.2
Life Cycle Inventory (LCI) and Hypothesis

The life cycle inventory (LCI) analysis provides a catalogue and quantification of the energy and material use as well as environmental releases associated with the processes included in the system boundaries.

The fundamental idea underlying the calculation of environmental inventories is simple. Any group of industrial operations can be regarded as a system by enclosing them within a system boundary. The domain surrounding this system boundary is known as the *system environment*. This system environment acts as a source of all the inputs (materials and fuels) to the system and, in a full life cycle, as a sink for all outputs from the system. This concept is shown schematically in Figure 12.5, where the system is represented by the box.

An inventory for this system is therefore simply a list of the quantities of all of the inputs that pass from the system environment, across the system boundary into the system and all of the outputs that pass from the system across the boundary to the environment.

When the inputs are all derived from raw materials in the earth and the final products are all waste materials returned to the earth, the inventory is referred to as a *LCI*.

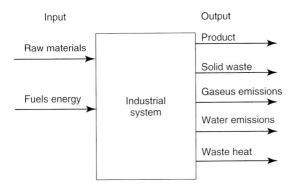

Figure 12.5 Diagram of an industrial system showing inputs and outputs.

It is no part of the inventory analysis to make value judgments about the relative significance of the different inputs and outputs. Instead, the analysis aims to provide the quantitative data upon which judgments can subsequently be made. It will, however, be clear from the above description of an industrial system, that the inputs to the system are the parameters involved in discussing conservation problems while the outputs are the parameters of interest when discussing pollution problems.

Because the inputs and outputs of any system are dependent on throughput, it is usual to normalize the data with respect to output so that all inputs and outputs refer to a unit output of product from the system.

It is also recognized that the performance of any plant is variable. Line closures, start-ups, maintenance, and so on all affect the overall efficiency of a processing operation. In an attempt to smooth out these variations, it is common practice to collect data for an extended production time period. The period chosen needs to be sufficiently long to smooth out such variations but sufficiently short so that it does not mask any improvements in practices. In general, most companies maintain records based on a 12-month period and for most purposes this period satisfies the practical requirements.

Thus the normalized data give a description of the average performance of the process during the period examined. This essentially represents a snapshot of the industry but gives no indication as to whether the period was typical or atypical, good or bad. The only way in which this can be determined is by repeating the exercise over a number of years until a general trend is established.

Since a system can be defined to include any collection of operations, there is no such thing as a correct system. Systems can range in size from a single factory or even a single machine, up to very large, extended systems which cover all operations from the extraction of raw materials from the earth through all production and use operations up to final disposal.

Thus the quantitative results of inventory calculations are specific to the system defined initially. It is therefore important to recognize that the definition of the system is an essential part of the results. Much of the confusion that arises when the results of life-cycle inventories are discussed is due to the imprecise way in which system definitions are reported.

When extended systems are chosen for analysis, the component operations will usually be carried out by a number of different companies. It is therefore important to analyze the overall system into its component sub-systems. Effectively this means setting up a detailed flow chart for the component operations. The level of detail in the flow chart should be such that it matches those physical operations for which data can be obtained. Analysis to a greater level of detail is unnecessary because the quantitative information is simply not available.

In general, the level of detail that should be included in the analysis can only be determined after an examination of the process itself. Before doing this, it is wiser to analyze to the greatest level of detail and subsequently recombine some of the operations. This is usually more simple than having to breakdown the original analysis to include further detail.

When drawing up flow charts, it is important to ensure that all packing and transport operations are included. Significant quantities of packaging are used in the transport of intermediate products and these can frequently be overlooked if not specifically identified at an early stage.

It is also tempting to simplify the flow chart by assuming that some operations are negligible and can be omitted. In the initial stages, no such assumptions should be made; if an operation exists in practice, it should be included in the analysis even though it may subsequently be demonstrated that its effect on the overall system is negligible. For those operations where no data are available it is better initially to find a surrogate – that is an operation that is thought to be similar to the required operation. Note that simply omitting an operation is essentially the same as treating all of its performance data as zero.

Data and information used in LCA studies can be divided into two main categories, *primary data* and *secondary data*:

- **Primary data** are collected directly from the plant and, therefore, guarantee a high level of accuracy. PVD process data have been collected in laboratory and refer to mass and energy flows of the LARC®-Platit ion plating technology. Chromium electroplating data were also supplied by electroplating centers and according to indications regarding food application chromium-plated coatings provided by Diad s.r.l., the end-user of functionalized surfaces. SiO_x plasma deposition data has been collected by using specific experimental data. Data regarding the process to obtain oleophobic/hydrophobic layers on PET and PET+cotton substrates have been collected from the ACTECO partners that perform plasma functionalization. In addition, data concerning those treatments using traditional techniques have been supplied by industrial partners of the ACTECO consortium and stem from current processes already in the market.
- **Secondary data** are obtained from databases, other analysis previously carried out or published reports. As far as the production of fuels, raw materials, and transports in terms of energy, resources consumption, and emissions to the environment are concerned, data come from the Boustead Model 5 and refer to: Italy, average Europe, and France energy mixes.

In addition, despite its completeness, the Boustead Model 5 library does not contain all the specified chemical compounds used in these processes. For instance, for the olephobic treatment considered it has been particularly difficult to find data regarding CF_4 (tetrafluoromethane) and similar perfluorocarbon (PFC) gases that were firstly investigated as precursors in plasma processes for oleophobic/hydrophobic treatments. These gases are not included as a possible input material in Boustead library, since commonly they are a by-product of several industrial processes, and not employed as input raw materials, also due to the regulations that ban the use of ozone depletion gases (such as CF_4). In order to cope with this situation, it has been decided to add the energy contribution of these gases to the total gross energy requirement (GER) of the whole process.

The GER represents the energy used when all materials and fuels are traced back to the extraction of raw materials from the earth and it is a measure of the total energy resource that must be extracted from the earth in order to support the system.

GER comprises the following elements: indirect energy, direct energy, transport energy, and feedstock energy.

The first element (indirect energy) is the description of the energy consumed by the fuel producing industries. Although variations in energy use in coal, oil, and gas extraction, and in the processing technology for these fuels are small, there are significant variations in the production and delivery of electricity. This arises because of the mix of fuels used in thermal generation, the age of the generation plant, and the efficiency of distribution. Materials processing operators have no control over the elements in this part of the table: it is a product of the infrastructure of the country in which a plant is located. This is why in our analysis we consider three different energy mixes (Italy, average Europe, and France).

The second element (direct energy) is the technology-dependent part of the table. This element represents the energy that is directly consumed by the process industries (i.e., the non-fuel producing industries). So, it depends upon the efficiency with which the materials handling and processing operations are carried out. It is the data that should be used if the technologies for two production systems are to be compared. This part of the inventory data is directly under the control of the process operators.

The third element is the transport. This, of course, depends upon the types of transport employed but, more importantly, is geography dependent: it depends upon the source of supply of the inputs to any process and the geographical location of the processing plant. It should be noted that the location of plants is often a historical accident. For example, early iron and steel plants were usually sited where there was a local supply of iron ore, limestone, and coal. As these local supplies were exhausted, raw materials had to be imported and so, what started out as an operation involving little transport, became an industry that increasingly relied on transport.

Finally there is the feedstock element in energy accounting. As noted earlier, the feedstock energy is kept separate from the other energy contributions because it is rolled up in the material of the product. Whereas fuels may often be used interchangeably, feedstock cannot. If a process demands a pure hydrocarbon feedstock, then only a pure hydrocarbon feedstock is acceptable.

An example of what is previously stated is provided by the case of the LCA comparative analysis to obtain anti-corrosion properties, in which three different scenarios have been taken into account to determine how the origin of the energy (production, transport, efficiencies) may affect the environmental burden of the same process. This is due to the choice of the energy mix. Indeed, the way in which primary electricity is produced in each country leads to different environmental consequences that are important to be evaluated. The scenarios that were chosen describe completely different options to produce electricity and represent a good way to evaluate how energy and its production affects to the environment:

- Italy energy mix, in which the fossil fuel source is predominant;
- Europe energy mix, which defines an averaged energy source configuration;
- France energy mix in which nuclear energy source is predominant.

For each scenario the computational model weighs the different energy sources depending on the selected scenario (Table 12.1).

However, some exceptions have to be considered. The process to produce titanium used in PVD process has only been considered using the Italy mix, since data regarding the Europe mix or France mix are not available; the same for production of CrO_3. This limitation affects the calculations minimally, since this level of detail is not necessary for this analysis.

Regarding the consumption of natural gas, the value always refers to European mix since the origin in terms of production of natural gas does not significantly vary for different European countries.

12.3.3
Inventory Data and Results

This section is dedicated to the interpretation of the environmental results to assign the equivalent values to the standard ISO 14040 four main parameters for impact categories above described, the so-called impact assessment: GWP, AP, EP, and photochemical smog (POCP).

Table 12.1 Summary of energy origin for each energy mix to produce 1 MJ of electricity.

	Italy mix (%)	Europe mix (%)	France mix (%)
Coal	12	27	7
Oil	34	8	2
Gas	34	16	1
Hydroelectrics	10	6	7
Nuclear	9	39	82
Other	1	2	1

Results of an LCA analysis can be split into the two following categories:

1) **Energy results**: describe the gross energy consumption for each process referred to the FU.
2) **Environmental results**: values indicating the natural resources consumption, air emissions, water emissions, and solid wastes normalized to the FU.

In particular, results concerning the GER, total raw material, air emissions, and water emissions are computed for all case studies listed at beginning of Section 12.3. LCA analysis has been carried out taking into account the following general hypothesis:

- Computational software: Boustead Model 5.
- Energy consumption values regarding electricity have been considered taking into account the Europe energy mix.
- Natural gas consumptions have been considered taking into account the general Italy mix (since no major differences could be found with respect to other choices and being equivalent to European one).
- All the results are referred to the established FU.
- This analysis contains the results coming from the data regarding the inputs and outputs of the process system and does not consider the contributions coming from the production of the substrates or the production of the machinery employed for each technology.
- No post processing such as water regeneration or abatement of emissions is considered in the overall LCA balance.

12.3.3.1 The Anti-corrosion Process

Considering the analysis to obtain anti-corrosion properties, three different technologies have been compared, obtaining significant differences as far as GER is concerned. As was predicted, changing the energy scenario does not affect the total energy consumption, but, as will be discussed later, strongly affects the environmental impact of the different solutions. In Figure 12.6 a schematic diagram is provided to compare GER, evaluated in MJ per FU, for three processes dependent on different scenarios, related to Italy, France, and average European energy mixes. From this partial result the overwhelming energy requirement of the microwave (MW) plasma treatment compared to the other two processes is evident. The latter two processes show almost comparable energy requirements. Using the European mix leads to a marginally lower energy requirement. These data show that plasma processing is generally energy intensive, though, as will be possible to assess from the following data and considerations, with less impact on environment.

Indeed to provide a representative assessment of the environmental impact, it is important to analyze the quality and origin of the energy employed for each process, focusing on the direct energy consumed, independently from the energy mix scenario. For this purpose we show the contributions of different direct energy sources in the total energy consumption of the three considered deposition processes, using the average European energy mix in Figure 12.7.

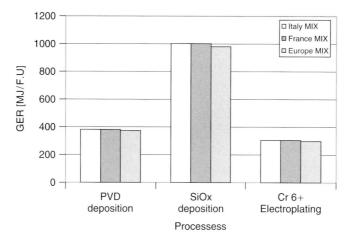

Figure 12.6 Diagram comparing GER for three processes:
PVD deposition, SiO$_x$ deposition (by MW plasma), Cr^{6+}
electroplating, which are dependent on different scenarios,
related to Italy, France, and average European energy mixes.

From Figure 12.7 comparisons among direct energy for PVD, MW plasma, and electroplating processes may be made, in particular direct energy for PVD treatment is mainly represented by electricity consumption. This highlights how the origin of the energy mix will certainly affect the PVD process emissions and therefore the environmental impact due to the production of electricity [11].

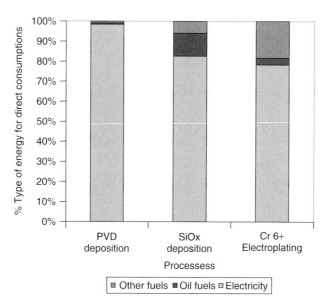

Figure 12.7 Energy source for direct energy (European mix).

Apart from the energy results, the LCA analysis provides data on the total raw material and energy consumption associated with each process. As an example, we present the differences in the consumption of water and main raw material (represented by chromium as the most significant raw material in the inventory) for the anti-corrosion processes. Plasma processes have a limited direct water consumption, which is negligible for PVD. It must be pointed out that since plasma processes are dry treatments, the water shown in the chart below is used for the raw materials or energy production. This is the reason why the SiO_x process uses this large quantity of water: it is indirectly used in energy and raw material production. The quantity of water of the electroplating process, a wet surface treatment, is considerable and can be split up in both direct and indirect consumption.

When considering the main substances that can be found in water emissions for each technology using the European energy mix it has to be considered that, since neither PVD nor SiO_x treatments use water as direct emissions, these emissions correspond to the indirect processes concerning the energy production or the extraction of raw materials. On the contrary, data concerning electroplating technology should be applied to both direct and indirect emissions.

As can be noticed in Figure 12.8 the electroplating process provokes the emission of water containing Cr^{6+} and resulting from direct emissions of the process itself. These emissions do not appear for the other two processes. Water coming from the electroplating process should then be treated adequately in order to neutralize the effect of these ions before being re-introduced in the biosphere. Some of the water treatments consist of Cr^{6+} neutralization in acid solutions, which is a heavy metals precipitation process, using NaOH and flocculants and then a sludge compacting process. Then the pH of the water for each post treatment should be permanently controlled and represents a critical point in the whole process. The

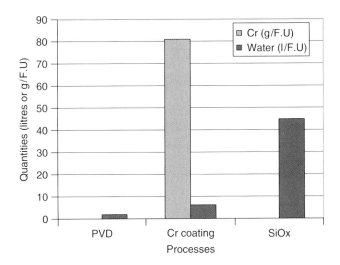

Figure 12.8 Water used and chromium emission for PVD, electroplating, and MW plasma coating.

Table 12.2 Substances in water emission.

Substances in water (mg FU^{-1})	PVD deposition	Cr^{6+} electroplating	SiOx deposition
COD	30	16	242
BOD	3	2	53
Na+	–	240	21 500
Suspended solids	825	39 200	7 000
Dissolved solids	–	170	54 400
Cr^{VI}	0	859	0
Zn+	0	217	1

environmental burden caused by these post treatments has not been taken into account at this stage, and would certainly change the values of the parameters in favor of plasma processing.

Different considerations should be made for the other substances found in water emissions, as reported in Table 12.2. As previously stated, PVD and SiO$_x$ emissions come from indirect processes, above all from the energy production. Contributions to the biochemical oxygen demand (BOD) and chemical oxygen demand (COD) mainly correspond to the emissions due to the production of energy, implying that the more energy is consumed, the higher values for these parameters. This is perfectly correlated to what has been stated for the energy consumption related to each process.

12.3.3.2 Textile Processes

The three different functionalization processes for textile application regarding oleophobic/hydrophobic properties considered in this chapter, require separate considerations. Actually, the processes to be compared are atmospheric pressure plasma processes and wet treatments. These technologies have different configurations with respect to the ones reported for the anti-corrosion treatment.

12.3.3.2.1 Total Energy Requirement

Regarding the energy consumptions, the following charts show the differences found for each process and each technology. Figure 12.9 and Table 12.3 show the GER concerning the oleophoby/hydrophoby treatments on PET and PET+cotton textiles. It is clear that atmospheric pressure plasma (APP) employed to obtain the same functionality requires less gross energy than a traditional wet process. Regarding the plasma process, the following hypothesis should be also considered:

- The plasma process requires a precursor gas for its activation. In the considered case this gas is R134 (CF$_4$). Values regarding the energy consumption to produce this gas are not available. This is why it has been hypothesized that the energy to produce this PFC can be assimilated to 5.9 MJ mol^{-1} [8, 12, 13]. Considering

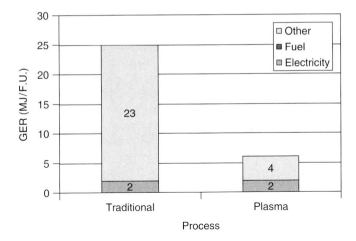

Figure 12.9 GER for the textile treatments considered. Note that the GER is the same for all considered textiles process (oleophobic PET, oleophobic PET+cotton, hydrophobic PET+cotton).

that the amount of gas used is 0.035 kg FU^{-1}, the energy contribution of this gas is 2.03 MJ FU^{-1}. This value has been added at the end to the total GER.

- PFC emissions (corresponding to the PFC that, during the process, have not been deposited on the surface) are treated as a generic PFC compound, which the Boustead model treats as a non GWP gas contributor.
- In addition to the PFC emissions, the process emits N_2 used during the treatment as a carrier gas. The Boustead model does not include this molecule as a possible parameter to calculate the environmental burden and therefore its emissions have been neglected.

In this case, the plasma treatment requires a much smaller overall energy input, only about 20% compared to the traditional process. Clearly, this has direct implications on the environmental impact. This effect is further enhanced by the fact that the share of the electrical energy in the total energy requirement is low, it is about 20% of the overall energy input.

Table 12.3 Energy requirements (GER) for traditional hydrophobic process on PET+cotton (values in MJ FU^{-1}).

Fuel type (MJ FU^{-1})	Production energy	Process energy	Transport energy	Feedstock energy	Total energy
Electricity	4	2	0	0	6
Fuel	0	0	0	0	0
Other	1	24	0	0	25
Total	5	26	0	0	31

12.3.3.2.2 Output of the Oleophobic PET Processes Data regarding the main raw material consumption and the emissions are reported in Tables 12.4–12.6. With respect to the raw material input, plasma processing is advantageous as it requires only about 40% of the input needed for the traditional process. Interesting is that 65% of the raw material input for plasma processing is simply air. Omitting this contribution, we find that the plasma process requires only 14% of raw material use compared to the traditional textile processing for attaining oleophobicity.

Concerning air emissions, traditional processes imply massive outputs of harmful substances such as PFCs, HCs, NO_x, CO, and CO_2 which all contribute to the GWP. The corresponding plasma values are orders of magnitude lower. Only for the emissions of SO_x, the levels are almost comparable. Dust emissions are five times higher for traditional processes with respect to the plasma ones.

Also for the substances emitted in water, the output of the traditional treatment is clearly larger. Substances related to COD exceed 10 times the ones resulting

Table 12.4 Raw material consumption for oleophobic processes on PET (data in mg FU^{-1}).

Raw material (mg FU^{-1})	Oleophoby traditional PET	Oleophoby plasma PET
Bauxite	0	50
Sodium chloride (NaCl)	27 900	1083
Fe	24	8
Limestone ($CaCO_3$)	19 029	15
S (elemental)	0	749
O_2	0	4 600
N_2	2	225
Air	2	12 590
Total	47 000	19 322
Water (total) (l)	35	5

Table 12.5 Air emissions for oleophobic processes on PET (data in mg FU^{-1}).

Substances	Oleophoby traditional PET	Oleophoby plasma PET
Dust (PM10)	1060	218
CO	1785	421
CO_2	1 388 492	342 863
SO_x as SO_2	758	689
NO_x as NO_2	2906	0
HCl	9	0
HF	0	779
Hydrocarbons	1245	10
Metals	0	317
CH_4	1 2081	0

Table 12.6 Water emissions for oleophobic processes on PET (data in mg FU^{-1}).

Substances	Oleophoby traditional PET	Oleophoby plasma PET
COD	17 200	1 751
BOD	4000	126
Na+	27	29
Cl$^-$	28	40
Suspended solids	4694	77
Dissolved solids	19	15

from plasma processing, while substances determining the BOD amount to more than 30 times the alternative plasma solution. Solid suspension in water due to traditional fluorination exceeds the water emissions by plasma treatment by a factor of about 60. Emission of sodium and chlorine ions into water and the output of dissolved solids are comparable in magnitude.

12.3.3.2.3 Output of the Hydrophobic PET/Cotton Processes Here, we compare the traditional and plasma process to render a PET+cotton textile hydrophobic. Tables 12.7–12.10 provide the raw material consumptions, the air emissions, the

Table 12.7 Raw materials consumption for hydrophobic processes on PET+cotton (values in mg FU^{-1}).

Raw material (mg FU^{-1})	Hydrophoby traditional PET+Co	Hydrophoby plasma PET+Co
Bauxite	0	50
Sodium chloride (NaCl)	40 160	1084
Fe	39	52
Pb	0	0
Limestone (CaCO$_3$)	464	24
Ni	0	0
Rutile	0	0
S (Elemental)	521	749
Dolomite	0	1
Cr	0	0
O$_2$	1	1
N$_2$	41	33 229
Air	7923	12 594
Olivine	0	0
Iron/steel scrap	0	0
Total	49 150	47 784
Water (total) (l)	22	5.0

Table 12.8 Air emission for hydrophobic processes on PET+cotton (values in mg FU^{-1}).

Substances	Hydrophoby traditional PET+co	Hydrophoby Plasma PET+co
Dust (PM10)	1051	579
CO	2148	965
CO_2	1 735 071	857 403
SO_x as SO_2	3327	3950
NO_x as NO_2	0	2246
HCl	0	78
HF	3889	3
Hydrocarbons	24	692
Organics	1	0
Metals	1835	1
CH_4	0	3516
Perfluorocarbons (PFC) not specified elsewhere	2	23 375

Table 12.9 Water emission for hydrophobic processes on PET+cotton (values in mg FU^{-1}).

Substances	Hydrophoby traditional PET+co	Hydrophoby plasma PET+co
COD	145 001	1 471
BOD	33 000	106
Na^+	1039	29
Acid as H^+	1	1
NH_4^+	1	1
Cl^-	1429	40
Suspended solids	6634	97
Hydrocarbons	0	0
Phenols	0	0
Dissolved solids	532	15

water emissions, and the waste production, respectively. The listed values always refer to required amount per FU.

In line with what has already been observed for the oleophobic process, also for the hydrophobic one the main contributions of raw material consumption for the plasma treatment is represented by air and nitrogen inputs. The sodium chloride input is in this case considerably lower and practically negligible (2.6%) compared to that required by the traditional process. Limestone input (mainly calcium carbonate) is larger for the traditional process by a factor of about 20. For plasma processing the only input exceeding the one of the traditional process is the elemental sulfur requirement, although it is comparable in magnitude. Values set to zero represent negligible contributions, although they need not be exactly zero.

Table 12.10 Wastes for hydrophobic processes on PET+cotton (values in mg FU^{-1}).

Substances	Hydrophoby traditional PET+co	Hydrophoby plasma PET+co
Unspecified refuse	506	343
Mineral waste	350	80
Slags and ash	4018	12 667

12.3.4
Impact Assessment

According to ISO 14040, the general framework of the assessment phase is composed of several mandatory elements that convert inventory results into environmental indicators. For this analysis the following impact categories are considered (such characterization factors are also recognized by the system for EPDs):

- GWP in 100 years (GWP100): index used to measure the global warming. This is the phenomenon whereby CO_2 in the atmosphere, along with other compounds, absorbs infra-red radiation emitted from the Earth's surface, thus giving rise to an increase in temperature. In other words, GWP is the measure, based on concentration and on exposure time, of the potential contribution that a substance causes to the greenhouse effect as to that caused by the same mass of CO_2. The standardization of global warming is made by reporting the amounts of the inventoried substances in grams of CO_2 equivalents.
- The AP is used to measure the acidification impact into the atmosphere and water courses caused by the release of hydrogen ions. The standardization of acidification is made by reporting the amounts of the inventoried substances in grams of H^+ equivalents.
- The EP is used to measure the nutrient enrichment (eutrophication), which in turn may result in algal blooms, caused by the release of sulfur nitrogen, phosphorous, and degradable organic substances into the atmosphere and water courses. The standardization of eutrophication is made by reporting the amounts of the inventoried substances in grams of O_2 equivalents.
- The photochemical pollution potential (POCP) is used to measure the breakdown of the stratospheric ozone layer, which should protect the earth from UV radiation, caused by the emissions of reactive substances that mainly originate from chlorofluorocarbons (CFCs). The standardization of ozone depletion is made by reporting the amounts of the inventoried substances in grams of CFC-11 equivalents.

Now, the main qualitative results are reported as a result of the comparative impact assessment study between plasma technology and conventional technology.

Starting from the previous inventory data of the considered case studies, impact assessment provides a quantification of environmental burden through impact parameters. Figure 12.10 provides an example of the GWP of plasma deposition vs. electroplating based on different energy mix scenarios and the following Table 12.11 lists the related other impact categories based on the same scenarios.

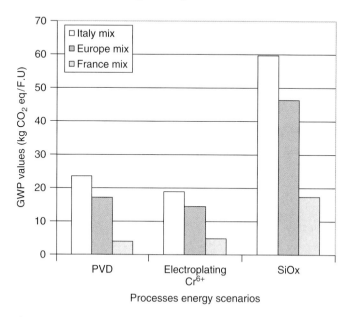

Figure 12.10 GWP comparison of plasma and electroplating processing based on different energy mix scenarios.

Table 12.11 Acidification, photochemical ozone depletion and eutrophication potentials: comparison among plasma processes and electroplating processes based on different energy mix scenarios: Italian, French, and average European energy mixes.

Treatment	AP (g eq SO_2)	POCP (g C_2H_4)	EP (g PO_4^{3-})
PVD Italy	245.47	32.38	8.71
PVD Europe	139.32	14.95	5.93
PVD France Mix	37.6	7.08	1.66
CrVI galvanic coating Italy mix	210.52	22.44	6.92
CrVI galvanic coating Europe mix	110.13	9.59	4.86
CrVI galvanic coating France mix	356.8	3.6	1.73
SiO_x plasma – Italy mix	718.75	75.99	22.92
SiO_x plasma – Europe mix	419.44	37.66	16.8
SiO_x plasma – France mix	197.35	20.54	7.46

12.3.5
Sensitivity Analysis

12.3.5.1 Managing Uncertainties

In general, it is important to consider that the validity of the LCA results, and in particular when comparing a plasma with a traditional process, the impact assessment considerations previously presented are only acceptable if the uncertainty range on the final impact parameters is determined. The total uncertainty of the computations may come from two different levels: on one side the uncertainty of the input data and on the other side uncertainty that comes from the computational model itself.

As far as the reference model (Boustead) is concerned, it can be stated that the possible error propagation due to multiple computations is within a very narrow range. However, even though difficult to quantify precisely, this is common to all the calculations performed. The use of another model will not eliminate this possible uncertainty because this kind of uncertainty is intrinsic to all known LCA models. Another point regarding the model is related to the database used. Even though it is very unlikely that data changes significantly in time, refinements of the data are common so that it is advisable to refer to updated databases to be more representative of actual situations.

The second level of uncertainty comes from the uncertainty of the input data and needs to be taken into account in the sensitivity analysis to allow a well-founded scientific comparison of impact average values. This brings the need of performing a sensitivity analysis on the input values. The main objective of the sensitivity analysis is to evaluate the precision of the input data collected (inventory) and to compare the input data uncertainty (variance) with the maximum uncertainty which allows discriminating among final average impact parameters of compared case studies. Only in this situation, it is fair to select a process as the one providing lower environmental impact for a specified impact category parameter (e.g., GWP). Hence, such a sensitivity analysis is required to qualify the comparative analysis.

When specifying the input data uncertainties it is possible to recalculate the environmental burden caused by the entire process by using (crossed) minimum and maximum input data in the uncertainty range in order to check for possible overlap of the impact parameter values.

These new results will then be compared in order to verify if the conclusions obtained from the previous studies are still valid. It can occur that due to initial parameter uncertainties, there is a possible overlap of the impact values. Consequently, the LCA study will not allow to discriminate among the average impact values, thus rendering the comparative analysis useless in such a case.

12.3.5.2 Example 1: General Sensitivity Analysis for the LCA Study of the Textile Processes

For the textile related case studies considered in this chapter, it is very interesting to perform a sensitivity analysis for the GWP. This parameter is mainly influenced by the emission of the greenhouse gas CO_2 and of PFCs. These emissions stem from the use of energy from fossil fuels (either the use of direct fossil fuel, or electricity coming from fossil sources), or from the use of specific chemical products during the processes.

Since the interest of the sensitivity analysis is to evaluate the maximum uncertainty allowed, the minimum variation of each plasma and traditional process GWP values have been calculated in order to reach the GWP overlapping threshold.

Calculation steps may be summarized as follows:

- Identify the average GWP value resulting from the analysis of average data concerning each process. This value comes from the results given by the Boustead model 5.
- Calculate the minimum sufficient percentage variation in the energy consumptions that mainly influences the GWP coming from plasma process (from now on GWP_p) and the GWP coming from traditional process (GWP_t), in order to have the same value of GWP for both processes (theoretical GWP).
- Once this percentage has been identified, recalculate the GWP values using the Boustead model 5 and validate the theoretical uncertainty.
- Compare the uncertainty calculated to the tolerance declared for each process.

The following case studies have been considered:

- Comparative LCA analysis between traditional and plasma processes for PET textile substrates to obtain oleophobic properties.
- Comparative LCA analysis between traditional and plasma processes for PET+cotton textile substrates to obtain hydrophobic properties.
- Comparative LCA analysis between traditional and plasma processes for PET+cotton textile substrates to obtain oleophobic properties.

For each of these cases a sensitivity analysis has been carried out in order to evaluate the maximum uncertainty allowed for the input data in order to maintain the differences found in the average analysis.

The case studies inventories reveal that most of the data processed during the analysis to get the average impact parameters are based on declared uncertainties spanning from 5% in a best case up to 30% in the worst case. Therefore, evaluating how this uncertainty affects the comparative analysis is necessary.

To provide a reliable comparison between the plasma and traditional processes, instead of re-calculating all the results using the minimum and the maximum input data values from the given uncertainty, it is more interesting, and more effective, to calculate the minimum variation of input data needed in order to have the same GWP result for each technology. In other words: the maximum allowed uncertainty to keep the average results as different values are calculated. This procedure puts a limit on the possible input data uncertainty for each process and at the same time checks the reliability of the LCA data interpretation.

One of the main difficulties that scientists and technicians come across while carrying out LCA analysis is data availability and its reliability. This is the case when dealing with confidentiality issues related to data supplied at the inventory step, in which data are incomplete or not precise enough. In other cases, processes use materials and chemicals whose composition is restricted or protected by patents. Another limitation occurs when (part of) the collected data stem from an experimental process, which typically leads to a considerable uncertainty with respect to the actual values of the industrial scaled-up process.

In the case studies considered, as far as the sensitivity analysis is concerned, the GWP values were determined by the energy consumption and the way it is produced. In the case of use of PFC gas not declared by the data provider, these contributions were treated as generic PFCs with no GWP influence. Thus, they do not have to be taken into account here.

The following table summarizes the variation values and the GWPs that were found, either by theoretical evaluation, and by using Boustead model. All the GWP values in Table 12.12 are expressed in kilograms of equivalent CO_2. The minimum theoretical relative (%) variations of GWP are calculated *a priori* starting from average GWP experimental values found for traditional and plasma processes, and define the minimal variation that determine the overlap between traditional process GWP and plasma process GWP. The theoretical GWP for plasma and traditional processes are referred to the absolute values corresponding to the overlap conditions. The calculated (extreme) GWP values are evaluated experimentally by providing as input data to the LCA code the average input data plus (minus) their declared uncertainty in LCI phase.

Note that the equivalence of the results between the oleophobic and hydrophobic processes for PET+cotton is due to the fact that the input values for the energy concerning these processes are the graphs (Figures 12.11 and 12.12).

The comparison between the calculated and theoretical GWP verifies the possibility of discriminating average values and, provided the required discrimination (as is the case), eventually define which one of the two compared processes has more impact in terms of GWP, on the basis of the declared assumptions.

Table 12.12 New GWP values applying min theoretical variations.

		Average GWP	Min theoretical variations in percentage for overlap (%)	Theoretical GWP for overlap	Calculated GWP
Plasma	Oleophoby PET	0.4	17	0.47	0.46
Traditional	Oleophoby PET	1.66	72	0.47	0.49
Plasma	Oleophoby PET+co	0.937	20	1.13	1.12
Traditional	Oleophoby PET+co	2.03	44	1.13	1.31
Plasma	Hydrophoby PET+co	0.937	20	1.13	1.12
Traditional	Hydrophoby PET+co	2.03	44	1.13	1.31

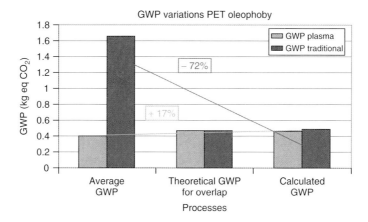

Figure 12.11 GWP variations for PET oleophobic treatments.

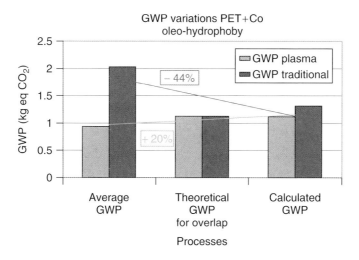

Figure 12.12 GWP variations for PET+cotton oleophobic and hydrophobic treatments.

The tolerance declared by the data providers for each process concerning the uncertainty of the experimental data is summarized in Table 12.13.

As can be noticed, one of the calculated tolerances is above 50%. This can be explained because most of the data were taken from experimental processes. Nevertheless, such large uncertainties (>50%) are not acceptable for a valid model and outcome as it implies that the industrial process is not very accurately documented. Clearly, for a more precise outcome of the LCA study, a more accurate study has to be carried out, and by refining the input data. However the *a priori* calculation of the minimum theoretical variations of the traditional and plasma processes GWP already sets a limit to the admissible uncertainties to the input data of investigated processes.

Table 12.13 Declared uncertainties vs. calculated tolerances.

Process	Energy declared uncertainty	Calculated tolerance (%)
Plasma PET oleophoby	5–20%	17
Traditional PET oleophoby	5–30%	72
Plasma PET+cotton oleophoby	Not declared	20
Traditional PET+cotton oleophoby	10–30%	44
Plasma PET+cotton hydrophoby	Not declared	20
Traditional PET+cotton hydrophoby	10–30%	44

12.3.5.3 Example 2: Design of Plasma Processes via LCA

In this part, we consider the specific possibility of using CF_4 gas for the plasma process employed to obtain the oleophobic properties on textile. Until now, in the analysis regarding textile plasma functionalization, it has been considered that the PFC gas emitted during the process was a generic PFC gas, which the Boustead model does not consider as a GWP contributor. This assumption, also confirmed by other plasma related LCA studies, is quite relevant as the effects of PFC gas are well known on the greenhouse effect.

In order to compare results deriving from employing a generic PFC and those deriving from a specific PFC forcing a higher environmental impact, CF_4 gas has been considered. As main result, it could be noticed how the chemistry composition of the gas is relevant for the impact assessment values, above all for the GWP parameter. CF_4 gas has a dramatic influence on the results and makes the GWP value for plasma process much higher than for the traditional process, reversing the analysis results. Therefore, it is necessary to emphasize that the exact chemical composition of the precursor gases has to be taken into account very carefully in this type of analysis as it results in dramatically different results for the exhaust gas composition.

Such an analysis and the resulting information are also quite precious for the design of suitable plasma processes having a lower environmental impact. Indeed, the appropriate choice of plasma precursor gases guarantees the environmental compliance of new solutions. The specific case considered here is a good example of how LCA study, if applied at processing design level, may significantly affect the surface treatment environmental compatibility.

12.3.6
Concluding Considerations on LCA Study

Based on the case studies presented, as far as average data are concerned, the innovative plasma processes proved to have a lower environmental impact compared to the traditional ones to obtain the same functionality.

In particular, it can be observed that the energy consumption has a direct influence on the GWP value. This parameter is calculated taking into account the quantity of CO_2, or equivalent, emitted to the air during the whole process, including the production processes to obtain raw materials and energy. The energy sources and their production origin determine the value of the GWP.

The sensitivity analysis proves the validity of the comparisons carried out showing no overlapping of the final impact parameters, even when accounting for the uncertainty ranges. Therefore, the LCA study performed enables to discrimination among average impact values.

The energy mix for electricity production has a major influence on the GWP value as the emissions of CO_2 vary enormously from one mix to another. In the case studies presented, the European energy mix was assumed. In that case, more than 61% of the total electricity production is obtained from fossil fuels or gas, implying an important emission of CO_2 that contributes to the greenhouse effect, and thus, to the GWP. Using renewable energy sources for the plasma process would certainly imply a further relevant decrease in the GWP parameter, even when using the same quantity of direct energy. In that case, the plasma process technology would be even cleaner, in spite of it already being the most environmental friendly technology nowadays available for the considered surface functionalization processes.

As stated before, the energy requirements of the traditional process is not only based on electricity consumption, but also on using a significant amount of natural gas employed to heat the water for the wet processes. This consumption causes the large value of the GWP value due to the CO_2 emissions.

Regarding the environmental burden generated by these processes in terms of GWP it is clear that the choice of the energy mix affects the results. Consequently, the choice of a suitable and more environmental friendly energy production and transport system represents the best option to reduce the global impact related to the greenhouse effect, and to emissions of pollutants that contribute to the eutrophication, acidification, and photochemical pollution.

Values regarding the specific consumption of raw materials, of emissions and of process by-products have also been included for each case study.

The traditional processes use significantly more water than plasma-based processes. In addition to the water consumption, the values for the COD and BOD, are also always higher for traditional processes and are responsible for the eutrophication of local water systems.

Moreover, the huge need for water in traditional processes requires that the waste water needs to be post treated before being emitted into the environment. Purification plants capable to neutralize all the polluted water coming from the production process are compulsory. Indeed, manufacturers that still use traditional wet processes are forced to invest in these types of installations by legislative constraint, thus avoiding massive environmental local pollution. Production plants equipped with these post-process treatments require consumption of extra process energy to clean all the waste water. This energy consumption as well as the energy and materials employed to produce and install such post process depuration systems have not been included in the performed LCA studies. Therefore, this additional source of environmental burden should be taken into account to set up a more complete balance between the environmental impact assessment of traditional and plasma solutions for surface functionalization.

Such considerations put in evidence that the case studies analysis is quite conservative, although results already show a net unbalanced situation for the two classes of compared processes (traditional vs. plasma), favoring plasma solutions as having an overall lower environmental impact.

It is extremely important to take into account all the issues inspected, above all in geographically dry regions where the availability of water is restricted, or regions where a suitable system to purify the waste water has not yet been established.

A further comment regarding water consumption should be added as all the considered case studies include a wet cleaning process. Plasma technology is nowadays nearly ready to provide a good alternative to substitute cleaning processes like wet de-sizing or de-oiling. Such substitutions would further reduce the water consumption to practically zero for most of the textile processing steps.

As far as air emissions are concerned, the results again show that values found for the plasma process come from the electricity generation and not from the process itself, due to the fact that pollutant substances emitted during the plasma process are negligible. This is interesting because in a fully renewable energy scenario, the plasma processes as a whole would provide negligible air emissions. Anyhow, pollutant emissions for plasma processes are lower compared to traditional process, especially for CO, CO_2, and NO_x. Also the dust emissions (PM10) are dramatically lower for plasma process.

In general, the uncertainty on the input data affects the results directly and can even change the balance in favor of one or the other technique. Therefore, it is crucial to be careful when collecting experimental input data. Underestimating the importance of this point could lead to a significant error that may invalidate the whole study. Despite the fact that from our analysis it can be stated that it is very unlikely to have the same GWP values for equivalent traditional and plasma processes, it can be deduced beyond doubt that all these analysis guarantee no overlap. The reason is that computations have been carried out taking into account several hypotheses that establish very specific initial conditions. Changing these hypotheses would certainly lead to differences, which have not been considered at this stage. This highlights how LCA, although being a powerful mean for quantifying environmental impact in a comparative way, should always be used with great caution, considering that environmental impact conclusions are very strongly related to the assumptions made. Therefore, for safety of scientific clarity, assumptions have to be explicitly detailed with great care.

In conclusion, plasma technologies offer a clear environmental advantage on traditional technologies, above all if considering the energy consumption and the use of water resources. All these results, as well as all the comments stated before, are valid within the framework of the initial hypotheses made.

12.4
Process Safety for the Working Environment

In the previous Sections we considered environmental issues that are mainly concerned with global impact. Here we focus on local impact and in particular on safety issues connected with the potential effect of plasma processes on the working environment.

When carrying out plasma processes for enhanced surface functionalization, it is sometimes necessary to work with precursors which, directly or indirectly, may constitute a hazard for the local environment. Actually, even when using a stable and not harmful chemical for plasma reactivity, when these molecules are introduced in the ionized medium they experience ion bombardment, ionization and often undergo molecular fragmentation and reaction with other plasma species and neutral molecules at different energy states. This results in the production of radicals and new molecules, which potentially may cause hazard for the working environment and for the operators. Given the short lifetime of radicals and metastable compounds this usually does not represent a real problem. However, stronger attention has to be given to long-lived compounds which can diffuse or even accumulate in the environment where plasma processing is carried out as well as to the use of precursors with relevant potential hazard to health and safety.

Since low pressure plasma treatments are by definition performed in confined reactor chambers with selective and monitored inlet and with outlet flows conveyed to the exhaust in a controlled way, our focus here is plasma processing in ambient conditions. This type of treatment is usually carried out at atmospheric pressure and temperature with open systems, which are free to exchange pollutants from the environment to the plasma processing volume and, conversely, to emit potentially unsafe compounds from the plasma discharge to operating environment.

The concerns for atmospheric plasma processing are not only related to the preservation of the working environment safety, complying with national regulations, but also to guarantee a controlled plasma-chemical processing environment to preserve process reliability, repeatability, and safety.

Regarding this issue, we report the main results of research activities within the framework of the ACTECO project which aim at assessing the environmental safety of the APP process and devising an innovative methodology for the recovery of safety conditions for process and environment, transferable to any processing unit configuration.

A particular technological challenging target in atmospheric plasma processing is to provide plasma fluorination; this implies use of fluorocarbon precursors. The study of their safety and impact on human health and environment is necessary and requires great attention. Moreover use of fluorocarbons includes some technological complications due to the generation of aggressive by-products which may determine the hazard for the processing unit components exposed to the reaction volume.

In particular here we focus on a processing methodology that:

- guarantees the safety of operators;
- guarantees a controlled atmosphere for process safety and reliability;

- avoids contaminations on mechanical parts of the system;
- provides feasibility of the process monitoring in terms of environmental compliance.

The prototype process developed was monitored for environmental and safety compliance providing testing through:

- differential pressure sampling;
- study of the flow pattern;
- chemical sampling of tracer, pollutants, and hazardous compounds.

12.4.1
Atmospheric Pressure Plasma Unit: Standard Configuration

The processing unit on which research has been realized is Environment Park's system based on APP with dielectric barrier discharge (DBD) planar modular electrodes. The unit is a lab scale version of an industrial production size unit and allows the development of innovative dedicated functionalization processes that can be directly up-scaled and transferred to industrial scale. The unit can be configured with dedicated and variable equipment so that different processes to obtain multi-functional surfaces can be implemented. In particular the unit has been devised with the due degree of flexibility: the equipment allows the injection of four different technical gasses and two vaporized liquid precursors inside the plasma region and a third precursor deposited by atomization on the substrate's surface, offering a high freedom in the choice of the plasma-chemistry of the process.

The possibility of using different liquid or gaseous precursors at the same time allows functionalization that would not be possible (and often incompatible) using traditional chemical process. Functionalization process schemes can be setup with in-line movement for nonflexible substrates or with a roll-to-roll configuration for continuous treatment of flexible substrates (e.g., polymer films, textiles). The unit provides processing on planar substrates and, depending on the configuration, treatment may be achieved on one side or both sides of the substrate material. The maximum surface roughness that allows a uniform functionalization to be obtained is in the order of a few tenths of millimeters. For dedicated surface shapes, with some modification of the cathode's configuration, treatment of not strictly planar surfaces can be achieved. In Figure 12.13 the DBD processing principle is represented, identifying the plasma volume, which is locally confined but not physically separated from the region outside the cathodes.

Figure 12.14 shows the APP unit having the following features:

- pre-heating stage (optional process step for different substrates)
- plasma treatment stage with planar plasma source for treatment of 60 cm width substrates of undefined length
- roll-to-roll speed from 1 to 10 m min^{-1} (industrial unit reaches 40 m min^{-1})
- flow control of four gasses for each process
- flow control of two vaporized liquid precursors

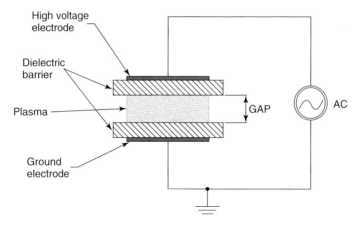

Figure 12.13 Diagram of the plasma reactor.

- evacuation hood
- two plasma sources
- two high voltage continuous/pulsed generators of 5 KW electrical power each;
- one atomizer for liquid precursors;
- manual and automatic control features through Programmable Logic Control (PLC) and PC (Microsoft Windows based OS) interface.

The unit is enclosed in a frame with poly(methyl methacrylate) (PMMA) window panels, which are not sealed with respect to the outer environment, therefore any gas exchange and mixing between the laboratory room and the unit box is possible, since no significant pressure differential is set between the two environments. The unit is designed in this way to be easily accessible for uploading the materials to be treated (typically as rolls) and to set up semi-continuous processing.

Figure 12.14 Atmospheric pressure plasma unit.

12.4.2
Devising Safe Processes for Industrial Applications Maintaining the Semi-continuous Feeding

In order to carry out critical processes safely, a new basic environmental design was devised and a prototype was realized by modifying the plasma unit in order to develop a confined APP system, still keeping the features of a roll-to-roll semi-continuous process. The system was partitioned into three domains having slightly different pressures, close to ambient pressure labeled as P1, P2, and P3 (see Figure 12.15). P1 is the zone where the plasma process is confined, P2 is the zone linked to the evacuation hood, P3 is the zone directly communicating with the outer environment, where an air inlet is provided.

The relationship between the different pressure domains are:

$$\begin{cases} p_1 > p_2 \\ p_2 < p_3 \\ p_3 > p_1 \end{cases} \tag{12.1}$$

p_1 is higher than p_2 to avoid mixture of the atmospheric gases (above all oxygen and water vapor) entering into the zone P1 because they may contaminate the plasma process. Indeed, in order to have process reliability it is important to work in a controlled atmosphere to avoid plasma pollution and dependence on meteorological conditions (such as humidity rate); p_2 is lower than p_3 because exhausted gases have to go to the evacuation hood and should not diffuse into the system. Finally p_3 is higher than p_1 to guarantee absolute safety.

In accordance with the material chosen for the plasma unit box windows, the three zones are separated by PMMA panels. Figure 12.16 shows the upgrade and modifications of the system. One can recognize the panels separating the electrodes zone (domain P1) from the intermediate pneumatic region P2.

In Figure 12.17 one can see the air inlet orifices, necessary to optimize the airflow in compliance with the hood suction capacity. The proper dimensioning of the system assures that the pneumatic requirements represented by the conditions in Equation 12.1 are met.

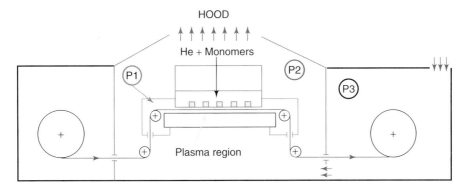

Figure 12.15 Diagram of the plasma unit equipped with different pneumatic domains.

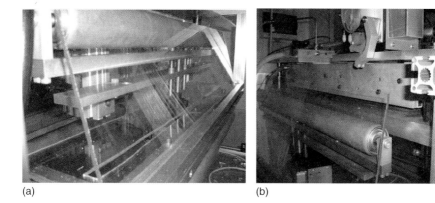

(a)   (b)

Figure 12.16 The inner pneumatic domains (a) P1 and (b) P2.

Figure 12.17 The panel separating the intermediate pneumatic domains P2 from the outer zone P3; the air inlet can be recognized by the round holes.

Pressure differentials are kept in stationary mode between the different pneumatic domains, however the pressure difference may be quantified in tenths of a millibar, which is enough to maintain the gas mixture separation. To assess differential stability in our specific case pressure sampling was performed to test the confined system functionality. The following data refer to different probe locations bridging the three pneumatic regions. A PCE-DM-30 digital differential pressure data acquisition system (measure range ± 30 mbar, resolution 0.001 mbar, precision $\pm 0.3\%$) was used. The graphs in Figures 12.18–12.20 report the differential pressure behavior in stationary mode and in transient mode (from off-mode to operating-mode). An abrupt change in the differential pressure is recognizable, corresponding to mode transition defining separate and well defined pressure regimes. As a matter of fact, fluctuations are always sufficiently small to allow a clear distinction between the two operating modes.

As it is possible to observe from Figures 12.18–12.20, the required relationships as of Equation (12.1) are experimentally satisfied, even in the largest fluctuating conditions. Indeed in Figure 12.20, although well discriminated, the pressure

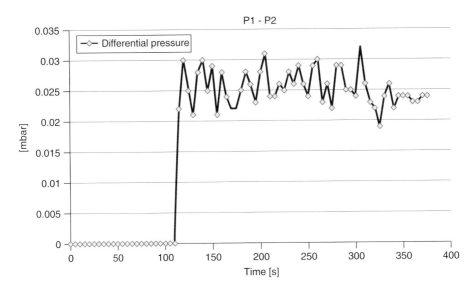

Figure 12.18 Time series of the differential pressure $p_1 - p_2$ in transition from off-mode to operating-mode.

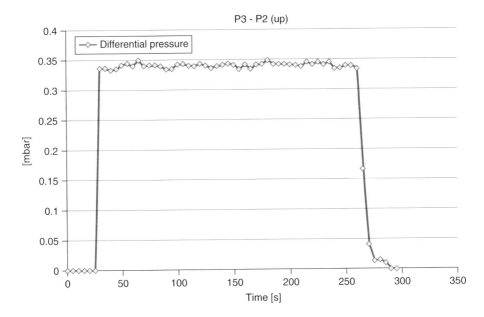

Figure 12.19 Time series of the differential pressure $p_3 - p_2$ in transition from off-mode to operating mode and off-mode again. Measurements were taken at the upper location in the zone sampling.

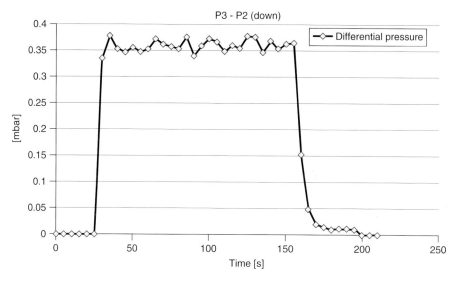

Figure 12.20 Time series of the differential pressure $p_3 - p_2$ in transition from off-mode to operating mode and off-mode again. Measurement were taken at the lower location in the zone sampling.

differential exhibits higher amplitude fluctuations due to the enhanced turbulent flow due to the inlet orifices shown in Figure 12.17, but never reaches zero.

A very important aspect for correct environmental monitoring is the analysis of the flow pattern, as shown in Figure 12.21. In order to account for transport phenomena due to the fluid dynamics, it is interesting to comment on some results obtained by providing a source of gas flow tracer to verify if any stagnation region or unexpected leakage from the different domains could be detected.

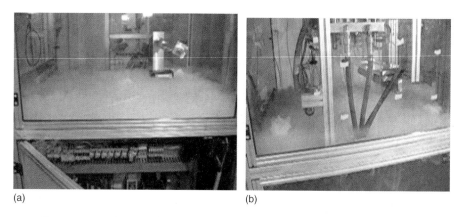

(a)   (b)

Figure 12.21 Side view of CO_2 tracer positioned in region (a) P3 and (b) P2 respectively, before the mode transition (system off-mode).

For this purpose, testing CO_2 vapor was chosen as tracer because it is a heavy molecule and tends to stagnate in the lower parts of pneumatic zones (Figure 12.15). Hence, it can give a significant proof of concept of the system functionality. Different tests were performed with the dry ice sources located in the regions P2 and P3. The system was switched from off- to operating-mode, while the hood was active and with process gas injection (in P1). After the transient, no traces of stagnation were detected and the flow pattern from P3 to P2 was proven as a guarantee that no leakage of substances evacuated from region P1 to region P2 could pass through the pressure barrier P2–P3 and be released to the working environment. Substances evacuated from region P1 are conveyed toward the hood, as specified, and air coming from the working environment (through P3) is injected into zone P2 (by the pressure differential) and finally evacuated to the hood. Hence, the set-up provides the right and safe exhaust admixture which complies with regulations.

Gas emission analysis was performed to monitor the plasma processes from a environmental point of view and to assess the effective presence of the gas tracer as well as the presence of CO and NO_x. The emissions of the precursor gases were evaluated by a detector tube system which combines an interchangeable detector tube and a portable pyrolyzer allowing for multiple gas sampling. This hand-held tool is quite flexible and may be positioned in different sampling locations with the possibility of mapping the process environment during operation (Figure 12.22). A direct-read scale is printed on each tube to provide easy access to values acquisition.

To gage, to check the system functionality and to measure its reaction time, it is recommended that the probe be set close to the tracer source and some initial testing be performed. The detection ranges of the probe used to monitor the plasma processes are provided in Table 12.14.

To provide a general assessment of the plasma reaction by-products some processes carried out within the framework of the ACTECO project were selected. In particular we report here some representative functionalization success cases related to the aminization process run with NH_3 precursors to attain hydrophilic properties used as textile pre-treatment (see also Chapter 6.5.2) and a process using

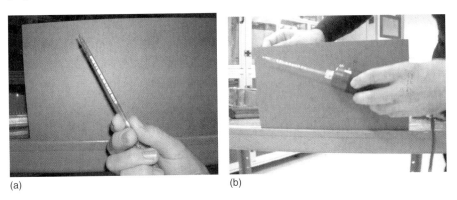

(a)   (b)

Figure 12.22 (a) Detector tube and (b) full assembly with pyrolyzer.

Table 12.14 Overview of the detector tubes measuring ranges.

Detector tube	Measuring range (ppm)
CO	5–50
CO_2	300–5000
$NO+NO_2$	5–625
$NO+NO_2$	0.04–16.5

Table 12.15 Gas emission sampling in region P1 to assess the intensity of the source.

Detector tube	Value (ppm)
CO	10
CO_2	1000
$NO+NO_2$	Tube is not colored
$NO+NO_2$	2

hexadecyltrimethoxysilane (HDTMS) as precursor to improve the hydrophobic properties of the fabrics.

In the first case, the chemistry involved comprises an admixture of NH_3 in solution (7%) with O_2 for a treatment duration of 30 seconds. CO, CO_2, and NO_x detector tubes were chosen to sample the by-products gases of the plasma process investigated. The sampling was done in the P1 zone to quantify the intensity of the 'pollutants' source. The results of gas emission analysis are showed in the Table 12.15. It can be noticed that the experimental data match consistently with plasma process chemistry.

In Zone P3 measurements were taken but the measured values were below the detectable threshold.

The second selected process was run using a much heavier and complex precursor molecule: HDTMS, see Figure 12.23. Larger molecules determine the possibility of many reaction channels involving fragmentation and recombination with the participation of other species present in the plasma (technical gases, other vaporized precursors, free radicals).

In this case the same detector tubes were used (i.e., for CO, CO_2, and NO_x) but four different zones where sampled, the three pneumatic domains P1, P2, P3

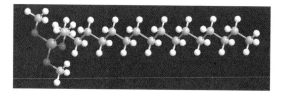

Figure 12.23 Chemical structure of hexadecyltrimethoxysilane (HDTMS).

Table 12.16 Gas emission analysis results in process monitoring: pneumatic domains P1, P2, and P3 were sampled including the working environment.

Detector tube	P1 zone (ppm)	P2 zone (ppm)	P3 zone (ppm)	atm (ppm)
$NO+NO_2$	0.2	0.15	0.1	0.1
CO_2	3750	2900	3250	3250
CO	10	≈ 5	0	0

inside the system and the working environment. Per zone, the average values are reported in Table 12.16.

From the reported data it is possible to verify that during plasma processing the amount of the investigated pollutants generated such as CO_x and NO_x is minimal and their concentration decreases in different pneumatic zones as expected: P2 has lower concentration than P1 and P3 has the lowest detected concentrations (even below the detection limit for CO). This is a very good result confirming the pressure measurements and the flow pattern investigations.

These results prove that the strategy of multipartitioning of the plasma system volume into different pneumatic domains provides the right diffusion barrier to pollutants injected/generated in the plasma working environment.

However, in order to devise a completely safe process, it is important to consider other critical aspects. As a matter of fact, condensation of vaporized chemicals on surfaces of the P2 zone of the plasma unit may be observed when the process is run for a long time (>30 minutes).

In order to face this inconvenience, the development of a new evacuation system to remove after each process the residual of the condensed chemicals is necessary to avoid the contamination of the system and to guarantee the safety of the operators while unloading/reloading the rolls.

The solution includes the use of a cleaning (purging) phase before completely opening the system. The cleaning phase is realized through a vacuum pump as shown in Figures 12.24 and 12.25. The specific protocol includes the standard plasma processing operation, in which the vacuum pump is OFF, the gas and precursors are injected through the inlet in P1, the hood is running in stationary mode and the target differential pressures between the chambers are set.

The second operation mode defined by the protocol is the purging mode, in which the vacuum pump is ON, the hood is off, the exhaust and the air inlet are closed, and the whole chamber has the same pressure. This pressure should be below the pressure of the environment, the target value is 0.5 bar (absolute) as this assures evaporation of the condensed residuals, thus recovering the safe conditions for opening the system and proceeding with new process set-up.

In practice, for industrial long term use, operation schedules of an atmospheric pressure process may require some time slots allocated for maintenance. This is to avoid possible accumulation of condensed residuals, thus assuring correct and safe operation of the system.

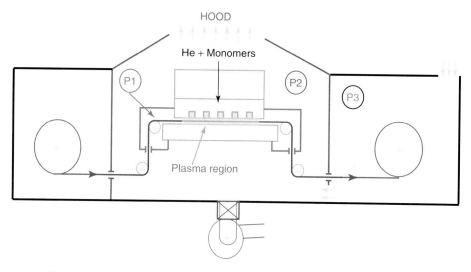

Figure 12.24 System with vacuum pump off-mode: unit in operation (plasma processing).

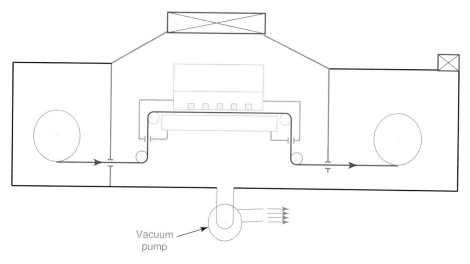

Figure 12.25 System with vacuum pump on-mode: purging phase, plasma process off-mode.

12.4.3
Final Considerations on Process Safety

In this Section, a method was introduced for safe operation of APP processes involving critical precursors. The devised system design, methodology, and maintenance protocols assure safe operating conditions.

Further, it was reported that in ordinary processes, the source intensity of compounds with a pollution potential are negligible.

References

1. Giudice, F., La Rosa, G., and Risitano, A. (2006) *Product Design for the Environment: A Life Cycle Approach*, CRC Press.
2. Chiellini, E. (ed.) (2008) *Environmentally Compatible Food Packaging*, CRC Press and Woodhead Publishing Ltd.
3. Yang, G. (2000) *Life Cycle Reliability Engineering*, John Wiley & Sons, Inc.
4. Imrell, A.M. and Sanne, K. (2003) EPD Key Interpretation, available at *.dantes.info*, (Accessed 3 November 2009).
5. *environdec.com*, (Accessed 3 November 2009).
6. Benveniste, G., Perucca, M., and Baldo, G. (2008) LCA comparative analysis of different technologies for functional coating in food applications. Proceedings of I-SUP-Conference, 22–25 April 2008, Bruges, Belgium.
7. Pinto Ferrerira, O.and Cuna, C. (2003) Eco-efficient solutions for innovative textile industry. Conference on Plasma Technologies – Turin 25th Sept 2003.
8. *http://eurostat.ec.europa.eu*, (Accessed 3 November 2009).
9. *boustead-consulting.co.uk*, (Accessed 3 November 2009).
10. European Commission Joint Research Centre, Life Cycle Thinking and Assessment, *http://lct.jrc.ec.europa.eu/*, (Accessed 3 November 2009).
11. Benveniste, G., Baldo, G.L., Perucca, M., and Ruggeri, B. (2007) LCA comparative analysis of different technologies for surface functionalisation. Life Cycle Management Conference 2007, Zurich Switzerland.
12. Ciantar, C. and Hadfield, M. (2000) An environmental evaluation of mechanical systems using environmental acceptable refrigerants, *J. Intern. LCA*, **5**, (4) 209–220.
13. Kim, S. and Overcash, M. (2003) Energy in chemical manufacturing processes: gate to gate information for life cycle assessment, *J. Chem. Technol. and Biotechnol.*, July 2004.

Further Reading

ANPA (2000) Banca dati italiana I-LCA.

Baldo, G.L., Marino, M., and Rossi, S. (2005) *Analisi del Ciclo di Vita LCA: Material, Prodotti, Processi*, Edizioni Ambiente.

ISO 14040 (1997) *Environmental Management – Life Cycle Assessment – Principle and Framework*.

Amato, I.and Montanaro, L. (2000) *Scienza e tecnologia dei materiali ceramici*, vol. III, Edizioni Libreria Cortino, Torino.

MSR 1999:2 (2000) Requirements for Environmental Product Declarations, EPD, Swedish Environmental Management Council, 27 May 2000.

Paracchini, L. (2003) *Manuale di Trattamenti e Finiture*, Tecniche Nuove.

Ciantar, C.and Hadfield, M. (2000) An environmental evaluation of mechanical systems using environmental acceptable refrigerants. *Int. J. LCA*, **5**(4), 209–220.

Kim, S.and Overcash, M. (2003) Energy in chemical manufacturing processes: gate to gate information for life cycle assessment. *J. Chem. Technol. Biotechnol.*, **78** (9), 995–1005.

Kesley, N. (2004) Chemical Energy Use in the Semiconductor Industry, University of California at Berkeley.

European Council (1999) Directive 1999/30/CE dated April 22nd 1999 relating to limit values for sulfur dioxide, nitrogen dioxide and oxides of nitrogen, particulate matter and lead in ambient air. *Official Journal of the European Communities*, **L 163**, 0041–0060.

Index

a

absorption measurements 86
accumulated power input time 70
acidification potential (AP) 349
acrylic acid (AA) 70
ACTECO project 48, 131, 235, 263, 269f., 353, 378, 385
actinometry technique 86f.
activated species 64
AcXys Technologies 57f.
adhesin 185
adiabatic compression 4
adiabatic electron sound speed 16
advancing contact angle 130
aging of textiles 144ff., 169f., 208
Ag NP/PEO-like coatings 211ff.
Ag/PEO-like coatings 201
Ahlbrand Coating Star 52, 54f., 152
Ahlbrandt System GmbH 52
Aldyne system 52, 170f., 53f.
Alfvén velocity 16
Alfvén waves 4
ambipolar diffusion coefficient 17
amplitude or diffusion contrast 126
Angora rabbit fibers 149
antibacterial drugs 197
antibiofilm 183, 197f.
antibiotic agents 193
anti-fog coatings 52
antifouling surface characteristics 195
antimicrobial activity, of polymeric QAS 196
antimicrobial-releasing biomaterials 193
antipathogenic drugs 197
anti-scratch coatings 52
antishrink treatment of wool 143f.
anti-wear coatings 52
– applications in food sector 268f.
– atoxicity and corrosion characterization 272, 280f.
– calotest and optical microscopy (OM), application of 275f., 277f.
– ceramic coatings 281
– chemical and structural characterization 273ff.
– coefficient of friction (COF) 271
– energy dispersive X-ray spectrometry (EDX), application of 274, 275f.
– for hammers for food scraps treatment 291, 340
– heavy metal release test 280
– mechanical characterization 276ff.
– need for wear resistance 270f.
– oxidation test 280
– pin-on-disk test 272, 279
– PVD ceramic coating 285ff.
– PVD coating potentiality for the food sector 264ff., 340ff.
– PVD coatings 267, 269f., 288ff.
– regulations 265
– requirements in food sector 269
– salt spray test 280f.
– saw blades used for cutting of food or packaging materials 288ff.
– SEM technique, application of 273ff.
– substrates and functionalities required for cutting applications 281f.
– technical analysis and selection of 282
– trends and market share 267f.
arc-physical vapor deposition (PVD) targets 16
Ar/n-hexane 83

astrophysical plasmas 4
atmospheric pressure nonthermal plasmas, physics and chemistry of
– gas discharges, modeling of 297*ff.*
– impurities, influence of 310*ff.*
– kinetic model for chemical reactions on a polypropylene surface 314*ff.*
atmospheric pressure plasma systems 17, 71
– corona-type surface systems 51*ff.*
– 'remote surface treatment' systems 54*ff.*
– source No1 56*f.*
– use 50
atomic force microscopy (AFM) 114*ff.*
– constant force mode 115*ff.*, 121*f.*
– resonant mode 117*f.*, 121*f.*
Aurora Borealis 3, 138
autoclaving 193
Avantes spectrometer (AvaSpec-2048-FT-RM) 305

b

backscattered electrons 122
bacterial adhesion process 184*f.*
bacteria–surface adhesions 190*ff.*
Baier curve 194
barrier coatings 52
binary coatings 266
biofilm formation 183*f.*, 351
– factors influencing 187*ff.*
– implications in biomedical field 184
– mechanism and different processes involved in 186*f.*
– PEO films and plasma deposition for preventing 202*ff.*
– prevention strategies of biofouling and microbial infections 192*ff.*
– process of bacterial adhesion 184*f.*
– role of plasma treatment in biofouling prevention 198*ff.*
biofouling 183
biofouling-resistant surfaces 198*f.*
Biomatech 213
Boltzmann constant 14, 29
Boltzmann equation (BE) 9, 298
Boltzmann statistics 27
Boltzmann transport equation 10
Bose–Einstein (boson gas) 15
Bragg's law 124
Brevundimonas diminuta 196
bronopol (2-bromo-2-nitropropane-1,3-diol) 201
Brownian motion 185

c

calotest and optical microscopy (OM), application of 275*ff.*
capacitively coupled systems 43*ff.*
capsule 190
catechol-containing zwitterionic polymers 196
cathodic arc evaporation 48
cathodic arc PVD systems 45*ff.*
Cathodic LARC® reaction chamber 48*f.*
caulobacter spp. 188
cefzolin 193
cellular adhesion tests 206*ff.*
CF_4 plasma 199
chain polymerization 68*ff.*
charged coupled devices (CCDs) 85, 125
chemical kinetics, of plasma surface interactions 68
chemical treatments, used in textile finishing industry
– chlorination of wool, pollution in 136*f.*
– cotton finishing, pollution in 136*f.*
– water usage 135
Child's law 31
chlorhexidine–silver sulfadiazine 193
ciprofloxacin 201
closed racetrack configuration 42
cloud–cloud lightning 3
cloud–ground lightning 3
co-aggregation process 188
coating process 18*f.*
coefficient of friction (COF) 271
Co^{60}-gamma radiation 196, 197*f.*
cold plasmas 6, 44, *See* Nonthermal plasmas (NTPs)
– waves in 12
colliding ice particles 4
collisional plasmas 21
collision operator, of Fokker–Planck model 10
completely ionized plasma 22
conditioning film 186*ff.*
contact angle (θ) measurement 66, 129*ff.*, 159, 164*f.*
contrast, on TEM images 126
COOH groups 70
coronas, plasma 17, 20
corona treatment, of cotton 148
corona-type surface systems
– controlled atmosphere corona treatment–aldyne treatment 52
– liquid precursor dosing system 52*ff.*
– standard corona system 51*f.*
Co–Sm magnets 41*ff.*

cotton   136*f.*, 148*f.*, 151, 352, 367*ff.*
Coulomb collisions   24
Coulomb forces   8
Coulomb interaction energy   24
Coulomb interactions   3, 9
Coulomb repulsion   189
100Cr6 blades   290
critical electron density   13
100Cr6 steel samples   269, 283, 286*f.*, 288*f.*
cryomicrotomy   117
cryo-ultramicrotomy   128
C_{1s}
– chemical shifts   97*f.*
– deconvolution of   74*f.*
– for PEO-like coatings   205*f.*
– XPS core level shifts   96
cut-off phenomenon   13, 40
cyclotron frequency   21
Czerny–Turner set-up   84

d

De Broglie wavelength   15
Debye length   14, 19, 28
Debye sphere   15
de-excitation process   80
desorption process   64
diallyl dimethylammonium chloride
    (DADMAC)   200
dielectric barrier discharge (DBD)
    17, 20*f.*
– configuration   71
dielectric tensor   12
– components   13
dielectric tensor, plasma   13
diethylene glycol vinyl ether (EO2V)   200
diffraction contrast   126
diffusion coefficients   17
direct-ionization rate coefficients   28
dispersion relation   12*f.*
dispersive interactions   130
dissociative ionization process   25*f.*
3D morphology, of a material surface   118*f.*,
    190
docking phase   184, 185*f.*, 187*f.*
Doppler broadening, of the line yields
    83, 88
2D plasma array   40
drift velocities   12
droplet deposition method   129
Duo-Plasmaline   36*ff.*, 249*ff.*
duty cycle (DC)   208
dynamic mode of SIMS   107
dynamic tensiometer, *See* Wilhelmy method

e

E × B drift motion   16
E × B drift velocity   16, 42
elastic scattering   24*f.*
electric field development, in clouds   4
electromagnetic energies   4
electron cyclotron resonance (ECR)
– condition   41
– discharges   4, 41*f.*
– heated plasmas   40*ff.*
– plasmas   16
electron cyclotron resonance ion sources
    (ECRISs)   21
electron distribution function   86
electron drift velocity   12
electronegative plasmas   23
electron energy distribution function (EEDF)
    23, 297*f.*
electron energy loss spectroscopy (EELS)
    126*f.*
electron gyrofrequency   16
electron–neutral collision frequency   13
electron spectroscopy for chemical analysis
    (ESCA), *See* X-ray photoelectron
    spectroscopy (XPS)
electron-to-mass ratio   16
electrostatic interaction chromatography
    (ESIC)   191
elementary reaction rate   23
energy conservation law, of photo electric
    effect   95
energy differences, between different atomic
    and molecular levels   81*f.*
energy-dispersive X-ray spectroscopy (EDS)
    123*f.*, 127
environmental product declarations (EPDs)
    349
Escherichia coli   83, 189, 193*f.*, 196
etching   202
ethylene oxide (EO) exposure   193
Eurocold Thermovide single chamber
    cleaning equipment   289*f.*
Europe, disease-related threat in   184
European Centre for Disease Prevention and
    Control   184
European textile industry   335
eutrophication potential (EP)   349
exopolysaccharides (EPSs)   187
extracellular polymer matrix (EPM)   183
extruded filaments, treatment of   148

f

F/C atomic ratio   100, 102*f.*
Fbe adhesin   190

Fermi–Dirac gas (fermion gas) 15
Fermi–Dirac statistics 15
fiber level, in textile manufacturing
    147*f.*
fibrinogen 188
fibrinogen-binding (Fbe) proteins 190
Fick's law 227, 237
field emission scanning electron microscopy
    (FESEM) 122*f.*
fingerprint, plasma 81
fluorination treatment 131
fluorocarbon atmospheric plasma finish
    154
fluorocarbon coatings 195
focusing mirror 85
Fokker–Planck equation 9
Fokker–Planck theory 10
food packaging, market analysis 339*f.*,
    See also Anti-wear coatings
fouling 183
Fourier-transform infra-red (FTIR)
    spectrometer 203, 239*ff.*
Frank–Condon principle 25
functionalization 144

g
gamma-ray irradiation 193
gas degeneracy 7
gas diffusion phenomenon, through
    polymers 225*ff.*
gas discharge modeling, of NTPs
– air-based 302*ff.*
– CF_4-based 309*f.*
– chemical reactions of neutral gas
    303*ff.*
– effective cross-sections 301*ff.*
– effects of electron and ion distributions
    298*f.*
– electron energy distribution function
    (EEDF) 297*f.*, 300
– electron kinetics 299*ff.*
– experimental UV, optical, and near infra-red
    emission spectra 302*ff.*
– full kinetic model 297*ff.*
– nitrogen-based 306*ff.*
– plasma chemical system 301*f.*
– radial distribution of active species and gas
    temperature in the plasma 308*f.*
– reduced model 297*ff.*
gaseous nebulae 3
gas flow rates 44
gas mixture ratio 236*ff.*, 243*f.*
Gauss theorem 30
GehD lipase 190

gendine 193
gentamicin 193
gigatron 36
global warming potential (GWP) 349
goniometry 159
grafting 66*f.*, 199*f.*
Gram-negative bacteria 188
Gram-positive bacteria 188
gyrofrequency 16

h
hardness, defined 276
harmonic time dependence 12
harmonix mode (VEECO) 121
He–Ar inert gases 153
hexamethyl disiloxane (HMDSO) 75*f.*
– coatings 201
– flow of oxygen-diluted in helium 102
– as precursor in SiO_x barrier films on PET
    foil samples 236*f.*, 243*ff.*
home-made stainless-steel reactor 44
homo-polymeric packaging materials 225
Hooke's law 116
Hostaphan RD 236
hydrophilic functionality of surface 67, 189,
    193*f.*
hydrophilic non-woven fabrics 154
hydrophilic textile treatment 36
hydrophobic fluoropolymer surfaces 191
hydrophobic functionality, of surface 67
hydrophobic textile treatment 36
2-hydroxyethyl methacrylate (HEMA) 72*ff.*

i
I_A of a XPS photoelectron peak 97
ideal gases 7*f.*
ideal gas law 8
implant-related infections 192
inelastic scattering 24*f.*
Institut für Plasmaforschung of Universität
    Stuttgart 36
integrated product policy (IPP) strategy 350
intense cosmic rays 3
interstellar gas 3
ion energy distribution function (IEDF) 23
ion gyrofrequencies 16
ionization degree 17
ionization process 25*f.*, 65
ionized plasmas 10
ionosphere 3
ion plasma frequency 15
IsaB heparin-binding protein 190
ISO 14040 standard 348*ff.*, 353

j

j-th electronically excited state   27
jacketing fiber protections   84
jets, plasma   17
Joule–Clausius formula   8
Joule's effect   45

k

K110 (AISI D2) steel   341*f.*
kinetic theory of gases   7*f.*
knife, plasma   57
knitted textiles, plasma treatment of   152*f.*
Krook model   11

l

Langmuir, Irving   3*f.*
Langmuir waves   4
lanthanum hexaboride (LaB$_6$) filament   122
LARC$^®$ PVD technique   48*f.*
laser induced fluorescence (LIF)   87*f.*
lateral force microscopy   115*f.*
lateral rotating cathodes   48
Lewis acid–base interactions   130
life cycle assessment (LCA), of a product 49, 138
– in accordance with the standard ISO   140, 349*ff.*
– anti-corrosion properties   360*ff.*
– application to plasma surface processing   353*ff.*
– benefits   349*f.*
– CF$_4$ gas, use of   375
– comparisons among direct energy for PVD, MW plasma, and electroplating processes   362*f.*
– concept   347*ff.*
– consumption of water and main raw material   363*ff.* 366*ff.*
– data and information used in   358
– design of plasma processes via   375
– functionalization processes for textile application   364*ff.*
– functional unit   354*f.*
– hydrophobic processes on PET+cotton   367*ff.*
– impact assessment   369*f.*
– impact category parameter (GWP), analysis   371*ff.*
– inputs and outputs of any system   357*f.*
– inventory data and results   357*f.*, 360*ff.*
– life cycle inventory (LCI) analysis and hypothesis   356*ff.*
– need for   350*ff.*
– oleophobic PET processes   366*f.*
– precursors used   359
– results of an   361
– role in decision making   349
– sensitivity analysis   371*ff.*
– strategies derived from   350
– system boundaries   354*ff.*
– system environment   356
– tolerances   374, 375*f.*
– total gross energy requirement (GER) of the whole process   359*f.*, 362*f.*, 365*f.*
– wastes for hydrophobic processes   369*f.*
Lifshitz–van der Waals interactions   130
lightning   3
l'Institut Français Textile-Habillement (IFTH) Lyon   36*f.*
Liouville theorem   9
liquid precursors dispensing system   58
locking phase   185*f.*, 188
Lorentz force   9
Loschmidt's number   17, 19, 300
low pressure cold plasma technology   35*ff.*
low-pressure microwave plasma processes   225
low pressure plasma (LPP) processing   6, 17, 19–20, 84, 111, 140, 148, 150, 154, 156, 158, 162, 167, 171, 174*f.*, 211, 378
– capacitively coupled systems   43*ff.*
– microwave systems   35*ff.*
– PVD systems   45*ff.*
low surface energy polymeric coatings   195*f.*
low temperature plasma   138
lysostaphin   193

m

macroscopic physical state   7
Martindale testing (ISO 12947/2)   171
mass spectrometry (MS)   65
matrix sheath   30
Maxwell–Boltzmann distribution   86
Maxwell–Boltzmann statistics   15
Maxwell–Boltzmann velocity   7
Maxwell distribution   14
mean free path   14*f.*
mean-free-transit time   10
mechanical energy   4
medical technology market   337*ff.*
metal-sputtered glass fiber fabrics   152
microbial surface components recognizing adhesive matrix molecules (MSCRAMMs)   190
micro/nano-structures   6
micro-organism characteristics   190*f.*
micro-plasmas   20

microporous polyethersulfone membranes 199
microscopic physical state 7
microtensile stage 119
microwave discharges 21
microwave energy 21
microwave plasmas 21
– Duo-Plasmaline 36*f.*
– electromagnetic wave propagation in 38*f.*
– electron cyclotron resonance heated 40*f.*
– surface-wave sustained 36*f.*
– for textile treatment 36*f.*
minocycline–rifampin 193
MKS mass-flow controllers 45
Mocon O$_2$ transition rate test system (OX-TRAN Model 2/61) 235
modeling, plasma sources used for 55*f.*
molecular ionization mechanisms 25*f.*
molecular static secondary ion mass spectrometry (SIMS) 107
monoethylene vinyl ether (EO1V) 200
monofilament textiles 146, 148
MultiPerm oxygen and water vapor permeability analyzer 235
Mylar window 127

n
nano-composite ceramic coatings 49
nanothermal microscopy 119
natural plasma state 3
Nd:YAG laser 87
N$_2$ ion analysis 88
Ni plating 153
noble gas-based plasmas 65
non-collisional homogeneous plasma 13
nondissociative ionization process 25*f.*
nonequilibrium chemistry 19
non-ideal gas regime 7
nonthermal plasmas (NTPs) 6, 17*f.*, 138, *See also* gas discharge modeling, of NTPs
– features of 22
nonthermal reactive plasmas 33
non-woven textiles, plasma treatment of 153*f.*
northern lights 3

o
O$_{1s}$
– deconvolution of 74*f.*
– XPS core level shifts 96
O$_2$: HMDSN gas mixtures 244, 249*ff.*
O$_2$/Ar/NH$_3$ mixture 162
O$_2$-containing plasma 73

Off-gas cleaning 19
O$_2$/HMDSO ratios 76
one atmosphere uniform glow discharge plasma (OAUGDP) 153
optical absorption spectroscopy 86*f.*
optical emission spectroscopy (OES) 65
– bench set-up 83*ff.*
– optical emission theory 80*ff.*
– spectroscopic technique 82*f.*
optical emission theory 80*ff.*
optical plasma diagnostics methods
– laser induced fluorescence (LIF) 87*f.*
– optical absorption spectroscopy 85*ff.*
– optical emission spectroscopy (OES) 79*ff.*
optimum roughness, concept of 189*f.*
organic precursor activation, in plasma 71*f.*
organic precursors 67
out-of-equilibrium thermodynamic regime 6
oxygen-based plasmas 67

p
Pascal principle 8
PEG coatings 194*f.*, 200
PEG oligomers 200
PEG polymer 194
Penning ionization 86
PEO films and plasma deposition, for preventing biofilm formation 202*ff.*
PEO-like coatings 202
PEO-like films 203*ff.*, 208*f.*
– silver, role of 211–215
– sterilization 210*f.*
– surface deposition/ablation effects on 209*f.*
perfluoroacrylate (AC8F17) 172
Perthometer C5D profilometer 236
PET fabric 163, 171
– hydrophilic properties of 167
– oil repellency levels 173*f.*
PET fiber 129*f.*
phase contrasts 118, 126
phosphonium salts (PSs) 196
photochemical ozone creation potential (POCP) 349
photodiode array detectors 85
photomultiplier tube 85
photon emission, associated with electronic states 81*f.*
physical vapor deposition-chemical vapor deposition (PVD-CVD) coatings 265
physical vapor deposition plasma technology 45*ff.*
physical vapor deposition (PVD) coatings, *See* Anti-wear coatings
pilin 190

pinhole-free films   203
pin-on-disk test   272, 279
pitting corrosion   280
Plank constant   15
plasma
– advantages   138
– applications   18*ff.*
– approximation   15
– definition   3, 137
– discharge treatment   193
– and electromagnetic interactions   8*ff.*
– electrostatic waves in   16
– (equilibrium) density   15
– and equilibrium regime   14
– finishing/coating   144
– as a fluid   11*f.*
– historical background   4*ff.*
– ideal gas regime for   7*f.*
– kinetic transport parameter   17
– long-range particle interactions in   8*ff.*
– off phase   209
– parameter   15
– on phase   209
– phase transitions   6
– physical quantities and parameters
     characterizing   14*ff.*
– physics   4
– quasi-neutrality conditions   11*f.*
– resistivity   16
– sterilization   201*f.*
– transport parameters   16*f.*
– velocity parameter   16
– waves in   12*ff.*, 16
plasma–chemical processes   22*f.*
plasma-enhanced chemical vapor deposition
     (PECVD)   71, 101, 200
plasma enhanced chemical vapor deposition
     (PECVD) reactors   17
plasma fluorination, of poly(butyl
     terephthalate) (PBT)   100*f.*
plasma polymerization   70*ff.*, 200*f.*
– of hexamethyl disiloxane (HMDSO)
     75*f.*
– of 2-hydroxyethyl methacrylate (HEMA)
     72*ff.*
– industrial application of   71*f.*
plasma–surface interactions
– chain polymerization   68*ff.*
– chemical kinetics   68
– grafting   66*f.*
– kinetics   64
– plasma polymerization   70*ff.*
– surface etching   65*f.*
– surface roughness, impact of   65

plasma surface processing, stages   295*f.*
plasma–wall interface charge neutrality   28
plasmodul   40
platelet adhesion   200
PMMA intra-ocular lenses   199
Poisson equation   11, 30, 299
'Polar' interactions   130
poly(diallyl dimethylammonium
     chloride)-coated glass surfaces   196
polydioxanone sutures   189
poly(ethylene glycol) (PEG)   189
poly(ethylene oxide) (PEO) coatings   189, 195
polyethylene (PE) films   66
polymer etching   65*f.*
polymer food packaging materials
– barrier concepts   233*f.*
– diffusion, solubility, and permeability
     coefficients   227*ff.*
– molecular diffusion in polymers, evaluation
     of   230*ff.*
– plasma deposition of SiO_x barrier films on
     polymer materials. *See* SiO_x barrier films,
     for polymer packaging
poly(methyl methacrylate) (PMMA)   189
polyolefins   65
polypropylene non-woven fabrics   153*f.*
polypropylene surface, kinetic model for
     chemical reactions on
– abstraction of H atoms from H-sites   320*ff.*
– alkoxy and peroxy radicals, surface
     concentrations of   322–325
– chemical reaction modeling   314*ff.*
– degree of the surface modification   322
– features of the time variation of the surface
     concentrations   322*ff.*
– O/C ratios   327*f.*
– surface concentration modeling   320*ff.*
polysaccharide intercellular adhesin (PIA)
     185
poly(sulfobetaine methacrylate) (pSBMA)
     196
poly(vinylbenzyl trimethylammonium
     chloride) (PVBT)   196
Porphyromonas gingivalis   191
PP food trays   247*ff.*
PP microfiltration membranes   19
$PP/SiN_y/SiO_xC_z/SiN$ coating   129*f.*
PP/SiO_x coating   128*f.*
PP tray surface   118
PP yarns   150
precursors, in plasma processing
     26, 58, 67, 359
presheath   30

primary radicals   68
protein adhesion   205*f.*
Pseudomonas aeruginosa   194
Pseudomonas aeruginosa biofilms   193
Pseudomonas putida mt-2   191
pseudomonas spp.   188
Purkinje, Johannes   4

q
QAS-grafted silicone rubber surfaces   200
quantum correlation   7
quartz crystal microbalance (QCM)   203
quasi-neutral plasmas   23
quaternary ammonium salts (QAS)   196
quorum sensing   187

r
radio frequency glow-discharge
   (RFGD)   200
radio frequency (RF) generator   44
Raoultella terrigena   196
reaction cross-sections   23
reactions rates (k)   23
reactive plasmas   22*ff.*
reactive species   64
regular Trichel pulses   298
relative directional mobility   17
'remote surface treatment' systems   54*ff.*
RF plasma glow discharges   86
Rockwell C hardness sampling   287
Rockwell hardness test   277, 279*f.*
Rutherford formula   24

s
S. pyogenes   192
safety issues, of plasma processes
– concerns for atmospheric plasma
   processing   379*f.*
– standards for industrial applications
   maintaining the semi-continuous
   feeding   380*ff.*
scanning capacitance microscopy (SCM)
   119
scanning electron microscopy (SEM)
– chemical analysis   123*f.*
– image obtained at P = 1 Pa and 3 kV   125*f.*
– imaging in   122
– LEO 1530VP   123
– new generation of   122*f.*
– principles   121*ff.*
– sample preparation and applications   124
scanning spreading resistance microscopy
   (SSRM)   119
secondary colonizers   188

secondary ion emission   107
secondary ion mass spectrometry (SIMS)
   measurements   106
self-consistent set   301
self-sterilizing surfaces   196
SFE inert surface   189
SFE surfaces   195*f.*
sheaths, plasma   28*ff.*
Si–CH$_3$ groups   240, 241*f.*
Si–CH$_2$–Si groups   240
SiCl$_x$ coatings, aging of   146
silver   193
SiO$_2$ sample, XPS survey spectrum of   95*f.*
Si–O–Si bond   75
Si–O–Si groups   241, 242*f.*, 245
SiO$_x$ barrier films, for polymer packaging
– deposited from O2: HMDSO gas mixtures
   243*f.*, 236*ff.*
– determination of diffusion coefficient   237*f.*
– on 2D substrates   236*f.*
– electron cyclotron resonance (ECR) plasma
   deposition   237*f.*, 247*ff.*, 251*ff.*
– FTIR analysis of chemical composition
   239*ff.*
– industrially relevant plasma deposition
   255*ff.*
– materials   234*f.*
– measurements   235*f.*
– O$_2$ permeation measurements   238*f.*
– on PET foil   236*ff.*
– on PP films   247*ff.*
– steady-state O$_2$ particle flux   235, 246*f.* 248*f.*
   250*f.* 253*f.*
– thickness of   235*f.*
SiO$_x$ coatings   58
– determination of growth of   101*ff.*
skin depth   40
Slime   184
Solarization resistant assemblies   84
solar UV radiation   3
solar wind radiation   3
spray test (EN 24920)   164*f.*
Staphylococcus aureus   189*f.*, 192,
   194, 196
Staphylococcus aureus AATCC 6538 bacteria
   213
Staphylococcus epidermidis   185, 189*ff.*, 194,
   201
Staphylococcus epidermidis RP62A (ATCC
   35984)   214
state of matter   3
static mode of SIMS   107
Stenotrophomonas maltophilia   191
stepwise ionization process   26*ff.*

stepwise ionization rate coefficient   28
sterilization   143
sterilization processes   201*f.*
stimulated emission   80
Streptococcus mutans   191
Streptococcus salivarius,   194
Streptococcus sanguis   195
super-hardness   266
surface activation   144
surface area estimation, of woven fabric   157*f.*
surface energy, of bacteria   191
surface energy calculation
– Good and Van Oss Model   131
– Owens and Wendt Model   130*f.*
surface-sensitive analytical methods
– atomic force microscopy (AFM)   114*ff.*
– characterizing plasma-treated surfaces
    91*ff.*
– contact angle measurement   129*ff.*
– scanning electron microscopy (SEM)   121*ff.*
– time of flight static secondary ion mass
    spectrometry (ToF-SSIMS)   106*ff.*
– tranmission electron microscopy (TEM)
    124*ff.*
– X-ray photoelectron spectroscopy (XPS) or
    electron spectroscopy for chemical
    analysis (ESCA)   94*ff.*
surface treatment, using plasma technologies
    352
– atmospheric pressure plasma systems   49*ff.*
– capacitively coupled system   43*ff.*
– low pressure cold plasma technology   34*ff.*
– physical vapor deposition plasma technology
    45*ff.*

t
'3D' character of textiles   154, 156*f.*
'tapping mode' atomic force microscopy
    (TMAFM)   118, 121, 120*f.*
technological plasmas
– cold   19*ff.*
– hot   18*f.*
teicoplanin   193
tensiometry   159
tetraethoxy silane (TEOS)   52
textile and clothing industries, market analysis
– general   335*f.*
– hydrophobic and oleophobic textile market
    336, 337*f.*
– medical textiles   338*f.*
– technical textiles   336*f.*
textile manufacturing process   351*f.*
– abrasion durability   173*f.*
– assessment of surface energy   158*ff.*

– atmospheric plasma sources for   141, 142*f.*
– and classification of plasmas   138*ff.*
– cotton, plasma treatments of   149, 151
– degummed silk (BombyxMori) yarns,
    plasma treatments of   149
– effect of air permeability on plasma
    treatment   153
– fabric level in   150*f.*
– fiber level in   147*f.*
– filament level in   148
– first screening test   168*f.*
– general flow chart   147*f.*
– hydrophilic properties imparted by plasma
    treatment   167*ff.*
– hydrophobic/oleophobic properties
    imparted by plasma treatment   171*ff.*
– impact of plasma treatment   135*ff.*
– industrial production process and related
    issues   175*ff.*
– integration of plasma processes into the
    textile manufacturing chain   146*ff.*
– intermediate/finished textile material   154*f.*
– knitted textiles, plasma treatment of   152*f.*
– linen, plasma treatments of   151
– low pressure experiments   167*ff.*
– moisture regain and air adsorption post
    treatment   158
– non-natural materials, plasma treatments of
    150
– non-woven textiles, plasma treatment of
    153*f.*
– oil repellency test (EN ISO 14419)   166*f.*
– physico-chemical reactions   138, 139*f.*
– plasma treatment for   137*f.*, 142*ff.*, 174*ff.*
– pressures encountered and their impact
    140*f.*
– silk, plasma treatments of   151*f.*
– specific properties of textile materials and
    plasma requirements   155*ff.*
– spray test (EN 24920)   164*f.*
– substrates and plasma effect on substrates
    141, 142*f.*, 143*f.*
– traditional *vs* plasma-based treatment   139*f.*
– two-dimensional textile objects, treatment of
    150*ff.*
– washing durability   172*f.*
– water/alcohol repellency test   165*f.*
– wet textile finishing processes   145*f.*
– wool, plasma treatments of   149, 151
– woven textiles, plasma treatment of   151*f.*
– yarn level in   149*f.*
thermal energy   4
Thermal equilibrium   14

thermal plasma processing technology, applications of   18
thermal plasmas (TPs)   14, 17
thermosphere   3
theta surface   195
Thompson scattering   83
Thomson ionization cross-section   24
Thornley–Everhart scintillator/photomultiplicator   122
3D Fourier transform   12
threshhold R_a 190
threshold roughness, concept of   190
Ti/hydrocarbon plasma polymer film deposition processes   83
time-of-flight secondary ion mass spectrometry (ToF-SIMS) images   54, 156
time-of-flight static secondary ion mass spectrometry (ToF-SSIMS)
– applications of   107*ff.*
– data treatment using ToF-SIMS datasets   108*f.*
– depth profiling   108
– ion imaging mode   108
– plasma functionalization of textiles and thin film coatings   109*ff.*
– principles   106*f.*
– spectrometry mode   108
– static and dynamic modes   106
titanium nitride (TiN) coatings   47*f.*, 264*f.*, 283*f.*, 285*f.*, 290
Tonks, Lewi   3*f.*
T-Peel test   150
transmission electron microscopy (TEM)
– applications   127
– chemical analysis   126*f.*
– image contrast   126
– principles   124*f.*
– resolution   126
– sample requirements   127*ff.*
transient electric discharge   3
transport parameters, of plasma   16
triclosan (2,4, 4P-trichloro-2P-hydroxydiphenylether)   201
3-(trimethoxysilyl)-propyldimethyloctadecyl-ammonium chloride coating   196
tufted carpets   154
tunneling atomic force microscopy (TUNA)   119

u
ultra-high modulus polyethylene (UHMPE) filament   157
ultramicrotomy   128

unspecific functionalization   67
usnic acid   193

v
vacuum electric permeability   14
vancomycin   193
Vickers microhardness test   277, 279*f.*

w
Washburn method   129, 160*ff.*
water/alcohol repellency test   165*f.*
wavelength-dispersive X-ray spectroscopy (WDS)   123*f.*
waves, in plasmas   12*ff.*
weakly ionized plasma   22
Wilhelmy method   129, 159*ff.*
wool   352
– antishrink treatment of   143*f.*
– Ar–CF₄ discharge plasma treatment   199
– chlorination of   136*f.*
– plasma treatments of   149, 151
woven textiles, plasma treatment of   151*f.*
W/ZrO field emission gun (FEG)   122

x
X-ray photoelectron spectroscopy (XPS)   65, 94*ff.*, 203
– angle-resolved   100*ff.*
– chemical characterization of plasma-treated surface using   98
– chemical mapping of plasma fluorination of carbon fibers   104*f.*
– concentration depth profiles   94
– core level chemical shifts   96*f.*
– determination of growth of SiO_x coatings   101*ff.*
– lateral surface imaging and analysis   94*f.*
– mapping of sites   104*f.*
– microanalysis of two selected areas   105*f.*
– nitrogen plasma-treated polypropylene (PP) film, quantitative analysis   98*ff.*
– principles   95*f.*
– quantitative analysis   97*f.*
– spectroscopy   94

y
Yasuda factor   67
Young–Dupré equation   66, 130